T0139049

Power for the World

Wolfgang Palz in front of the PV façade of the public library of Mataró, Spain, in 1992. It was one of the first semitransparent solar façades in Europe, co-financed by an EU programme.

Power for the World

The Emergence of Electricity from the Sun

Wolfgang Palz
World Council Renewable Energy

With Contributions from 41 International Solar Pioneers

Published by

Pan Stanford Publishing Pte. Ltd.
Penthouse Level, Suntec Tower 3
8 Temasek Boulevard
Singapore 038988

Email: editorial@panstanford.com
Web: www.panstanford.com

British Library Cataloguing-in-Publication Data
A catalogue record for this book is available from the British Library.

POWER FOR THE WORLD
The Emergence of Electricity from the Sun

For photocopying of material in this volume, please pay a copying fee through the Copyright Clearance Center, Inc., 222 Rosewood Drive, Danvers, MA 01923, USA. In this case permission to photocopy is not required from the publisher.
ISBN 978-981-430-337-8

Printed in Singapore.

Contents

About The Author

Wolfgang Palz

Currently Chairman, World Council Renewable Energy, Bonn/Paris

- Member French National Aid Commission, Demonstration Fund for Energy Research, Paris
- Honorary Chairman, 5th World PV Conference, Sept. 2010 in Valencia, Spain
- Board Member, ISFOC for PV Concentrators, Puertollano, Spain
- Member of the Board of REN21, Paris
- Advisory Board Member, American Council On RE, ACORE, Washington DC.
- Member of Global Bio-energy Partnership GBEP, Rome
- Honorary Board Member of ISES, Freiburg, Germany

Palz holds a PhD in physics (Dr.rer.nat.) of University of Karlsruhe, Germany, from 1965. In 1965–70 he was a Professor for semiconductor physics in Nancy, France. 1970–76 he was in charge of power systems development at the French National Space Agency CNES in Paris. In 1973 he was co-organiser of the UNESCO Congress 'The Sun in the Service of Mankind' in Paris. In 1976/78 UNESCO published his book 'Solar Electricity' in seven languages.

1977–2002 he was an official of the EU Commission in Brussels, the executive body of the European Union. 1977-1997 he managed the development programme of the Renewable Energies; it included policy development and contracting to European industry and academia of the Commission's budget (almost $1 billion over that period). The R&D programme comprised the sectors of Solar Architecture, Solar Energy, Wind Energy, Biomass, Ocean Energy.

In 1997 he became an EU Commission Counsel for renewable energy deployment in Africa and, besides, advised the EU Commissioner for Energy on the EU White Paper RE issued that year 1997. From 2000 to 2002, Palz was member of an Energy Committee ('Enquête Commission') of the German Parliament, the Bundestag in Berlin, to establish an energy strategy for

Germany on the time horizon 2050. In 2005/2010 he was consultant of the EU Commission for PV programmes in Latin America. Palz is bearer of an Order of Merit of Germany (Bundesverdienstkreuz), has been recognised a wind energy pioneer in Britain, and received the European Prizes for Biomass, Wind Energy and Solar Photovoltaics respectively.

Contact
w@palz.be

List of Contributors

Hermann Scheer, Germany
Franz Alt, Germany
Yves Bamberger, France
Stefan Behling, UK
Joachim Benemann, Germany
Karl Wolfgang Böer, USA
Dieter Bonnet, Germany
Anil Cabraal, World Bank
Gallus Cadonau, Switzerland
Dominique Campana, France
Denis Curtin, USA
Michael Eckhart, USA
Hans-Josef Fell MP, Germany
Americo Forestieri, USA
Gao Hu, China
Biswajit Ghosh, India
Adolf Goetzberger, Germany
Giuliano Grassi, Italy
Michael Grätzel, Switzerland
Wolfgang Hein, Austria
Osamui Ikki/Izumi Kaizuka, Japan
Helmut Kiess, Switzerland

Stefan Krauter, Germany
Harry Lehmann, Germany
Alain Liebard/Yves-Bruno Civel,
    France
Daniel Lincot, France
Antonio Luque, Spain
Bernard McNelis, UK
Bernd Melchior, Germany
Ricardo Melchior Navarro/Manuel
    Cendagorta, Spain
Thomas Nordmann, Switzerland
Monica Oliphant, Australia
Morton Prince, USA
Haiyan Qin, China
Bunker Roy, India
Walter Sandtner, Germany
Fred Schmid, USA
Richard Swanson, USA
Peter Varadi, USA
Philippe Verburgh, Switzerland
Neville Williams, USA

Hymn to the Sun

Your rays feed the fields
You shine and they live
They are abundant for you
You have created the seasons
So that all you have created can be alive
The winter for cooling,
The heat

How numerous are your actions
Mysterious in our eyes!
Only God, you who has no likeness
You have created the Earth as per your heart
When you were alone,
Man, all animals, domestic and wild,
Everything on Earth marching on feet
Everything in the sky flying with wings
The foreign countries, Syria and Nubia
And the land of Egypt

Akhenaton Pharaoh of Egypt 1378–1362 before our time

Nofretete, Echnaton and family, 14th century BC, Neues Museum, Berlin.

Foreword

Hermann Scheer

Member of the German Bundestag, Berlin; President of the Association EUROSO-LAR, Bonn and General Chairman of WCRE

Emancipation means liberation from mental and physical dependencies so that people can go their own free way. Unless this concerns merely one individual goal, emancipation must be possible for everybody. This means: emancipation must be possible for everybody — along the lines of the philosopher Kant's central directive, his "categorical imperative": "Act only according to that maxim whereby you can at the same time will that it should become a universal law." This imperative is the basic value for any humane civilization. Many believe that the ideal state this implies, involving enhanced individual freedom coupled with an ori-
entation to the common weal, is desirable, yet never attainable. In the energy area, though, it is largely attainable, namely by using photovoltaic technology.

This is a unique gift for humanity — a gift whose far-reaching significance many have still to appreciate. With photovoltaics, it will be possible for roofs, façades and — soon — windows as well to generate power; electricity will be produced in equipment cases, in the bodies of cars, on the outer hulls of ships, on the noise-protection walls of roads, on the outer surfaces of greenhouses, or on outdoor terrain with vertical PV modules, under which grass cultures can grow that remain suitable for grazing. Villages or even whole towns can thus be supplied, either completely or very largely, with electricity. The dual functions of the various usable surfaces mean that the cost of power generation will become minimal: everywhere on Earth and affordable for everyone — enabled by the unique possibility of direct conversion of photons into electrons. We will then need fewer and fewer power lines and no longer require gigawatt power plants — thanks to electricity production without emissions, without noise and with no need for water.

Over a century ago, the general emergence of electricity proved that it is a cultural power. With PV technology, it is now turning into an emancipating cultural power. Yet the fact is that hardly anybody recognised this cultural potential in the early days of PV technology, although the world's best-known founding father of electrical engineering — Thomas Alva Edison — had already described his vision of power being produced in every house, i.e. on a decentralised basis. In the first century of the history of electricity, however, it was the opposite approach that practically won the day: ever larger power plants were built with ever longer power lines, i.e. more and more centralisation. In Italian usage, power production itself even became synonymous with major power stations, and a power plant is called *centrale*.

Major power stations became the focus of thinking and action, as if their evolution had been based on a natural law of economic energy. This was true even of most PV specialists back in the early days of the 1950s, 60s and 70s. They developed PV technology for specific applications — like the power supply for satellites sent into space. They failed to realise that they were working on a technology that harbours the potential to revolutionise energy supply. Any technology is only ever worth as much as there are people with enough imagination to appreciate its actual utility and with a will to launch it into the real world. This does not happen by itself, as is shown by the numerous examples of ignored and suppressed inventions. The reasons for ignoring them usually lie in an underestimation of their value and in mental barriers erected by conventional ways of thinking and by past experience. Those who have power and markets to lose when a new development is sighted generally have grounds for suppressing it.

Later, when wide-spread application of the new development is taken for granted, people will ask why it took so long. Their questions will include: Who ignored this technology, and why? Who got it on the road, and what were the human sources of the emancipation from energy which is provided by big nuclear and fossil-based power plants? As for who has been stonewalling, the answer is clear: the losers in this development, i.e. the big electricity companies and primary-energy suppliers who will inevitably lose their supply monopoly. Besides them, however, there are many others who lack the imagination to appreciate that autonomous producers with countless PV systems could replace the few major producers. And there were, and still are, many decision-makers who are all too closely involved with the energy companies. For PV technology, what is needed is social imagination and, above all, emancipatory motivation.

So, if society were to be mobilised for PV, the public would have to be informed, thus laying bare the motives behind the stalling tactics. This required courage and powers of persuasion, since the mere articulation of comprehensive and long-term perspectives, and the refutation of the abundant disinformation about PV were considered an attack on established energy interests. Hence, political concepts had to be developed and put in place, smashing through the manifold privileges granted to nuclear and fossil energy. The breakthrough for PV application called for all-round activities. It took pioneering work on many fronts. A new idea, after all, will only get off the ground if it has active supporters. An aphorism by the satirist Stanislaw Lec runs: "Those who are ahead of their time often have to wait for it in uncomfortable quarters." In the technocratic discussion about potentials, the only questions usually asked are those about the natural, technological and economic potentials of PV. But the most important potential issue is that of the human potential: Who was, and who is motivated to mobilise the technological potential and to open and enter the channels for its dissemination? Who spotted and seized the opportunity to enable PV to power the emancipation from existential energy dependence?

The driving forces taking PV from lab to roof — in other words, the first big steps on the road to practical applications — were local action groups, committed to environmental protection and self-determination. They were no longer willing to wait around for government initiatives. They had a justified distrust of the big energy companies who were talking the public and governments into believing that there are no alternatives to nuclear or coal-fired power stations. The action groups wanted to prove that life can be different. They not only provided information about the alternative, but also set landmark examples with practical projects. They called upon and urged the political authorities to take initiatives of their own. In a nutshell: they declared PV to be a public task that should no longer be left to the big energy companies. That's how it started in Germany: first of all in just one city, and then in more and more cities. This trickle turned into a swelling civil movement in the 1990s that placed the ethical demand for zero-emission and independently produced electricity before the question of cost. These people were thinking not only of the present, but also of the future, of individual freedom and the common weal, of autonomy instead of heteronomy.

The spirit behind this emerged in the 1960s — in the movement that is generally referred to in Germany as the "68er". It was an emancipation movement, demanding equal civil rights for all races and genders, and rejecting bureaucratic and technocratic patronage, war and the exploitation of people. The protests against the Vietnam war were the cement in this movement, from the US to Western Europe. Many movements with different focuses emerged from this spirit: the civil-rights movement, the peace movement, the feminist movement, the third-world movement, the environmental movement. The jointly shared experience was that, even in countries with democratic constitutions, governments were not willing to do what had to be expected of them. Without this background, it is not possible to explain the process that led in Germany to a popular movement for renewable energies that helped create the democratic basis for legislation to mobilise PV. But it was necessary to have players who would channel the spirit of this new civil movement into PV — and renewables in general.

I am part of this emancipation movement from the 1960s. In 1969, I was one of the leaders of the student movement at the University of Heidelberg where I was president of the student parliament. Heidelberg was — alongside Berlin and Frankfurt — one of the hotspots of student unrest in Germany, like the University of Berkeley in California. This spirit marked my personal and social attitudes.

Nor did it forsake me in 1980 when I was elected to the German parliament — the Bundestag — to which I have belonged ever since without interruption. In my early years in parliament, I was a committed backer of nuclear disarmament. As early as 1982, I became spokesman of my parliamentary group, and was later chairman of the Subcommittee on Disarmament and Arms Control. I was a member of the Foreign Affairs Committee and had a reputation as a great foreign-policy talent. In 1986, I published my book "Die Befreiung von der Bombe" (Liberation from the Bomb), in which I described a possible route toward a world free of nuclear weapons. It met with a spectacular response; Germany's leading political magazine *Der Spiegel* — the German equivalent of the *TIME Magazine* — called it an "icebreaker".

Still, I was not only against nuclear weapons, but against nuclear-power plants as well. I was also increasingly concerned with the issue of alternatives, since I was aware that a future with fossil energy was no option either. In 1981, I had read one of the earliest books on the looming CO_2 climate disaster. My immediate conclusion was: we must focus our efforts on renewable energies. But here were the energy experts saying that the potentials would not suffice to allow us to do without nuclear power and/or fossil energy. My critical spirit, honed back in the 1960s, told me not to believe this. I became curious about renewables. And a mere second glance showed me: a world without nuclear power and fossil energies is an option.

I asked politicians — in my party, in parliament or in the government — who were responsible for energy policy why they did not focus their policy on this perspective. I asked environmental organisations, but they, too, thought this perspective was unrealistic. I contacted the few solar scientists around, and my impression was that they as well were underestimating the potential — or that they did not have the courage to formulate a more comprehensive perspective. One told me: "If we manage to cover 10% of our energy needs from renewables by 2050, you can consider yourself lucky." I felt that all of this was faint-hearted. So, I wrote my first article on solar energy in the mid-1980s, at a time when there was talk everywhere of the "Star Wars" programme — the Strategic Defense Initiative (SDI) of US president Reagan. My article bore the title "For an ecological SDI — the Solar Development Initiative": just as new armament technologies were being driven forward costing billions, so, too, must solar energy be driven forward — as a political mission to defend our environment. Now I was accused of morphing from realist to dreamer by embracing such an unrealistic cause. One friend told me: "Why are you suddenly concerned with such an unimportant issue?"

I increasingly came to realise that the precondition for a successful political initiative to promote solar energy was to create a public climate that would enable it to tear down the walls in people's minds. And it also became clear that this subject must not simply be left to the big energy companies, the energy politicians and conventional energy science. This is why I set up EUROSOLAR in 1988 — the European Association for Renewable Energy — and became its president, which

I still am today. The founding document talked of the "reality-based vision" of a solar age: the vision of a complete substitution of nuclear and fossil energies with renewables. Politicians, scientists and entrepreneurs became members of EUROSOLAR. The two most important tasks we set ourselves were to enlighten the public about this reality-based vision, and to work on political concepts for action as well as draft legislation to promote renewable energies. A short time later, I withdrew from my foreign-policy offices in parliament and focused entirely on being a source of public and political inspiration for the historic energy turnaround. The year 1989 brought the first political success: I authored a resolution for parliament which was to prioritise the promotion of solar energy in research and development policy. This resolution was passed unanimously by the Bundestag. Although it was never implemented in its entirety, one immediate practical consequence at least was the 1000 Roofs Photovoltaic Programme of the German Federal Government — the first systematic PV market-launch programme — as well as the 250-MW wind energy programme. I launched an in-depth round of lectures with about 200 talks a year — at party assemblies and for action groups, environmental organizations and universities. Media attention increased, with Franz Alt, a TV journalist, and his programmes becoming pioneers. Among the environmental organisations, too, calls for political promotion of solar energy grew louder. Also, the number of my colleagues in parliament — especially from my own party — who had become curious now grew, and I was asked to give talks in their constituencies. In this way, the support base expanded in public and in my party, the Social Democrats (SPD), one of the two major parties in Germany.

The year 1990 saw the second success: in an effort mounted together with several members of parliament from the other major party (the Christian Democratic Union/Christian Social Union, CDU/CSU) and one member from the Greens (Wolfgang Daniels) — who had likewise become members of EUROSOLAR — the first feed-in tariff law for renewable energies came into being. In September 1990, the bill was adopted by parliament. It regulated guaranteed grid access for electricity from renewables supplied by independent producers and guaranteed a feed-in price of between 75% and 90% of the underlying electricity price. We had persistently urged that this law be passed — in the Bundestag and in the Bundesrat, the legislative chamber representing Germany's federal states. Important support came from Reimut Jochimsen, the economics minister of North Rhine-Westphalia, Germany's largest federal state. At the time, he was chairman of the conference of economics ministers, and I had persuaded him in several talks to support this law. This opened the road for a steady rise in wind-power investment. For PV electricity, though, the feed-in tariff was not yet attractive enough. Still, it was not possible to achieve any more in those days. Up to the very last minute, the big electricity companies tried to frustrate the legislation. It was later said that, had they known how important this law was to become, their resistance would have been insurmountable.

The next step was a pinpointed mobilization for PV's market launch. To this end, I presented for the first time in 1993 a draft for a 100 000 Roofs Photovoltaic Programme. It seemed utopian even to the representatives of the still very small PV industry. My argument was: if we wanted to gain more support for the solar perspective in politics and society, big steps would have to be demanded. After all, only few people are interested in small steps, and the underestimation of solar energy could not be overcome that way. This convinced my party's youth organization, to start with, and it mounted a solar campaign and declared my book "A Solar Manifesto" published in 1993 to be "required reading".

It dealt, not with technologies in renewable energies, but with a policy for such energies, the declared aim being a complete transformation of the energy supply toward the use of renewables: 100%! This book became a bestseller in Germany with a total of eight print runs. It gave a political voice to the commitment for renewable energy. It defined ignorance about renewable energies as a "mistake of the century", and revealed the disproportion between the political promotion of renewables at the lowest political level and the political promotion of nuclear power and fossil energies at the highest level. And it disclosed the reasons and methods behind the resistance and the obstruction of renewables, including the technological pessimism in their regard, which was in stark contrast to the technological optimism in favour of any other technology. This book evolved into a reference work for many players in political parties, for policy-makers at local-government level, for companies and also for scientists, all of whom were inspired by it. Above all, it offered stimulus for a solar movement. Such a movement had existed in California in the 1970s, but had largely withered away in the 1980s in the wake of the roll-back of renewable energies during the Reagan administration. In California, too, it had been marked by the spirit of the 1960s. What had died away there was now revived in Germany. And the most concrete, practical medium at grassroots level is photovoltaic technology.

When I brought the 100 000 Roofs Photovoltaic Programme into the public debate in order to open the doors for industrial mass PV production, the Federal Government's 1989 1000 Roofs Photovoltaic Programme had long since come to an end. But there was still no chance of a political implementation of this more far-reaching concept. Even Siemens-Solar, at that time the best-known German manufacturer of PV cells and modules, declared the concept to be immoderate. At local-government level, however, more and more activities were getting under way. People were setting up solar clubs. One of these was the solar-promotion club in Aachen in North Rhine-Westphalia run by Wolf von Fabeck, a former army officer, as the driving force. He called for a "cost-covering price" according to the feed-in tariff principle for people who had installed a PV system. The Aachen city council took this up in 1994 and decided that its municipal utility should pay DM 1.80 per kilowatt hour of solar power fed into the city's grid. But the new economics minister of North Rhine-Westphalia initially refused to grant his approval. I then talked to his prime minister, Johannes Rau, and was able to win him over for an approval. In the course of the next four years, more than 40 towns and cities followed Aachen's example. One of the towns was Hammelburg in Bavaria, where a local action group had formed. Like many others, it had invited me to give a talk in their town. This is where I met Hans-Josef Fell in 1995. He was councillor there, a member of the Green party and the engine behind cost-covering prices in his community. He became a member of EUROSOLAR. These local initiatives made a crucial contribution toward launching a public wave of support for PV. They created a local base for a PV market. Without them, it is unlikely that the still-small PV industry of the 1990s would have survived in Germany. The "baby" had become a toddler.

1998 was the key year for the next big steps that were to make Germany a force to be reckoned with in PV. The outcome of the general elections brought a majority for the SPD and the Greens who together formed the new Federal Government. Previously — in April 1998 — I had succeeded in anchoring both the 100 000 Roofs Photovoltaic Programme and the feed-in law for renewable energies in the SPD's election manifesto. This nearly faltered because, just a few days before the meeting of the SPD executive that was crucial for this thrust, Greenpeace published a demand for a 50 000 roofs PV programme. So I was asked at the meeting of SPD leaders why the SPD should demand more than even Greenpeace, but I won the day with the argument that we should be thinking of industrial mobilization for such a very big step. Although I prevailed with my demand for 100 000 PV roofs, it was missing initially in the government programme drawn up by the SPD and the Greens after the election. I then demanded subsequent negotiations and was able to push this point through at the last minute. I realised that this was the time to act, and that the programme had to be translated into practice at once. But this, too, was extremely difficult, since the new government had resolved not to pass its new budget until June 1999, i.e. not until eight months later. That was dangerous for the PV industry. All those purchasing PV systems would then have an eight-month wait for the programme to start, so that there would be no orders for PV systems during that time. This is why I contacted the state-run KfW development loan corporation to ask whether it might be willing to pre-finance the programme so that it could get started straight away. This was accepted. Within a few days, I drafted the programme, together with the KfW, the finance ministry — Oskar Lafontaine, who supported the programme personally, had become the new finance minister — and two Green members of parliament. One of these was Hans-Josef Fell, who had been elected to the Bundestag in 1998 and who has since become my most important fellow campaigner in parliament. The programme started on 1 January 1999. This is the date of birth worldwide of industrial mass PV production. The first system under this programme was installed on my garage on 2 January, next to a system that I already had on my roof. By contrast, all other mass programmes that had been previously proposed remained on paper. My first proposal for a 100 000 Roofs Photovoltaic Programme from the year 1993 had been taken up in 1995 by Peter Michael Mombaur, MEP, who in a report for the European Parliament recommended a 100 000 roofs programme for the EU. In 1997, US President Bill Clinton spoke out in favour of a one-million roofs programme, and Mechtild Rothe, MEP, demanded this for the EU as well. But none of these initiatives were ever implemented in practice.

The 100 000 Roofs Photovoltaic Programme was a zero-interest scheme, with the government paying the difference between zero interest and a loan's market interest rate. In the first two years, nothing had to be repaid, and repayments were to be made between years three and ten, at 12.5% each. The final instalment was to be waived. But this by itself was not yet enough to generate a wide movement on the market. To provide a further boost, we also wanted to increase the feed-in tariff. This needed legislation, and our aim here was to extend the existing Electricity Feed Act. From the government, there was no initiative to be expected for this. Oskar Lafontaine had resigned as finance minister. So, we had to take the initiative ourselves. Starting in the summer of 1999, four Bundestag members worked out a draft bill: Dietmar Schütz and I for the SPD parliamentary group and Michaele Hustedt and Hans-Josef Fell for the Green parliamentary group. This was how the draft Renewable Energy Act (*EEG*) came about for which we were able to win over our parliamentary groups in the autumn of 1999. The final vote on the bill was scheduled for 25 February 2000. For PV — in addition to the grants for the 100 000 Roofs Photovoltaic Programme — we pencilled in a guaranteed price of 99 pfennigs per kilowatt hour. Even this seemed utopian to nearly everyone. To help ensure a majority in both government parliamentary groups, we limited this to the quantity of the already running 100 000 Roofs Photovoltaic Programme. This got us a majority. But the biggest political hurdle to be taken came out of the blue two weeks before the final vote in parliament: the cabinet called upon us to seek the approval of the EU Commissioner for Competition before we adopted the Programme. The reason given was that the Renewable Energy Act would violate the EU's market rules, and it was not compatible with European law. This was a last-minute attempt to torpedo the law.

The days that followed were like a High Noon situation in a Western. The tie-breaker was how the SPD parliamentary group would respond, meaning the larger parliamentary group in the government. In this situation, the economics minister and the new finance minister suggested postponing the vote on the bill and negotiating with the EU Commission first. I knew only too well that this would spell the end of the bill. Things came to verbal blows between ministers and myself in the parliamentary-group session three days ahead of the envisaged final vote in parliament. I moved to bring the bill to the final vote unchanged and without delay, and won the vote by about 90%. The Act became effective on 1 April 2000. But the conflict was far from over. As early as 7 April, the EU Commission filed a lawsuit against the Act on which the European Court of Justice then had to pass a judgement. Eleven months later, on 13 March 2001, the Court of Justice dismissed the lawsuit. This meant that the Renewable Energy Act was off the ground — along with PV.

When the Act was up and running, there was total installed PV capacity of 50 MW in Germany. By comparison: new installations in the year 2009 alone were more than 3000 MW. Not only the German PV sector, but the worldwide PV industry, including America's, China's and Japan's, owe their upswing to the German market — and all of them are indebted to the Act for the last decade's rapid falls in prices, which are now already in the neighbourhood of grid parity. To get this far — always by way of parliamentary initiatives with ever-new joint thrusts along with Hans-Josef Fell and a growing number of other MPs, and always against the resistance of the electricity industry and one-dimensional economic-science institutions — we successively lifted the restrictions on PV in the Renewable Energy Act: first from 300 MW to 750 MW and then — starting in 2009 — without any quantity limitations. In the Bundestag, there was a EUROSO-LAR group of parliamentarians.

This also helped get the PV industry off the ground. I remember the European photovoltaic conference I chaired in Glasgow in May 2000. This conference was held after the start of the 100 000 Roofs Photovoltaic Programme and the Renewable Energy Act. Some of the PV firms attending the event showed initial reserve. For years they had been accustomed to producing for market niches, and they were used to that. Now they, too, were being called upon to act: with new investment in a market that was now widening up. But all too often, they had experienced political "go-and-stop" programmes. They didn't trust the new opportunities and suspected that another political stop was just around the corner. There were also many political attempts being made to halt this development again. They are still being made — even in Germany.

Political advocates in parliament have again and again successfully countered such attempts. For this, action was needed on all fronts of the conflict: at the political front in parliament, in the media, and also on the legal front. The energy companies tried for years to bypass the legal provisions. There were countless lawsuits before the courts. For this reason, it had become important to change the prevailing opinion among legal scholars and in legal literature. This prompted me in 1997, along with several lawyers and law professors, to found the law periodical *Zeitschrift für Neues Energierecht* (Magazine for New Energy Law), which has been appearing ever since. It has made a crucial contribution toward changing the legal culture in favour of renewable energies.

Our most important ally in the course of time was public opinion, which was now listening more to the many grassroots initiatives than to the warnings of the big electricity companies. The latter spent many millions on their disinformation campaigns to scare people away from renewable energies. We fought them with a better idea: with the higher humanitarian value of renewable energies for society. The PV systems on countless roofs that all of us see everywhere convey a new hope. It is the hope that many people can take their energy future into their own hands — and that they need not wait for the results of global climate negotiations. It is the now tangible opportunity for energy emancipation from big, anonymous structures. PV as energy for the people, by the people. This struggle is not yet over. Now that electricity companies have failed with their original strategy of ignoring or ridiculing PV, they are now busy developing a new strategy. Large solar power plants and large wind farms — wherever possible off-shore — are set to replace the big nuclear and fossil power stations. These companies want to preserve their supply monopoly. The future will be marked by the conflict between centralised and decentralised systems. The companies were able to prevail for a century with their big centralised power plants because they could generate electricity at lower cost thanks to the large quantities generated. But these economies of scale cannot be ported to PV: for PV, the economies come from increasing quantities in material and cell production — irrespective of whether the modules are for decentralised installation on roofs or centralised on large surfaces. Indeed, a large-scale industry for solar cells cannot prevent solar modules being installed on a decentralised basis. This means: neither technical nor economic barriers exist to energy emancipation, and any barrier that exists is a neutral one. Decentralised or centralised: that is the offset between autonomy or dependence, freedom or compulsion. This is why PV technology and decentralization cannot be stopped. It will come because the need for freedom will win the upper hand — at all events when its implementation does not harm our fellow beings, but is also for their benefit. That is why PV is an opportunity for humanity.

Introduction

This book is about power, electric power. Electricity is synonymous of light, it is something precious. Our whole civilisation is built on it.

Electricity for lighting became in general use only in the 20th century — when industrialisation started in earnest in the 19th century it took place in the 'dark'.

After the electricity revolution that was started at the very end of the 19th century by Thomas Edison in New York, our 21st century that we just started will bring us another revolution: in the course of this very century our electricity will have become 'Solar Power'. The 'oil age' that started in 1848 in Baku will be coming to its definitive end in the foreseeable future. It will be displaced by the 'Solar Age' for a more sustainable World.

The coming World of Solar Electricity will be dominated by 'Photovoltaics, PV'. Just some years ago you would have immediately stopped going further on in this text: understanding electricity is already difficult enough — PV must then be something for the very specialised engineer.

By now things have changed. Millions of ordinary people from around the World have become involved in PV as investors, producers or buyers; over a hundred thousand new jobs were newly created in a global market in just a few years' time. Some people even got very rich with it.

And this is just the beginning: the prospects for PV are tremendous; global business will soon reach 100 billion USD per year and more.

The Photovoltaic effect was already discovered in Paris in 1839 long before conventional electricity started to conquer the World. But it remained a dream of some enthusiasts. By now in 2010, with PV's extraordinary market success, the time of the PV pioneers — in the technical, ethical, political, financial, industrial, commercial, or social sense — is coming to an end. Hence, it was timely to make the effort to keep at least some of it on record in this book.

But this book cannot more than to shed a little light on something tremendously big. You need an encyclopaedia to keep a full record of all the things that happened in PV and that are more than ever in progress: PV Conferences almost every day somewhere around the globe, newsletters, and meetings — the 'CO_2 footprint', as it is called by the environmentalists, is not so good for PV these days.

I was personally involved in PV since 1961 and never gave up my enthusiasm for it and never stopped working for its development. Already my thesis work as

Baku where the 'Oil Age' began in 1848. Picture taken in 2009 with W.Palz on the left, Walter Sandtner on the right and colleagues.

a physicist at Karlsruhe University in Germany was on PV. So today I have become one of the oldest experts still active in the PV business.

This book has been written for the interested layman; a complicated story made simple and entertaining. Obviously I had to include my own experience on PV — but this is by no means the story of my life; that would again be another story...

The PV experts, too, can learn in this book some facts and figures they cannot find on the internet or in the specialised meetings they are used to attend. I hope they will enjoy reading and find themselves more than gratified to have chosen PV for their professional life.

A great asset of the book are the original contributions from many pioneers around the World who were and still are convinced actors and fighters for our all ethical endeavour: Solar Power.

These texts are not edited and were contributed under the authors' own responsibility. They provide a lot of colour to our book, not only because of the many nice pictures that illustrate their own experience.

I am more than grateful that all these colleagues and friends accepted my invitation to make the effort to remember all their 'deeds'. I am sure you will have fun reading these articles that give a taste about PV also from angles you would never have thought of: what is the relation of PV with the cathedral in Cologne; or what is the relation of PV with piano builders in Cincinatti?

I also take this opportunity to thank the publisher in Singapore to put all that together. Without his insistence I would not have done it. And eventually I had great fun doing it.

Chapter 1

Part I

The Rising Sun in a Developing World

1. Electric Power, a Pillar of Modern Society

1.1 *Electricity in Today's Life*

We are so used to the electricity around us that we could hardly imagine a life without it. Powering communications, lighting, appliances and industrial processes, electricity is everywhere. Even in heating and transport sectors, electricity is making inroads: electric heat pumps and electric or hybrid automobiles combining a combustion engine and an electric motor are currently attracting enormous interest.

It is true, however, that without electric power we would not go back to the Stone Age, but rather to the end of the 19th century — before then, nobody had electricity (well, as we will see later, just a little for communication purposes). This is only 130 years ago. The fountains of Versailles had to work without electricity; even the elevators of the Eiffel Tower in Paris can run without electric power because it did not exist at the end of the 19th century when it was built.

Modern civilisation would not be what it is today without the availability of electric power; it makes our lives a lot more convenient and comfortable. And electricity is also one of the reasons we live longer and healthier lives — just consider the electric equipment used in hospitals and dental clinics for instance.

1.2 *The Conventional World of Electricity*

Electric energy is a good and has to be paid for by every user. The bill comes in units of kilowatt hours (kWh); the unit cost varies from country to country and ranges between 10 and 35 US cents, including taxes.

Power for the World by W. Palz
Copyright © 2011 by Pan Stanford Publishing Pte Ltd
www.panstanford.com
978-981-4303-37-8

Electric power is the capacity installed to deliver this energy. In a home it amounts to a few kilowatts (kW), depending on the size and the consumption level of the family. Electricity is currently produced in central power plants far away from its consumers. The power capacity of such plants may reach up to 5 million kW, with up to 1 million kW per single block.

If the nameplate capacity of a power plant is 1 million kW, it will generate, when operated, 1 million kWh of electricity in one hour. The layman is often confused about the difference between kW and kWh. Consider a cow and the milk it produces: there are cows that can produce a lot of milk, and others that have a smaller capacity; that capacity can be compared with the capacity of a power plant to produce electricity, measured in kW. The electric energy produced corresponds to the milk; that quantity is measured in kWh.

Most of the conventional fuels that are at present employed to generate electric energy are lost in the conversion process, not only because of the "Carnot efficiency laws" but also because they are not optimised. Currently, half of global greenhouse gas emissions are produced by power plants. Most of the electricity produced is lost or wasted, too, and only a fraction is actually used.

Electricity transmission is a major source of the wastage. Transmission lines are costly to build, adversely impact the landscape tremendously, affect the health of people living nearby, and occupy huge areas of land. Electricity is presently transported back and forth in huge quantities over long distances across the globe's international networks.

Trade in electricity depends on the prices quoted on specialised exchanges for electric energy, and these may vary hour by hour. One example is Switzerland, where hydroelectricity is available to cover 60 percent of national consumption — but half of that is exported. The result is that consumers in Germany, say, may claim that they only consume electricity from renewable sources, while Swiss consumers, without knowing what is going on, are getting imported electricity generated by nuclear and coal-fired power stations.

You may want to call this the absurdity of the liberalised energy market. However, in the conventional energy system we have today, it makes sense since it may be better to transport great distances electricity produced from a nuclear source — which cannot be modulated easily for technical reasons — to spare stored hydroelectricity at a particular point in time instead of wasting it. We will see later why it is so important to promote decentralised solar power to reduce the need for power lines.

At the consumer level, much progress has been made to reduce energy wastage and inefficiency in electric appliances. This has been achieved both through technological advancements and by encouraging energy efficiency through labelling on products. From everything from light bulbs to entire buildings, the trend has been on increasing awareness and reducing energy consumption.

Today, some 4500 million kW of electric power capacity is installed worldwide. The US, China and the European Union are the largest consumers, accounting for more than 20 percent each of the total electric energy produced. The majority of today's power plants are fuelled by coal, oil or gas; equal shares from large hydro and atomic power plants make up the remainder. Wind power capacity installed worldwide amounts to 150 million kW; all of it was installed during the last 15 years. The current total PV capacity — another newcomer and the subject of this book — has reached just 20 million kW to date. There is a clear trend: over the last few years PV and wind power have enjoyed the highest growth rates globally, and nuclear has expanded the least.

More than 99 percent of global electricity is presently produced via mechanical energy conversion. As a rule, generators are driven by turbines, which is even the case with the cleaner renewable power systems such as solar thermal plants, geothermal power plants, wind farms and bio-power plants of various types.

One should note that mechanical power was available to man long before electricity was "invented". Even before the development of the steam engine, which eventually led to the industrialisation of the 19th century in Britain, Germany and the US, and beyond, mechanical power was used in wind — and water mills. And let us not forget the power generated by horses: even now, mechanical power is often measured in horsepower.

By comparison, electric energy was the "nobility" among energies, and it took time for it to become accepted by the market. Railway trains were powered for almost a hundred years by mechanical engines before they became electrical — and a lot cleaner. Displacing the fuel-driven engine in cars is proving even more difficult and may never succeed completely. Indeed, one thing is already clear: direct solar PV will never make it into the aircraft sector, while derived forms of energy will.

1.3 *Solar PV: A Part of the New Semiconductor World*

We should emphasise at this point that photovoltaic solar power is the only source of electrical energy generation that can do without a mechanical engine: it employs a semiconductor and the Sun's rays as the only fuel. Is it the action of light, "the visible hands" of the Sun, the sustainer of life, that convinced the Vatican in Rome to put as much PV power on its buildings as possible?

It's clear that PV stands firmly at the intersection of modern electric power and semiconductor technology. Semiconductors have developed, next to electricity, into another pillar of modern society. As a semiconductor physicist, I observe this with satisfaction; when I was a student, semiconductors occupied niche markets only. Eventually, amplifiers and television sets did away with their vacuum tubes; mechanical switches and relays were displaced by electronic chips; monster-sized computers were miniaturised; telephone connections became cheap. Semiconductors were at the forefront of the social revolution that occurred worldwide with the emergence of cellular phones and the Internet.

Semiconductors are now everywhere: in LED lamps, displays, cameras, TVs, computers and solid-state lasers.

PV is an integral part of the semiconductor world. Today's global production volume of silicon solar cells has already equalled the amount of silicon employed for electronic chips. Optoelectronic devices that employ compounds different from silicon, as far as they are concerned, belong to the same family of semiconductor materials as the high-efficiency solar cells that are needed for applications in generators for space satellites and for PV concentrator systems. It is also worth pointing out that all the many space applications we have today — satellites for communication, television, the Internet, meteorology, GPS, not to mention spy satellites — would not be possible without PV. In space, for a long time, PV has had no competition.

2. Looking Back to Light the Future

2.1 *The Emergence of Electricity*

The modern age of electricity started on September 4, 1882. That day in New York, Thomas Edison switched on the world's first power plant, with a capacity of 600 kW. Not only had Edison's company created a power plant, but he had put together everything that goes with it: a cable network for distribution, electricity counters, fuses, and a network of consumers. The electricity was exclusively utilised to power some electric light bulbs. That was what Edison was really interested in promoting. He had spent years developing a commercially viable incandescent bulb, something that was completely new for the time. The event became the subject of enormous public interest: ordinary people began to understand the future possibilities of electric power.

To this day, it is not uncommon for electricity providers to have "power and light" and "Edison" in their names. When people refer to electricity in some countries, they speak about light; after all, electricity is something abstract, something one cannot see, while light is the first application people are interested in. Before that momentous day, petroleum lamps were employed for most indoor uses, and gas lamps provided street lighting. After Edison unveiled his first power plant, electricity went global very quickly: Paris had its first power plant in 1888; by 1900, the Parisian Metro had already gone electric.

It is important to realise how recent the availability of electricity to us really is. When my own father was born, at the end of the 19th century, the family had no electricity — because nobody anywhere had electricity! Even today, with nearly 7 billion of us, one in four still have no electricity, neither for lighting nor comfort — not even for basic survival. We will return to this subject later in this book.

Edison's electric generator was powered by a steam engine. He built this generator, called a dynamo, himself. It was the central piece of the whole enterprise. But it was not his invention. The generator had many fathers, starting with Michael Faraday, who built a small prototype in 1831 in England. Later, a generator was developed in Belgium that was used in a limited way as early as 1849 for the lights of lighthouses and to produce a hydrogen fuel.

An important milestone in this development was the invention of the dynamo-electric effect in 1866 by Werner von Siemens. It was he who built the first dynamo. It did not work well enough, however, and it enjoyed some commercial success in the 1870s in France only after the Belgian Gramme had improved it. It is interesting to note that, up until the present, the companies formed by Werner von Siemens in Germany under his own name and the General Electric company (GE) created by Edison in the US have maintained a dominant position in the global electricity sector.

Before electricity revolutionised the lighting sector, it laid the basis for another strategic sector: communication. In concrete terms, it was the telegraph, running on "low-power" electricity, that brought countries and people closer together, from the mid-1800s onwards. Electric signals were transmitted through cables. (Wireless communication became possible only after Heinrich Hertz discovered electromagnetic waves in 1887. I am proud to say that when studying physics in Karlsruhe, Germany, the auditorium we used was Hertz's former laboratory.) The source of electricity for that purpose was a Galvanic cell, or "Voltaic pile". The electrochemical battery, put together around the year 1800, was actually the first

Figure 1. The beginning of electricity in the service of mankind: Volta's electrochemical pile of zinc and silver pallets as shown in 1800 to Napoleon, a big event in those days. The picture illustrates why it is called a "pile". Accordingly, some French people still refer to a solar cell as a "photopile".

invention of electricity for man's needs. The inventor was the Italian physicist Alessandro Volta.

Thus, the first important use of electricity was associated with the development of cable networks for the telegraph. The immediate transmission of information was a revolution at that time. Siemens, who had created a telegraph company in 1847, first had to connect Berlin, the residence of the Prussian Government, with Frankfurt, the seat of the German National Assembly. Later, he was involved in laying cables from London via Berlin all the way to India, and then underwater from Europe to the US. Edison also started his career in the telegraph business. As early as the 1870s, the telegraph was welcomed by stock market speculators in the US, in particular the gold market.

2.2 *From the "Voltaic Pile" to the Photovoltaic Cell*

The photovoltaic effect was first observed in Paris in 1839 when a young man played around with a Voltaic cell. That year, just 19 years old, Edmond Becquerel reported in *Les Comptes Rendus de l'Académie des Sciences* "the production of an electric current when two plates of platinum or gold diving in an acid, neutral, or alkaline solution are exposed in an uneven way to solar radiation."

At that time, Edmond Becquerel was an assistant to his father, Antoine, who had been appointed, the year before, to the chair of applied physics at the Muséum National d'Histoire Naturelle in Paris. (Becquerel is also a famous name in nuclear physics — in association not with Edmond, but with his son Henry, who received a Nobel Prize with the Curies.) The building still exists today and I am currently living within walking distance from it in central Paris. In 1868, Edmond Becquerel published the book *La lumière, ses causes et ses effets* (Light, its origins and its effects).

Figure 2.   Edmond Becquerel, the discoverer of the PV effect in 1839.

The BBC in London — surprisingly, not French TV — produced a programme to mark the 150th anniversary of Becquerel's discovery that was inspired by the late Prof. Bob Hill, a pioneer of PV research. That same year, on his suggestion, I established, as an official of the EU Commission in Brussels, the "European Becquerel Prize for Merits in Photovoltaics". Since then, the prize has been awarded, on a yearly basis, to the VIPs in European PV research — though not exclusively: among the winners there are also two Americans, a Japanese, and a Russian.

I should mention just a few of the outstanding winners: the late Roger van Overstraeten from Leuven in Belgium, a wonderful friend of mine and chief adviser of the EU's R&D programme on PV under my responsibility; the late Werner Bloss from Stuttgart, a lifelong advocate of thin-film cells; Antonio Luque from Madrid, a great scientist working on solar cells and solar concentrators; Morton Prince from the US, a member of the team that developed the first silicon solar cells in 1954 and long-time manager of the PV programme of the US Administration in Washington; the late Karlheinz Krebs, an EU official from the Joint Research Centre in Italy, Europe's key promoter of PV certification; Adolf Goetzberger, the founder of the Fraunhofer Institute for Solar Energy Systems (ISE), the first solar energy institute in Germany and an important promoter of the International Solar Energy Society (ISES), the world's oldest solar energy association; Walter Sandtner, long-time head of the PV programme in the German Ministry of Research and Technology and initiator of the 1000 PV Roof Programme; Viacheslav Andreev from St. Petersburg, Russia, a world leader in high-efficiency cells; Masafumi Yamaguchi from Toyota in Japan, a record holder in solar cell efficiency; Richard Swanson from the US, a silicon solar cell pioneer and record holder in silicon cell efficiency; and Mechtild Rothe, a prominent Member of the European Parliament and great fighter for PV and renewable energies in the political arena.

The Becquerel Prize is solemnly awarded at the Opening Ceremony of the European Photovoltaic Solar Energy Conferences. These yearly conferences are currently the largest specialist meetings on PV in the world. With more than 4000 registered participants from 75 nations and 50 000 visitors to the associated exhibition attending, it is bigger than events like the America's Cup! This series of conferences was set up in 1977 by a group from the EU Commission in Brussels: Dr. Günter Schuster, the Director General for R&D; Albert Strub, the Director for Energy R&D; Roger van Overstraeten, our Chief Advisor; and myself. Later, Peter Helm became the driving force behind developing the conference into the world's leading event in PV. The field owes much to Becquerel, the Becquerel Prize, the Prize winners, and the European Conferences that serve as a forum for them.

In 1849, ten years after its discovery by Becquerel, Alfred Smee coined the term "photovoltaic" effect in London: "Upon exposing the apparatus to intense light, the galvanometer was instantly deflected showing that the light had set in motion a voltaic current which I propose to call a photovoltaic circuit".

The photovoltaic effect in solid-state matter, the semiconductor selenium, was first found by Charles Fritts in New York in 1883. He produced the first solar cell by coating the selenium with a transparent gold layer. His work had been preceded in the 1870s by the work of W. Smith, W. G. Adams and R. E. Day in Britain, who discovered photoelectricity in selenium.

The big question that remained was understanding and interpreting photo-electricity and photovoltaics. The knowledge we have acquired since those days is that photo-electricity is due to the action of the photons of light: each incident and absorbed photon creates one mobile electron that contributes to electric conductivity in the solid. In addition, a "barrier" layer is necessary for charge separation to create a voltage. The barrier layer can be "built in" by doping the semiconductor, a process that is employed not only in solar cells but also in all the active elements of a silicon or other type of semiconductor chip. The barrier layer appears also through contact with a metal or some other type of solid matter. In Fritts' experiment it was gold.

Only in 1897 was the existence of the electron proven by Joseph John Thomson. But previously, various authors, such as Heinrich Hertz or Wilhelm Hallwacks in Germany, had worked on the "external photo-effect", the emission of electrons by the action of light. In 1900, Philip Lenard found that the speed of emitted electrons is a direct function of the frequency of the light. That same year brought a discovery of fundamental importance: that light is composed of individual particles or quanta, called photons, and that their energy is directly proportional to the frequency of the radiation. That discovery was due to another German, Max Planck.

One only had to combine these discoveries to understand photoconductivity. One reason why Albert Einstein got the Nobel Prize in 1921 was his interpretation of the external photo-effect, the "emission of electrons" by the action of radiation. But this was not yet the interpretation of photovoltaics. As already mentioned, that also has to do with charge separation of the light-induced electrons by an internal field effect inside the semiconductor.

There was a time when electron emission found important industrial applications: one should keep in mind that long before semiconductor chips acquired the role they have today — the transistor was developed only in 1954 — the key component of electronics was the vacuum tube. Its operation is based on the cathode emission of electrons, though not by radiation but by heat.

An interpretation of the photovoltaic effect in semiconductor/metal contacts was given by Walter Schottky at Siemens in Germany in 1930. Everything is connected: Schottky was a doctoral student of Max Planck. "Schottky diodes" are still extensively used today. The preferred semiconductors of those days were selenium and cuprous oxide. Silicon, which currently dominates the solar cell market as well as most of the semiconductor chip market, was not yet available in sufficiently high purity at that time.

The understanding of the p-n junction, the classical cell produced by the doping of the pure semiconductor, is due to Russell Ohl, working at Bell

Laboratories in New Jersey, US. In 1941, he filed the patent for the modern solar cell; the patent was granted in the US five years later. He produced some silicon cells himself, but the efficiency was too low to be of any practical use. Achieving silicon crystals of sufficient purity had to wait for the work of Calvin Fuller, also at Bell Labs. Doping them with boron led to the first practical silicon solar cell in 1953. Associated with his work were Daryl Chapin, Gerald Pearson, Walter Brattain, Morton Prince, and William Shockley, all at Bell Labs.

Public opinion in those days understood well the importance of the emergence of the first practical solar cell: when the invention became known in 1954, the *New York Times* had it on the front page.

2.3 Photovoltaic Power: The First Steps

Some people claim that Werner von Siemens was the first to understand the importance of PV for electricity generation on a large scale. I glanced through Siemens' autobiography myself and could not find any mention of PV. Siemens was interested in the dynamo generator, of which he discovered the principle, in the Stirling engine, and then oil exploration when coming to Baku — but not PV. In his day, the solid-state PV battery had not even been invented.

Still, it is true that in the second half of the 19th century, as coal and oil has just begun to displace wood and conquer the energy markets, people expressed concern that these fossil resources were limited. This may explain why the Frenchman Mouchot's solar steam engine attracted enormous public interest: in 1866 he was invited to show his invention to the French Emperor, and 12 years later at the Paris Universal Exhibition. After Hermann Scheer founded the EUROSOLAR association in Bonn, he established the "Mouchot Prize", and in 1991 I became the first to receive it. It should also be emphasised that Mouchot's steam engine produced only mechanical power, not electricity.

My friend Prof. Biswajit Ghosh from Calcutta (Kolkata) discovered a reference from 1891 that was one of the first visions of PV for power generation. R. Appleyard had written in the *Telegraphic Journal*: "The blessed vision of the Sun ... by means of photo-electric cells ... these powers gathered into electrical storehouses to the total extinction of the steam engine and the utter repression of smoke". Even before then, in 1886, Charles Fritts, the inventor of the first "solid-state" solar cell mentioned above, was quoted as saying: "Solar energy will pour down on us long after we have run out of fossil fuel".

Albert Einstein is sometimes cited as the originator of PV power. The fact is that he contributed to the understanding of the effect of light on solids, but the energy he was interested in was not that of everyday life. Rather, he was one of the fathers of atomic energy and weapons: his discoveries explained the annihilation of matter and its conversion into hot energy.

PV became a serious business only after the pioneers at Bell Labs had produced silicon solar cells of a revolutionary 6-percent conversion efficiency.

Applications headed in two very different directions: a small market for toys in the US, and a much more significant market for space satellites. In the race against the Soviet Union, the US launched the first satellite with solar panels in March 1958, Vanguard I. The silicon cells came from Hoffman Electronics.

3. Solar Power for Space Satellites

PV generators have a unique advantage for space applications: their low weight. Weight, rather than cost, is the determining factor for satellites. PV generators get the Sun's fuel for free; hence, all Earth-orbiting satellites employ PV associated with a storage battery. This is also true for the inhabited International Space Station. Only for interplanetary missions are generators employing radioactive materials preferred; only 1 percent of the unmanned space vehicles of that time employed radioisotope thermoelectric generators. The Apollo missions to the Moon were powered with fuel cells.

Most of the first satellites in space had scientific research objectives. Correspondingly, the President of the National Space Administration in France (CNES), with which I was associated from 1966 until 1976, was a pure scientist. At that time, I was in charge of the energy department at the Programme Directorate at the CNES Headquarters in Paris.

One scientific programme we worked on was Eole: 500 balloons equipped with tiny solar-powered radio emitters were to be released in the southern hemisphere and their journey on the wind monitored by satellite. We had to develop thin-film solar cells on plastic for the balloons; we chose a flexible cuprous sulphide (CdS/Cu2S) cell on a plastic sheet called Kapton. We will say more about these cells later.

One of the first important commercial applications of satellites emerged during the 1960s: intercontinental telecommunications. The capacity of the transmission through the cables under the sea was insufficient for TV programmes. The transmission of the Olympic Games became a primary reason for the installation of telecommunication satellites in geostationary orbit, that is, an orbit such that the satellite appears in a fixed position from locations on the surface of the Earth.

An inter-governmental agency was created for that purpose, INTELSAT, with its headquarters in Washington DC. Comsat, also in Washington DC, was selected as the first company to implement programmes, i.e., to install and operate the satellites, initially over the Atlantic between Europe and the US. They were used for TV and telephone transmission.

I succeeded in establishing a cooperation agreement between Comsat and CNES, my employer. It concerned the development of a solar generator that could be deployed in space once the satellite reached its orbit. The deployment structure was a pantograph like the one currently used on the solar generators of the

Something New Under the Sun. It's the Bell Solar Battery, made of thin discs of specially treated silicon, an ingredient of common sand. It converts the sun's rays directly into usable amounts of electricity. Simple and trouble-free. (The storage batteries beside the solar battery store up its electricity for night use.)

Bell System Solar Battery Converts Sun's Rays into Electricity!

Bell Telephone Laboratories invention has great possibilities for telephone service and for all mankind

Ever since Archimedes, men have been searching for the secret of the sun.

For it is known that the same kindly rays that help the flowers and the grains and the fruits to grow also send us almost limitless power. It is nearly as much every three days as in all known reserves of coal, oil and uranium.

If this energy could be put to use — there would be enough to turn every wheel and light every lamp that mankind would ever need.

The dream of ages has been brought closer by the Bell System Solar Battery. It was invented at the Bell Telephone Laboratories after long research and first announced in 1954. Since then its efficiency has been doubled and its usefulness extended.

There's still much to be done before the battery's possibilities in telephony and for other uses are fully developed. But a good and pioneering start has been made.

The progress so far is like the opening of a door through which we can glimpse exciting new things for the future. Great benefits for telephone users and for all mankind may come from this forward step in putting the energy of the sun to practical use.

BELL TELEPHONE SYSTEM

An advertisement published in the US National Geographic Magazine, Washington DC, August 1956

International Space Station. A 1-kW generator was built and tested jointly at Aérospatiale (now part of the European Aeronautic Defence and Space Company, EADS) in Cannes on the Côte d'Azur.

This was intended to prepare for the larger power needs of a new generation of three-axis stabilised satellites. To stabilise satellites in their orbits, most of the first vehicles in space were permanently rotating: the spinning satellites. Larger structures, which are too difficult to have spinning, are kept stable with an internal flywheel that has the same effect. Spinning satellites with the PV generator on the envelope were expected to get too big to accommodate a generator of sufficient power on their limited surface area. This was thought to be an issue particularly for the direct TV satellite broadcasts picked up by dish antennas fixed to people's houses. At the time, that was only a vision, but one wanted to be well prepared for the future. At CNES, we had a working group to prepare for the three-axis satellites. It was coordinated by Maurice Fournet, who had previously developed the laser reflector on the Soviet Moon vehicle Lunokhod. (This optical reflector retransmitted a laser beam sent from Earth exactly to the origin, whatever the position of the vehicle on the Moon.)

It turned out that the need for ever-larger power capacity in TV satellites was by and large compensated by progress in electronics; the space sector in the 1960s was populated by the pioneering developers of the age of digital electronics and communication. As a result, over time the need for power diminished, as digital electronics became considerably more energy efficient than analogue systems. What has ultimately been achieved is quite extraordinary: with just a few Watts — not even a kilowatt — of electronic emission, modern satellites, positioned 36 000 kilometres above the equator, can provide direct TV coverage for a whole continent of consumers.

Such satellites are also suitable for interactive Internet transmission. Their application is not yet widespread, as operators want first to amortise the dense cable networks they have already installed. But by switching to satellite transmission, it is possible to save a lot of expensive copper cables. Correspondingly, spinning satellites stayed in use for much longer than we had initially thought; three-axis stabilised satellites with large deployable solar paddles were needed only later with the arrival of big projects.

With the end of the US Apollo programme to go on the Moon in the early 1970s, there was a degree of disillusionment in the space business in the US and Europe. CNES' marketing people asked all of its potential clients in France whether they saw a need for satellite services. The answer was a resounding "no"; the French meteorological service claimed that, if necessary, aeroplanes were sufficient to get the job done. The public TV broadcasting authority suggested that French people — their customers — did not want more than the single channel that they had, or perhaps at most just a second regional one. The situation was similar to what is encountered in the energy sector worldwide: when it comes to the future prospects of a sector for which such people are responsible, the response is inadequate. Politicians hate any kind of vision and professional managers prefer to stick to what they are used to; they are too conservative and get it wrong.

As we now know, there are so many applications for satellites that space is crowded with satellites and debris, in particular geostationary/synchronous orbits. But things developed slowly — even in the US. By 1971, 600 US and 400 Soviet spacecraft had been placed in orbit; of those, only 50 US satellites with no more than a combined power of 7 kW had been placed in synchronous orbit.

Virtually all the satellites launched during those years were equipped with silicon solar cells. But they are actually not ideally suited for the harsh environment outside the Earth's atmosphere. There, the cells are bombarded with particles from the solar wind and the Van Allen radiation belt, which affects the active layer of the cells, degrading efficiency. Silicon solar cells are just five times thicker than a human hair. But thin-film cells like the cuprous sulphide mentioned above are even a great deal thinner — actually 50 times thinner than a human hair — and accordingly are much less degraded in space. In the 1960s and 70s, like many other groups around the world, we worked on such thin cells in France. As mentioned, people like Schottky had already worked in the 1930s on a similar material, cuprous oxide.

I went to NASA to see the late Bill Cherry — the same man to whom the American IEEE Conferences on PV would pay their greatest respects by creating the Cherry Award after his passing. I asked him whether he could try using some of our Cu_2S cells in his satellites. His polite reply was that there was no need for cells other than silicon ones unless they were better and cheaper. I was not one to give up easily, and we eventually succeeded in getting a special small test satellite for our Cu_2S cells, SRET, which was put in orbit by the Soviets.

For most satellites in the US and Europe, silicon cells were displaced by cells of the GaAs family. Their active layer is also very thin, a fraction of that of a hair, and their efficiency is higher than that of silicon cells, in particular at the end-of-mission of a satellite, since they degrade much less readily. Why were they not employed earlier? The technology was less mature and they were more expensive.

4 First Ideas about Lighting with Solar Power

4.1 *Mutations of the Societies in the US and Europe*

Traditionally conservative, societies in the "West" adhered to the Christian Bible's rallying cry to dominate the world and to exploit nature to the fullest extent possible. And that was certainly the guiding principle in matters concerning energy. In the decades after World War II, electricity production and consumption grew by 7 percent a year; it doubled every 10 years. When the think tank The Club of Rome produced the book *The Limits to Growth* in the early 1970s, it was a real shock to the "establishment".

Those times are over. Most countries, and the world community as a whole, generally accept that ecological sustainability is necessary for man's wellbeing and survival. Nature must be seen as our partner, not our slave. As the disastrous experience with totalitarian regimes was fresh in everybody's mind, the freedom of the individual against the dominance of society became an important issue after WWII.

In 1949, George Orwell published *Nineteen Eighty-Four* about the risks of a totalitarian society masking itself as democratic. When West German rearmament was being debated in the 1950s, the country's intellectuals, unions, Christians and women's groups expressed their opposition in a protest movement called *ohne mich* (count me out). "Easter peace marches" have been held in Germany since 1960. In those early days, they were — unlike today — quite important popular movements.

Traditional authorities in the West have been challenged since the mid-60s, with the youth as the driving force. People protested against the Vietnam War; the hippy subculture began to spread; the Woodstock Festival of 1969 made an indelible mark on history. Also, Indian philosophies gained in popularity; the Beatles travelled to India to receive spiritual guidance; Hare Krishnas marched through the streets.

The events of 1968 in France, Germany, and many other countries defined a whole generation — actually, my generation. In May 1968, my home in the Latin Quarter of Paris was awash with street battles and barricades. The intellectual inspiration for those events came, inter alia, from Jean-Paul Sartre and the Jewish-American-German philosopher with the French name, Herbert Marcuse. Marcuse's book *One-Dimensional Man* became the basis for Germany's student revolt. From 1966 onwards, Marcuse addressed thousands of students in Heidelberg and Frankfurt. His subjects were the authority of resistance over industrial society, the need for the autonomy of the individual, the right to happiness. He reported to Germans about the student revolt in the US, which was triggered by the Vietnam War: "... a spontaneous unity ... the rebellion in attitudes, language, sexual morality, clothing ...". In France, the "May events", as they are called there, challenged the whole of society: President de Gaulle had to go; in the Rue de Grenelle in Paris, representatives of the government and trade unions negotiated the Grenelle Agreements in an effort to put an end to the strikes and solve the social crisis.

Power and energy were not issues during the fundamental changes in society at that time. But the debate can fairly be translated to the present day as a movement for energy autonomy. The trend is for people to master their electricity supply themselves, be it in their homes, in towns and cities, or in regions. The traditional notion that electricity as provided by big, anonymous companies simply comes from the wall socket is widely challenged today. And this is where solar power comes in: everybody has access to it, free of charge. It is the route to individual power supply.

Atomic energy is not a subject of this book; it is a big issue on its own, beyond our purpose here. But it is instructive to recall one particular situation in Germany in the 1980s concerning plans to build a nuclear reprocessing plant in the town of

Wackersdorf in Bavaria. So aggressive were the clashes between protesters and police guards that the project's chairman decided to shelve the plans as there was a danger of major civil unrest.

There is no risk of civil unrest with solar power.

4.2 A New Awareness for Solar Power

I am a PV specialist myself but I had never been familiar with the purpose of providing PV power for everybody's electricity needs until I met the late Martin Wolf in 1969. He had emigrated from Germany in the 1950s and became a key player in the early stages of academic and industrial development of PV in the US. He invited me to his home in Princeton — and opened my eyes. He and his friends had worked out the entire concept of PV power for all; I always thought the Americans knew the little secret of how to stimulate one's enthusiasm.

In June 1971, the US President issued a message calling for programmes to ensure adequate supplies of "clean" energy for the years ahead. Under the authority of the President's Office of Science and Technology, a Solar Energy Panel was jointly organised in 1972 by NASA and the National Science Foundation. Professor Martin Wolf was to chair the Sub-panel on Photovoltaics. The goal that was anticipated was a million-fold expansion of PV production rates and the attendant automation.

At the time, the US' official projection was of a five-fold increase in energy demand between 1969 and 2020, half of which would be provided by nuclear energy, the other half being provided by fossil energy. William Cherry, mentioned above, and Fred Morse presented the overall conclusions of the US panel at the following congress in Paris in 1973. Contrary to the existing projections regarding energy supply, the Panel recommended to the US Government a target of 20 percent solar energy contribution to the US' total energy needs by 2020, equivalent to the total energy consumed in the US in 1970. Twenty percent of the electricity needs should also be met by solar energy. Sounds familiar?

In July 1973, in Paris, UNESCO hosted a congress entitled "The Sun in the Service of Mankind". I was the organiser of the PV section of the meeting. The notion that the Sun was at man's service rather than his partner was typical for the time. More than 1000 delegates from all over the world, including a strong US delegation, made this congress an important milestone in solar power development. It was opened by a "Hymn to the Sun" presented in a speech by Pierre Auger, a famous semiconductor physicist.

In his message to the congress, Wernher von Braun, father of the Apollo Programme, declared, "I believe we are at the dawn of a new age, one which might be called the 'Solar Age'." A working party chaired by Felix Trombe, with prominent members from France, the US, India, Chile, Canada, Australia, Niger, Israel, Japan and the USSR, concluded: "Recent events in many countries make it apparent that a take-off point for large-scale solar energy development may be at hand. There is a widespread appreciation of the limited lifetime of fossil fuel supplies and in many countries there is also concern about the pollution engendered by the use of fossil and nuclear fuels. There is therefore a growing recognition that solar energy, as a renewable and non-polluting source, may play a major role in meeting future worldwide power needs." That clear vision was set out in 1973.

Three months after the congress in Paris, the prestigious California-based Jet Propulsion Laboratory, better known for their pioneering interplanetary missions in space, held a PV workshop at Cherry Hill, New Jersey. It became another important milestone leading up to the "honeymoon" of modern PV. Viewed from our present perspective in 2010, it is rather surprising that the projections of large-scale PV implementation expressed at that meeting of some 150 American experts were already so realistic. Those pioneers — among them, Paul Rappaport, who later became the founder of what is now the prestigious National Renewable Energy Laboratory (NREL) in Golden, Co.; Karl Böer; and Joseph Loferski — already understood several important things: in particular, that future cost decreases are dependent on production volumes.

The Cherry Hill workshop projected that for solar cells US\$0.50/Watt could be achieved at a manufacturing volume of 500 000 kW (0.5 GW), and US\$0.10/Watt at 50 million kW (50 GW). Taking inflation into account over the more than 37 years that elapsed since the time of that analysis, we are not far from our current understanding of the cost problem. In addition to silicon, the workshop recommended thin-film solar cells, such as $CdTe$, $CuInSe_2$, among others that became very successful in the 21st century. Currently, in 2010, producer of $CdTe$ modules First Solar, Inc. has become the global market leader for PV. $GaAs$ and optical concentration for PV were also on the workshop's list of recommendations.

US\$0.50/Watt per PV array (or US\$500/kW) was an important objective as, in those days, it was thought to be the price of nuclear power that alternative energies were expected to compete with. The role of nuclear power, which was then still favoured by energy engineers, illustrates the magnitude of the challenge ahead for PV. In 1971, Alvin Weinberg from Oak Ridge National Laboratory anticipated the future electrified world thus: 30 years after programme start, one would have to build a 5-GW nuclear fast-breeder reactor every day, at a cost of US\$2500 million. (Imagine that: the breeder was supposed to cost only US\$500/kW!) After 30 more years, one would need to build two reactors at US\$5000 million per day. The mountain of problems the first pioneers of PV saw before them did indeed appear sky high.

4.3 *The Oil-Price Shocks, and the Nuclear Disaster of 1986*

Looking back, it is amazing to realise that the first major international call for the mobilisation of photovoltaics, culminating in the initiatives of 1973 described above, actually came before the first oil price shock — which happened in the autumn of that year! Just like those that followed in 1979 and 1990, it was triggered by events in the Middle East.

One could claim that around 1970 the awareness of "Western" societies about the world's environmental and climate problems was already much developed; views then were actually not very different from those today. At a major 1974 press conference we gave with CNES in Paris about the need to push forward solar PV, climate change was indeed a big issue. In particular, journalists were keen to know how high sea levels would rise if ice masses melted in Greenland and elsewhere. The oil price shocks had enormous impacts on governments, industry, and society at large. One of their many outcomes was stimulating interest in renewable energies in general, and PV in particular.

The oil price shocks had political origins: they were not directly related to the limited global supplies of hydrocarbons but rather to their geopolitical concentration in the Middle East. Irrespective of that, however, right from the first oil exploitations in the mid-1800s, experts have always had underlying concerns about the limits of the world's fossil resources.

In the meantime, experts in oil and gas exploration have presented no consensus view about how much of our fossil resources are left for future needs. The question of "peak oil" — the maximum rate of global petroleum extraction — has great political and strategic implications; experts have been arguing for years whether exhausting half of the planet's total recoverable hydrocarbons has already occurred or if it will happen in five years. In the US, peak oil occurred 40 years ago. In any case, the discussion remains largely academic, since the general trend of the coming exhaustion of all of the world's oil and gas reserves in the course of this century is beyond question.

The International Energy Agency (IEA) in Paris, created in 1974 after the first oil price shock under the umbrella of the Organisation for Economic Co-operation and Development (OECD), for years made no mention in their publications — and in particular their annual "World Energy Outlook" — that fossil and nuclear fuels were not infinite. In optimistic projections of consumption towards the middle of this century, they saw that global energy demand would grow steadily and that this demand would primarily be met by the classical fossil energies; only marginal roles were given to renewable resources. Only when confronted with mounting evidence that this could not be true did the IEA start to lament the limited availability of fossil resources. Not so the US energy experts in 1973. They feared that a long-term fossil fuel shortage could start as early as 1985, and to some extent they were right: in particular in the US, most of the hydrocarbons in the ground had indeed been consumed by then. This was one of the important motivations for stressing the need to immediately begin looking for alternatives.

After the fallout from the second oil price shock in 1979 subsided, worldwide political interest in energy fizzled out. Instead, the driving force for sustainable alternative energies since the early 1980s became the global peace movements and the growing opposition to atomic energy, largely as a result of the lack of public appetite for atomic weaponry. More than anything else, however, the meltdown at the Chernobyl plant in Ukraine in 1986 underscored the realisation that new ways of generating energy must be sought. The disaster saved many renewable energy programmes from extinction at the last moment.

The 1990s began with the fall of the Berlin Wall, a new oil price shock, and a blossoming awareness of the need for a sustainable civilisation (the 1992 Earth Summit in Rio de Janeiro); among many others, these events witnessed the emergence of solar PV in commercial markets. More on all this later.

5. After the Vision: A Mountain of Challenges

5.1 *PV in the Starting Blocks in 1973*

What we had achieved by 1973 at the beginning of terrestrial PV development can be summarised thus: we did not have a great deal more than the know-how about the market of applications for satellites; the yearly global production by this small specialised industry did not exceed 10 kW a year (global production volumes are a hundred thousand times bigger today). Silicon solar cells of high quality and an efficiency of 14 percent (today's commercially available cells are not very much higher) could be mass-produced. In those days, cells were circular and much smaller in size than today.

As for the prices of silicon cells, the literature quotes numbers that sound rather fantastical to me. I personally telephoned the Sharp company in 1974, and I got an offer of US$100/W. But at the same time the new Solarex company (now BP) in Washington offered us at CNES in Paris cells for US$10/W — in writing. In those early days of development, we often heard that the price had to be reduced by three orders of magnitude, i.e., a thousand times. Wrong. As we now know, reducing it by a factor of 10, down to US$1/W, was good enough.

What ultimately determines the cost of solar electricity is the total system cost: the solar modules themselves, the cabling, the electronics and the control strategy, the mechanical support structures, possibly also some storage, and the interface with the user. At the time, however, there was not yet any relevant experience with complete systems, and the total cost was just speculation.

Thin-film CdS/Cu_2S, CdTe or CIS solar cells existed only as laboratory prototypes. The 6–8 percent conversion efficiencies achieved were already encouraging and are actually not much worse than those offered by today's commercial thin films. One should mention that amorphous silicon solar cells or microcrystalline silicon cells had not yet been invented.

5.2 *The Cost Problem: Technological Challenges*

Technological development is obviously a key for cost reduction. Most efforts targeted the crystalline silicon solar cell, which had always occupied the centre stage of PV interest since its first appearance in 1954. Electronically pure silicon was readily available, as it was the same material that was being used in the electronics industry in semiconductor "chips". Relatively big industrial plants are needed to convert the impure metallurgical silicon that is derived from sand; the main constituent of beach sand is silicon in its oxidised form.

For the purpose of solar conversion, the silicon purity is less stringent than for chips, and consequently, one could go for a less costly "solar grade" silicon. But that effort has been made only fairly recently, when the PV market took off from 2004 onwards. Initially, the chemical industry, such as Wacker in Germany, which was primarily interested in the chip market, was not prepared to meet the new demand. The supply of silicon feedstock on global markets became so tight that its price on the spot market in 2007 could reach US$500/kg, while production cost lay between US$10–20/kg. It took time for the necessary developments to take place before new plants around the world brought silicon of electronic and solar grades to market in sufficiently large quantities.

Solar cell technology also had to be developed further in order to simplify production and reduce cost. Silicon crystals were pulled and sliced; a wire saw was created to produce thinner slices without too much "kerf loss". "Doping" of the material had to be optimised. Optical absorption inside the crystal and optical surface properties were improved. Anti-reflection coating was an important topic of research. For the electrical contacts, technologies with cheaper metals were needed; the option of screen printing was explored.

Automation was also needed to reduce the cost of the whole cell-production process. More recently, this has developed into a key business in the global PV market. Automation is also important in countries with high labour cost when it comes to module production, as cells must be interconnected electrically with metal strips. In a solar module, one must also optimise the mechanical properties of the glass support, the optical properties of the glass cover, and the plastic materials that glue the whole thing together.

For their part, thin-film solar cells employ deposition on a substrate; they are too thin to be cut from a solid crystal in the way it is done for silicon. Various deposition methods were studied, including chemical vapour deposition (CVD), screen printing, evaporation, and spraying. The choice of substrate, the metal contacts, and encapsulation are also important for thin-film solar cells.

As the technology of thin-film cells is, as a rule, somewhat simpler than that of mono- or semi-crystalline silicon cells, prospects for reducing cost were always considered more promising. Accordingly, despite their lower conversion efficiency, cost in terms of US$/W for a module using thin-film cells can be lower. Still, this cost advantage can be partly eroded by the need for more substantial support structures: as a result of their lower efficiency, a larger area is required to generate a unit of power.

As far as concentrating PV is concerned, the family of GaAs solar cells was developed further from the 1976 laboratory prototypes with their efficiencies of 22 percent. By 2009, an impressive efficiency of 42 percent was announced by

Spectrolab — a record for this kind of material. It is interesting to note that among the hundreds of PV manufacturers today that particular company in the US is one of the very oldest.

What limiting factors exist for producing inexpensive PV? The principal one is the cost of the glass used in the modules; there is a lot of glass employed in mechanical support and encapsulation. The invention of the float glass process for flat glass manufacturing — the Pilkington process — has helped to minimise that particular cost.

Once, when I gave a presentation on the future of PV in the 1970s, an electrical engineer from a power utility in France remarked that even if the cost of the solar cells were zero, PV could not compete with other sources of electricity because of the cost of the rest of the PV system — cabling, electronics and structural support. Others pointed out that the cost of a greenhouse structure — something similar is needed for PV installed in the field — is already prohibitively high. In response, we suggested that there were mechanical structures, such as those used in vineyards, that could be a lot cheaper. My colleague Giuliano Grassi at the EU Commission actually built and tried out a small PV test field on the vineyard he owned in Chianti, Tuscany in Italy.

The inverter that converts the DC of a PV array into AC, as is now used everywhere, was another unknown parameter in PV development. A new initiative came later from Werner Kleinkauf in Kassel, Germany, who founded the firm SMA to develop the technology needed for modern inverters. Before the PV market was established, SMA grew through its business with the German railway company, creating inverters for rail coach air-conditioning systems. Today, it is one of the global market leaders in PV.

For optical concentrating PV systems, Fresnel lenses, mirrors and structural orientation devices were needed; these kinds of systems were the last to get the necessary attention in PV development.

As a rule of thumb, it turns out that in terms of the cost of a complete PV system ready to produce electricity, half goes to the solar PV modules and the other half goes to the rest, or the "balance of system" (BOS) as the Americans call it.

5.3 *The Chicken and Egg Problem: Mass Production*

Experience shows that even with the best technology one cannot achieve the low costs that are needed to serve large markets where competition is high without mass production. The "learning curve" describes how prices come down as the market volume of a particular product increases. It occurs naturally when you have a novel product everybody wants to have; that was the case, for instance, with laptop computers and Internet access. The situation is different for PV in a market that is already populated with much cheaper providers of the same product — electricity.

In off-grid applications in many cases, PV providers could compete economically even before mass production because the conventional providers can meet those particular needs only at high cost. Unfortunately for PV, electricity from the grid is currently already provided almost everywhere in the industrialised countries, as well as in China and Russia, and so on. This particular market turned out to be insufficient to initiate the mass production of PV and lower production costs.

At the same time, World Bank and United Nations estimates have suggested that more than one-and-a-half billion people in the developing world currently still lack access to conventional electricity through the grid. In most of these cases, PV would be — even without considering the cost advantages of mass production — the smartest and most cost-effective way of solving this problem. We have attempted to put the case for PV — a major part of this book will illustrate in detail what various PV pioneers have tried to achieve — but we have had only moderate success so far. The problem is structural: because of a paucity of information about solar power, both the people in need and potential donors are not keen to embrace PV. Some communities have said that they would prefer to wait until they get full access to the grid like anybody else; they do not realise that this simply may never materialise.

The only viable path that remained open was promoting the notion of attaching PV to the grid — despite the unfavourable economics for PV in this configuration. To that end, at the meeting in Cherry Hill in 1973, Bill Cherry said, "… the Government has got to do some pump priming. The semiconductor industry got started the same way." Going this route would immediately bring the conventional power utilities into the picture, since they run the electricity networks. In the US, the utilities were not totally opposed to this idea, but they were reluctant to enter into a programme that involved high costs. On the other hand, Europe's state-owned utilities were not traditionally known to be enthusiastic about PV. For them, there were many reasons to avoid PV contributions to their networks, particularly the challenge of accommodating many small, distributed and intermittently operating solar plants. In any case, with taxpayers having to bear the costs associated with any significant programmes, the question, as ever, was not one of grand visions of the future but of political expediency.

Generally speaking, the private sector has not risen to the challenge either. Companies tend to need to be profitable on a very short-term basis, and only a few with deep pockets have taken a longer view. The oil industry belongs to that category: it works within longer time frames and is in that respect similar to the electric utilities. In the case of the US again, the oil majors have turned out to be some of the biggest investors in PV right from the beginning; they have more cash to spend than the utilities and do not have the same constraints to follow.

The vicious cycle was ultimately broken by pioneers in government who succeeded in putting in place the necessary regulation to make the PV business profitable for everybody (a subject we will return to in later sections of this book).

The establishment of PV was also supported significantly by international competition in a liberalised marketplace. In the semiconductor markets of the 1950s, it was the financial intervention of governments that kept prices high: only when that support dropped off and competition kicked in did market prices come down. Indeed, once a mass market is established, the engine of development is set in motion: commerce grows; sources of finance become available, eventually including the stock market; and human capacity increases (engineers, scientists, installers, etc.) — until suddenly political opposition appears from nowhere and questions the whole enterprise!

5.4 *Entrenched Energy Strategies and Politics*

As we have discussed, unless it establishes a suitable position on the global energy markets where it can compete, PV power depends strongly on political support. However, in energy matters, both national and international policies follow what is considered to be the strategic mainstream — and that was never really solar power. The fact is that the key energy strategists got it wrong most of the time. In the 1970s, the international energy agenda was dominated by the development of nuclear breeder reactors and coal gasification. Needless to say, in hindsight neither of these options was realistic.

Interest in solar energy was placed at a similar level to the development of fusion reactors. In 1995, Emanuele Negro presented a study he had conducted on behalf of the European Parliament on the profitability of financing research on PV and on "thermonuclear fusion." His conclusion was that nuclear fusion research "has no longer to be regarded as an alternative source of electricity, but as a basic research domain. Its expectations have not to be compared with the ones of other sources of energy, but for their scientific interest". Today, PV is booming, while fusion technology will likely play a role, if at all, only in 50 years and beyond.

An additional challenge with these mainstream strategies is that they are not frequently changed; for practical reasons, this would not be possible. Hence, the same options, however obsolete they may actually be, have dominated the energy debate for decades, making it very difficult for new technologies to get a foothold. Even the sacred cow of today's energy policy — carbon capture and storage (CCS), or so-called "clean coal" — may too one day be relegated to the scrap heap of ill-conceived energy policy decisions.

This is the energy world in which we must evolve. Right from the beginning in the early 1970s, it has meant a terrible uphill battle for PV to get its share of attention.

5.5 *Against Dominant Allocations of State Budgets*

As we have seen, PV is very much dependent on a share of public financial support for research and technology development (RTD) and promotion. In reality, the financial needs for PV development are lower than those of most competing technologies in the energy sector. The reason is that PV technology is comparatively uncomplicated. To make PV happen, securing at least the minimum budgets to realise the options at hand is really just a question of fair treatment. As the overall volume of state budgets is obviously limited in size, PV has to compete with all the other budget expenditures, be they regional, national or international.

Within the energy sector, PV's chances were slim: we have seen that PV fell into the category of "other options". By comparison, development budgets for nuclear power have been a lot larger than those for PV over the years. And not only for nuclear: between 1918 and 1978, the US Government spent US$150 billion to stimulate oil, gas, coal, nuclear and other forms of electrical energy production.

The nuclear programme in France had a dramatic effect on the RTD of PV: in the mid-1980s, the government took PV off its agenda and reduced its budget to zero. Rolf Linkohr, himself a supporter of nuclear in the EU Parliament, promised me that he would push for at least equal RTD budgets for nuclear and renewable energies for the EU.

In the US, conservative Presidents like Ronald Reagan dramatically reduced energy budgets as a whole, as military spending became a big priority. Europe also contributes its share to the global military expenditure — to the tune of roughly US$1 trillion per year. Military budgets have barely changed, perhaps surprisingly, even though stateless terrorists have become the main enemy. In effect, eliminating a single terrorist costs a colossal amount of money; the debate goes on as to whether there are not cheaper and more peaceful ways of approaching the problem.

Another big consumer of public money is the social welfare sector, especially, many claim, in Europe. And these welfare systems are far from optimal. It is an established fact that a significant proportion of people in industrialised countries remains extremely poor; a figure of 3 percent of the population, or over 30 million people, is a typical estimate. Many of them, including many children, depend on welfare for survival, often by private charities — Germany's "soup kitchens", the "Restaurants du Coeur" in France, or via "food stamps" in the US. There must be something wrong with the way national budgets are allocated!

As far as the energy budgets are concerned, they are relatively modest in comparison with the giant mastodons mentioned above. But expenditures on energy are considerable as well. Globally, new energy investments are counted in the hundreds of billions of US-dollars per annum.

I have always felt that the energy stakeholders — those in charge of public budgets and those in the electric utilities who do little more than administer their clients' money — have a particular responsibility to follow the "right" strategies and to avoid wastage through development failures. It is not really acceptable that public money is wasted just because the strategies of the energy establishment proved to be wrong.

5.6 Administrations

Generally speaking, detailed budgeting is in the hands of public administrations. The people in charge have a huge responsibility and take the job very seriously. As far as PV is concerned, a certain expertise is necessary. But it is often the case that managers barely learn how to spell the word *photovoltaics* before they are moved to another position, to avoid any conflict of interest with their contractors.

Official policy in many cases is to organise budget allocations rationally. The first to introduce a very "objective method of budget allocation" was the late Robert McNamara, US Secretary of Defence from 1961 to 1968 during the Vietnam War, when there was plenty of budget to be managed. The method divides projects up into different evaluation parameters and associates figures of merit to them. In the administrations in which I worked, we used it regularly. However, most of the time the resulting recommendations were wrong, as the parameters had been tuned unrealistically.

I worked in administration in a position of responsibility for 37 years and PV was always a major part of the programmes I had to manage. We could overcome the challenges described above. No public money was wasted. PV has now become important, and we are proud to say that we were leading the charge.

5.7 Energy Pay-Back Time and Module Lifetime

Neither the "energy pay-back time" (the time it takes for a PV module to generate as much energy as it required to manufacture the PV module) nor the lifetime of modules was a particular concern for the pioneers who imagined PV strategies in the early days. They knew from their own hands-on experience that these were not real problems. It was only when PV started to attract a wider interest that opponents of the new technology started to argue against it. It was claimed then that it would take 15 years or longer to recoup the cost of building a PV power plant. That would have dealt PV a deadly blow — if that had been true. In practice, the energy invested in producing a module depends on its technology. It has often been demonstrated that the energy pay-back time is, on average, just one year of operation. For thin-film cells it is lower; for crystalline silicon, it is slightly higher.

In the late 1970s, the president of the Solarex company, the late Joseph Lindmeyer, was keen to demonstrate that PV cell production may not even need an external source of energy at all. At his facility near Washington DC, he built a "solar breeder" and ran it for a period time. Consisting of a PV manufacturing plant with a solar array on the roof, the solar breeder showed that all the energy the plant required could be generated by the Sun from its own PV array.

The issue of lifetime has quickly become obsolete as well. From the early PV applications on space vehicles, it was already clear that even in a harsh environment, solar cells are tremendously robust. If anything failed in a satellite's solar generator, it was the battery, not the PV array. Today, many manufacturers have a 20-year guarantee on the PV modules they sell.

In countries like Spain, the "feed-in tariff" is paid for a period of 25 years. As modern PV is only a little more than 50 years old, there is a hesitance to make commitments beyond a 25-year time frame. However, there are indications that PV modules with a lifetime of 50 years, or even longer, will become the standard. We actually have relevant experience from an 80-kW plant built in 1983 by the EU Commission with the Italian energy provider Enel on Vulcano Island, Italy. A detailed analysis after 22 years of continuous operation revealed a 6-percent degradation of the modules; it may be even less since the accuracy of the measurements was only 5 percent. Moreover, the quality of today's PV technology is even higher than that of those modules produced back in 1983.

Clearly, however, it is possible that compared to fixed arrays that do not move, the reliability is lower for arrays that follow the Sun to maximise energy collection or because they employ optical concentration.

5.8 Intermittency of Supply

The electricity generation of a PV array can be predicted to some extent, as it is associated with the weather forecast. In that respect, intermittency of supply is less of a problem for PV than it is for wind generators, for instance.

For grid-connected systems with an obligation by the grid to buy the solar electricity whenever it is available, disruption of solar supply is not a problem, at least not for the owner. It does not matter then if the system is run as part of the "feed-in" regulation that is popular in many countries or the "net metering" that is often employed in the US. The electric utilities in the US and in countries that have to meet a large peak load around noon because of widespread air-conditioning usage, solar electricity is all the more desirable, as meeting peak load by conventional means is costly.

When no grid is available, electricity storage is the first idea that comes to mind. There has been a recent renewed interest in storage technologies, which originated from the emergence of hybrid and electric automobiles. Lithium battery technology has developed rapidly, already monopolising the market in laptop applications and with larger sizes in the offing for use in electric cars. However, even today, people have a reserved attitude when it comes to PV applications.

The need for seasonal storage is a special challenge for PV, as the intensity of solar radiation drops in winter; this applies to most climates, even to some extent subtropical ones. In this case, storage is, as a rule, too expensive; a system that is used only once a year is difficult to amortise.

Eventually, the solution for autonomous systems is hybridisation of PV power with other generators of renewable electricity together with small battery storage to bridge short-term needs. Interesting options in that respect are small wind turbines, water turbines, and turbines driven by biofuels. We will return later to a hybrid project combining PV with wind that we organised in Latin America.

5.9 Environmental Challenges

Among the renewable energies, PV has received less opposition from environmentalists than wind turbines or liquid biofuels, for example. But there are a few concerns.

One issue was thought to be the biosphere. The President of CNES, at the time my own "big boss", expressed a note of caution when he addressed the audience at the UNESCO Congress in Paris in 1973 mentioned previously. He warned that the large-scale collection of energy from the Sun might affect the biosphere's equilibrium. Later, when I had assumed responsibility at the EU Commission in Brussels, Dr. G. Schuster, the father of European RTD and its first Director General, confronted me with similar considerations. I should, however, immediately add that Dr. Schuster was, from the start, a big supporter of PV and of my work at the Commission, though his traditional field of interest was nuclear fission and fusion energy.

It is now clear that the capture of solar radiation for energy generation plays a negligible role in the Earth's equilibrium. Certainly, it is infinitesimally smaller than, for instance, the impact of agricultural practices and other conversion of land for buildings or roads, which have serious detrimental effects on soils, water resources, and ecosystems — not to mention the disastrous consequences associated with increased global temperatures and atmospheric pollution. In 2008, the Max Planck Institute in Hamburg published a study on the change in land cover in Europe over the last millennium: about 5 million square-kilometres of natural vegetation were transformed into agricultural land between the years 800 to 1700. According to the study, historical events such as the Black Death "led to considerable dynamics in land cover change on a regional scale". In the last three centuries, the population explosion and rapid industrialisation have led to change on a massive scale.

A more serious concern was, and remains, cadmium — a major ingredient of the CdTe solar cells that have now attained a prominent position in global PV markets. US company First Solar manufactures and markets this technology, primarily in Germany, recently becoming a leader in the global PV business.

The problem is that cadmium is highly toxic and presents significant dangers to human health and the environment. The use of cadmium has raised serious concerns since the 1960s, when the PV favourite was CdS/Cu_2S, the "cuprous sulphide" solar cell that also contains the element. The argument goes that compounds of cadmium are less toxic than the pure metal; besides, it is also used in many other applications, such as in paints.

My personal belief was always that CdTe could not be neglected in PV development, as its theoretical efficiency is one of the highest of all possible materials. By the 1980s, when public support had all but disappeared, I made sure that there was, at the very least, enough financial support for cadmium-containing PV in our EU RTD programmes. Thus, its survival in Europe was assured, although industrialisation on a large scale would ultimately be undertaken by US firms. Not surprisingly, when First Solar was eventually listed on the stock market and a lot of people benefitted financially, the arguments against cadmium conveniently evaporated.

6. Leadership

6.1 *The Pioneering Role of the US*

There can be no doubt that the world's PV stands on the shoulders of American pioneers. We have seen previously how the US took the leadership role in PV by inventing the first commercial solar cells and by promoting them through their space programmes. They were also the first, as early as 1973, to imagine an aggressive PV promotion as a mainstream power provider in the US and around the world.

After the oil price shock of 1973, renewable energies and in particular PV began to garner overwhelming popular support. The American people became interested in solar energy in a profound way, the significance of which is hard to appreciate today — a "solar tsunami" with shock waves around the globe. It would not be unfair to say that what has followed since pales by comparison to the ardent enthusiasm of those days.

The effort culminated during Jimmy Carter's Presidency (1977–81). His Secretary of Energy was James Schlesinger. In California, Governor Edmond Brown Jr. led the charge for solar. The Director of the Federal PV programme was Paul Maycock, an outstanding pioneer who devoted his whole life to PV. Indeed, some of the information below is taken from his 1981 book *Sunlight to Electricity in One Step*, published by Brick House.

It's no exaggeration to say that in the latter half of the 1970s many people already saw already themselves living in the "solar age". They were committed to a better "quality of life". In May, they celebrated "Sun Day". In 1978, the US Congress passed an RD&D (research, development and demonstration) Act on PV. It called for a doubling of PV generation capacity every year until 1988, with an objective to reach a 4-GW level. The budget commitment was US$1.5 billion, an enormous sum at the time (even though it was tiny when compared with what had previously gone into nuclear research). I was invited to speak at the opening of one of the IEEE Photovoltaic Specialists Conferences, on the podium with a Congressman who elaborated on their objectives and with more than a thousand experts in the room. It was another occasion that helped me to set my mind unconditionally to fighting for PV.

The Public Utility Regulatory Policy Act (PURPA) of 1978 provided regulation that meant that utility rate structures could not discriminate against small producers of electricity and that utilities must buy solar electricity at fair prices. That same year, the White House Council on Environmental Quality projected an installed PV capacity of at least 500 GW, or 500 million kW, by 2020. From 1977 to 1980 more than 1,300 small PV systems (residences, water pumps for irrigation, and so on) were built under the framework of the Federal PV Utilization Programme (FPUP), which encouraged federal agencies to get involved. Since 1980, a tax credit of 40 percent of the purchase price has been offered for PV. Low interest loans were proposed and the establishment of a "solar bank" was considered.

The Federal budget for R&D on PV reached US$157 million in 1980. Hundreds of solar PV-powered buildings were developed, both stand alone and grid-connected, including Lord House in Maine; Olympic Natatorium in Atlanta, Georgia; PV Pioneers in Sacramento, California; recycling centres in New York; Georgetown University in Washington State; APS factory in Fairfield, California; a solar townhouse in Bowie, Maryland; and Liss House in Fairbanks, Alaska.

The first community relying entirely on solar power was the Schuchuli Indian village in Arizona, in 1978. The world's first PV-powered neighbourhood came about in 1985 when New England Electric installed 100 kW of distributed rooftop PV systems in the central Massachusetts town of Gardner. The first larger stand-alone PV systems of up to a hundred kW per unit were installed, among other places, in the National Bridges National Monument Park in Utah, Mount Laguna in California, and — for the purpose of irrigation — in Mead, Nebraska.

Among the American states, California had already emerged as a leader; it offered even better tax credits than Washington. In some regions or counties, people voted to shut down nuclear plants; almost every dollar coming out of nuclear went into PV. People were gratified to see that, contrary to the nuclear industry's prior warnings, electricity rates remained stable after nuclear plants were closed.

US industry followed suit. New production capacity was built and soon the country had overall 25 MW manufacturing capacity on stream, a five-fold increase achieved in just a few years. Even the energy "establishment" like General Electric (GE) and most of the power utilities, who were used to doing all their business with the "hard energies", started to wonder whether they should not get involved in PV as well. GE would build a five-acre PV installation for SeaWorld in Florida.

One particular aspect of what took place may in retrospect appear a bit strange to us now: the enthusiasm surrounding "tracking" or "optical" concentration. These technologies were not yet mature at that time.

In 1977, a massive "block buy" programme was launched by the Federal Government: almost 2000 kW were bought and installed by 1980; 10 000 kW more had followed by 1987. A 500-kW concentrating system was also installed in Saudi Arabia. All of them followed the Sun by tracking. The largest plant, built in 1983 at Carrisa Plains, California, had a nameplate of 6,500 kW and two-axis tracking. From a technical viewpoint, though, this system was not optimal. The solar PV modules that had been purchased from Siemens had a measured efficiency that was up to 30 percent lower than their label stated.

But there was a non-negligible practical issue to consider. This concerned the "capacity factor" of grid-connected plants, that is, the ratio of the actual output over a period of time and the output if the plant had operated at full nameplate capacity over that period. The measured capacity factor of the plants in California, Texas and Arizona turned out to be as high as 35 percent. The reason was the correlation of the Sun's availability with the peak air-conditioning load.

As we know, after Carter's tenure as President came to a close, the US stepped back from their solar commitment. One of the first actions of the newly installed President Reagan was dismantling the PV panel from the White House roof. For all US Presidents since, it's been a case of "business as usual". Even the Clinton-Gore Administration fared no better.

It was the same in California. John Geesman, formerly with the California Energy Commission, stated on September 2, 2009: "When the Renewable Portofolio Standard was signed into law in 2002, California derived 11 percent of its electricity from renewable sources. In 2008 that number was 10.6 percent. Every school child in California knows that most of that comes from policies enacted when Jerry Brown was Governor some 30 years ago."

It was not long before global leadership shifted to Germany, Japan and others.

6.2 France: A European Solar Pioneer

The French had already established themselves as leaders in solar energy in the 19th century, as exemplified by Becquerel's discovery of the PV effect in 1839. Mouchot's early work on solar concentration was followed in the 20th century by the establishment of a generation of solar furnaces by Felix Trombe: one in Algeria that belonged to France for a time, another at the Fort of Mont Louis, and a third at Odeillo in the Pyrenees. Those located in France are still in operation today: producing temperatures of over 3000°C, they can even be employed to simulate atomic explosions.

In the late 1970s, France built a solar thermal plant called Themis, with concentrating mirrors and a tower, and participated in a similar European project, Eurhelios, in Sicily. Centre National de la Recherche Scientifique (CNRS), France's main science organisation, has also operated a solar centre on thermal power on Corsica. After France's liberation in 1944, PV activities were concentrated at the semiconductor laboratories of CNRS at Bellevue, a suburb of Paris. In 1967, its director, the late Michel Rodot, recommended me as the manager of the PV programme in the newly created French National Space Agency (CNES). Shortly before, President de Gaulle had decided that he wanted to have his own satellites and his own PV programme — and not to depend on the US for it.

Figure 3. A French PV generator of CdS from 1973.

The SAT company in Paris, the official provider of silicon cells for French and for some European satellites, had a licence from Spectrolab in California. Our own contractual development work with French industry concentrated on thin-film cells: we developed CdTe cells on molybdenum sheets at Radiotechnique, a subsidiary of Philips in France, and CdS/Cu_2S cells on Kapton plastic at SAT.

I became part of a committed group at SAT with Besson, Ngyen and Prof. Vedel, a specialist in electrochemistry from the École de Chimie de Paris who worked for almost 10 years to develop these cells into a viable product. For me, it was a good opportunity to learn how industry works and to get a breather from my administrative duties. We achieved over 8-percent efficiency with both types of cells. We also thought we had solved the stability problem of cuprous sulphide. We were, at least, able to convince the late Paul Rappaport, the founder of what is now NREL in Colorado, who congratulated me personally on this.

But this was not the end of my engagement with CdS. In 1973, I was approached by John Francis Jordan, Senior Vice President of the Baldwin company in Cincinatti. They were just moving into the banking business and had bought 10 banks in Denver. The lady who owned the company spent her last years at the Ritz Hotel in Paris. They were originally piano builders, and John Francis had come across CdS when developing electronic musical instruments. We became very close friends.

John Francis had the unique vision of generating all of the US' electricity in the deserts of the southwest. From his house on a hill in El Paso, Texas, all the land one could see — which stretched for more than 100 miles — would have been sufficient for this. His idea was to produce CdS cells continuously in a float glass plant; modern glass is produced in such plants, where glass sheets are ejected from a liquid tin bath at high speed when they are still at a temperature of 500°C. Huge areas could then be produced by spraying the semiconductors on the hot glass: 500°C does indeed correspond to the temperature needed for CdS to crystallise on the substrate. A company was created to realise this idea: Photon Power.

Independently, the oil company Total approached me in Paris on the subject of CdS solar cells. I was introduced to them by my long-time friend Bernard Devin from the CEA, France's powerful Atomic Energy Commission. At that time, François Fiatte was Total's Director for Diversification into solar power, coal, etc., and I became his consultant and friend. Total wanted my advice about whether to develop a spray process that was offered to them by the company PA in Britain. Eventually, in 1975, a decision was made to build a plant in El Paso employing the spray process; the US glass company Owens Ford decided to enter into the joint venture and to finance the plant with the French. There was no public money of any kind involved.

The US$20-million plant was built and began operation. Unfortunately, it was a failure for two reasons, one technical and one human. As the CdS films were only 1 micron (1/1000 mm) thick, the slightest impurity in the glass substrate led to a short circuit. While the cells looked wonderful — and the prototypes in the laboratory had been shown to have an efficiency of 8 percent — the mass-produced cells did not work. The human factor was that the plant manager, who was Total's manager in London, had personal issues: uprooting from London to El Paso turned out to be such a shock for his wife that after a while he had to go back home to be with his family. By then, John Jordan was already well over 70 years old and was no longer in a position to fix things. A wonderful endeavour finished prematurely.

When I came to France as a young German scientist in the mid-1960s, the main platform for French interests in solar energy was the AFEDES association. It was closely linked with another one, COMPLES, whose goal was to promote solar energy around the Mediterranean Sea. These groups initiated and organised the solar congress at UNESCO House in Paris in 1973; we have already mentioned its important role in the worldwide promotion of solar energy.

I was introduced to this group by Pierre Vasseur, my superior at CNES and a good friend. He was also a director of the laboratories at École Polytechnique, France's prestigious technical college in Paris. Vasseur explained to me the background of the French motivation to promote solar energy: an important inspiration for them was the philosophy of Pierre Teilhard de Chardin (1881–1955), who was a Jesuit priest, often in conflict with the Vatican and the Pope, and a palaeontologist working much of the time in China. In his book *The Future of Mankind*, he wrote, "Since the Palaeolithic and the Neolithic age, mankind could always expand: growing and proliferating was the same for him. And now all of a sudden, emerges in front of us at great speed the wall of saturation. What to do to avoid that human concentration — although social unification is a favourable trend — passes an optimum beyond which all increase of numbers means only famine and suffocation."

Investigating the roots of French interest in solar energy, one also has to mention Frédéric Joliot-Curie, a professor at the Collège de France in Paris (100 metres from where I am currently living). Along with his wife, he received a Nobel Prize in Physics and worked on nuclear technology and a synchrotron in occupied Paris during WWII — while at the same time being connected to the "résistance". In 1948, he built France's first nuclear reactor.

In May 1956, at a nuclear committee in Paris, he declared, "I think that we must very seriously and immediately get involved in the utilisation of solar energy. It would not be reasonable to see in nuclear energy the only source for meeting the considerable future energy needs of our country." It is the same man who, in 1950, initiated the "Call of Stockholm" against nuclear weapons: "We demand the absolute interdiction of atomic arms, arms of terror and massive extermination of the people." (Information kindly provided by Jacques Dupin, CERN, Geneva.)

The oil embargo by the Organization of Arab Petroleum Exporting Countries (OAPEC) in 1973 was a shock for France. It suddenly saw its precious national independence in great danger. French stakeholders in energy, science and technology were asked to consider what their contributions to "energy independence" could be. CNES and myself were called upon to represent the potential of PV. I became directly associated with the President of CNES and played a role as the French "Monsieur Energie Solaire". After some months of brainstorming, the government decided that the solution had to be nuclear energy, and in 1974, the country embarked on a massive construction programme, with 58 reactors on stream today; solar energy was asked to leave the mainstream. I do not need to comment here on this decision but only recall what the French President said in 2009: for him the decision to give a monopolistic role to nuclear was wrong; he wanted from then on to give equal emphasis to nuclear and to solar energy. At least that is what he said.

But French interest in solar energy did not disappear altogether. The Director General of CNRS, Robert Chabbal, held regular discussions in his own home with four or five people, of which I was one. In 1978, the late Henry Durand, another heavyweight in solar energy, was appointed by the government as president of a specialised agency for renewable energy: from the mid-1970s until today France is one of the few countries with a specialised government agency for renewables (later combined with another one in charge of environmental matters).

The French government's growing respect for solar energy was also demonstrated by two other events. In 1975, Jean-Claude Colli, the "Solar Delegate" of the French government, led a delegation to Japan — of which I was a member — to sign the first co-operation agreement between the two countries on science and technology. A year later, another French delegation, with Chabbal at the top and me as a member, returned to Japan to sign another governmental agreement on solar energy co-operation.

In a 1976 report issued by the Prime Minister's office, Jean-Claude Colli expressed his belief that sooner or later PV power plants would become a reality; uncertainty

Figure 4. A French governmental delegation visiting Japan in 1975. Lower row: Chabbal, the DG of CNRS, second from right; W. Palz, standing, third from the right.

concerned only the time horizon not the eventual success, he claimed. But France was not willing to support solar energy with its own major budgets and instead suggested the EU Commission in Brussels take over. The first European renewable energy programme was decided by the EU's Council of Ministers in 1975; I myself had been a French delegate when it was drafted by official experts a year before. I was then the draftsman for PV.

I took the helm of the European programme in 1977. And here in Brussels I again found my French friends: Chabbal as Chairman of the advisory committee of the programme and the late Monsieur Pheline as a French delegate. I had become friends with Pheline, who had twice become frustrated with his French administration. Before we got to know each other, he was in charge at the French Embassy for Brazil and Argentina, tasked with selling French nuclear reactors to those countries. He had almost succeeded when France decided to give up its own nuclear technology at CEA and to adopt a US licence instead. His second frustration was that French budgets for PV were miserable in those days. But he did not know what would happen later in the 1980s: Monsieur S., the then-DG for Energy in the French government, decided that the support for PV had to be reduced to zero.

France's pioneering role endured nevertheless. One illustration of this was the study "ALTER, Study of an energy future for France based on renewable energy" from "the group de Bellevue". The study was implemented in the early 1980s by a group of French energy experts from CNRS, Électricité de France (EdF), Collège de France, and INRA (the National Institute for Agricultural Research). It was demonstrated in that study that: "in the perspective of 2050, it would be possible to change progressively in France from an energy system dominated by the fossil

Figure 5. A French delegation in Japan visiting a solar concentrator in 1975. Chabal, CNRS, 7th from the right; W. Palz, 6th from the right.

energies to a stable autonomous energy system based exclusively on renewable resources, i.e., solar energy in all its forms". This study was a world's first!

A key author of this report was my good friend Philippe Chartier, the long-time science director of the French renewable energy agency. He is still active today as a promoter of PV inside the Syndicat des Énergies Renouvelables, an important lobby group for solar interests in France.

In 1994, the French Government organised a national debate on "energy and the environment". Meetings were held for six months all over the country. The final report of December 1994 highlights concerns about radioactivity (mentioning large numbers of miscarriages after the Chernobyl accident) and electromagnetic field effects. French people wanted the R&D effort to be less concentrated on nuclear; they wanted energy diversification. The report stated that France spends 15 times less on renewables than Germany. Focus needed to be put on PV. Unfortunately, this endeavour did little to help: today that disparity between France and Germany in renewables investment is as wide as ever.

In May 2007, the French President started something similar to the 1994 initiatives, the Grenelle de l'Environnement — Grenelle being the name of a street in Paris with some significance since the social movements of 1968. Meetings were also held all over France for several months. My dear friend Alain Liebard was the chairman of the working group on renewables; see his contribution in this book.

The conclusions of the "Grenelle" are progressively being translated into law. A highlight of this new interest in renewables is a vigorous feed-in tariff (FIT) regulation for PV promotion in the country, putting the right emphasis on building integration. More details are given later in this text.

6.3 PV Start-up in Germany

Since the 1960s, Germany has been actively involved in PV for applications in space. AEG and Telefunken in Hamburg and Heilbronn looked after silicon cells, while Battelle in Frankfurt developed CdS/Cu_2S thin-film cells. There was traditionally also a small production of selenium photoconductors for light metering in cameras in Marburg.

The interest in terrestrial applications of PV started in 1973 when the Germans sent a delegation to the UNESCO Congress in Paris. A year later, Dr. Karius from AEG organised a national PV conference in Hamburg at which I served as a co-organiser. Shortly after, the German government in Bonn began its first R&D programme on solar energy. The first official to run it was Dr. Klein; when we met at a meeting in Paris, where he came with his Secretary of State, Volker Hauff, he told me proudly that they had not only thermal applications on the agenda but also PV. The programme has continued since then: after Klein, came Gerd Eisenbeiss, Walter Sandtner, and others. A particularity was that the Ministry delegated detailed management and financing to a special group that was installed at the Nuclear Research Centre (KFA) at Jülich, a little town close to Bonn. As a result, applicants for contracts had to clear two hurdles: after having convinced Bonn, they had to negotiate with Jülich.

There was good co-operation between our European programme on solar energy and the German one: German representatives were sitting on our Consultative Committee in Brussels together with those from all other EU member countries.

The interest in solar energy in Germany in those days paled by comparison with the enormous enthusiasm in the US at the time. Nevertheless, the sector gathered momentum thanks to a few pioneers. Adolf Goetzberger was the first to create a solar institute in Freiburg in 1981: the Institute for Solar Energy Systems (ISE) became part of the Fraunhofer Society and was supported by the regional government of Baden-Württemberg. Today, it is a leading institute in Germany and the world. Six or seven other institutes followed. The late Werner Bloss created the Centre for Solar Energy and Hydrogen Research Baden-Württemberg (ZSW) in Stuttgart as an extension of his Institute of Physics at the University. Then came the Institut für Solare Energieversorgungstechnik (ISET) in Kassel, and others in Jülich, Hanover, Berlin, and so on. I was invited as a member of the founding advisory councils of some of these institutes.

The main protagonists of solar cell development in the late 1970s and early 1980s in Germany were Dr. Eckehard Schmid from AEG in Hamburg for crystalline silicon and Prof. Werner Bloss from Stuttgart for thin-film cells. Around the mid-1970s there was some disenchantment with thin-film cells of the CdS/Cu_2S type: in Germany, like in France and the US, activities were largely discontinued at that time. The ZSW in Stuttgart switched its interest towards CIS, the copper-indium-diselenide family of semiconductor materials.

In the industrial world of PV in those days, Germany saw a lot of ups and downs. Solar AEG disappeared from the map in the mid-1980s when the whole company was sold and later disappeared altogether. Siemens in Munich, at that time a newcomer acting with much arrogance, threw in the towel after its investments in PV cost the company almost US$1 billion over three years — and after the acquisition of ARCO Solar in California turned out to be a disaster.

A notable personality of the industrial PV world in the Germany of the mid-1980 was Werner Freiesleben. He was then a director of Wacker Chemie in Burghausen in Bavaria. Wacker was, and is, the recognised world leader in the production of purified silicon of semiconductor-grade quality. Freiesleben is a very cultured man who lived in a medieval castle in the old town of Burghausen, where he also happened to have regular chamber music evenings.

I had many interesting discussions with him. I remember once I had seen a BBC report on climate change some 12 000 years ago: the ice sheets melted, the Sahara became a desert, and there was heavy precipitation (traces were visible on the Sphinx at Giza, which in those days had the head of a lion); we agreed that all this was due to a change in Earth's axial inclination, though not why the axis changed.

In those days, Freiesleben had a vision of starting a major development programme for PV with Wacker. But the family-owned Wacker did not have the appetite for the risks involved, and Freiesleben had to leave. I met him again in Brussels in 1988 as a representative of the German chemical industry to the EU. At his house, for the first time I met with Hermann Scheer, who wanted to see me after having just created the EUROSOLAR association in Bonn. I invited Werner Freiesleben to become the General Chairman of one of our big PV conferences — which he kindly accepted.

There are only four major PV companies in Germany that have survived from those days. The first is Wacker, which continues to be a global market leader for pure silicon material. The second is SMA, the world leader in inverters that was founded by Prof. Kleinkauf in parallel with the ISET research centre, which was integrated into a new Fraunhofer Institute for wind energy in 2009. The third is Schott Solar, which produces silicon cells and modules. It survived in Alzenau despite many changes and consolidations after Werner Bloss had tried to implant his thin-film technology into Nukem, a forerunner of the Schott Solar of today. The fourth and final company is Würth Solar in Schwäbisch Hall, which has produced CIS thin-film cells since 2008. Mr. Würth had become very rich marketing wall plugs. A few years ago, he made a smart decision to invest in PV and bought the CIS technology from ZSW.

What also remain from those days are the unique visions of two outstanding German individuals, Dr. Reinhard Dahlberg and Prof. Ludwig Bölkow. In 1986, Dahlberg, who was previously with AEG, proposed — right after the Chernobyl disaster — to deploy PV on a huge scale in the desert areas of Northern Africa. He estimated that 45 000 square kilometres would be enough to generate all of the primary energy needed for West Germany. His idea was that the electricity might be converted into hydrogen for transport overseas. While such a proposal was certainly not very realistic in those days, it had the merit of triggering awareness about the enormous potential of PV to become a mainstream energy provider. Indirectly, it contributed to the establishment of the huge domestic PV programmes that have become a reality in Germany since then.

Ludwig Bölkow (1912–2003) was a great man. His main achievement came after WWII when he founded his own aerospace company in Ottobrunn near Munich. It merged with other companies in Germany and France into EADS, which today produces Airbus airliners. When I visited him in Ottobrunn in 1994, he received me like a friend, offering me a copy of his book of personal memories (Ludwig Bölkow, *Erinnerungen*, 1994, F. A. Herbig, Munich). In it, he writes: "One of my first understandings was that many officials and politicians had heard of the problems of energy supply and its environmental constraints, but that they were not aware that we must start immediately to change things."

After retiring from his aeronautics company, he created the consultancy firm Ludwig Bölkow Systemtechnik GmbH. He financed it with his own fortune and earned substantial fees giving speeches all around the country. His consultancy firm is still active today, after his passing, and plays inter alia a prominent role in discussions about "peak oil".

In the late 1980s and early 1990s, Ludwig Bölkow did three things. First, he financially assisted various struggling companies, without requiring benefits in return. Second, he lent his weight behind Dahlberg's ideas on the large-scale deployment of PV. His consultancy firm conducted a detailed study together with other research institutes: besides conversion into hydrogen, electricity transmission via high-voltage DC was also considered. He writes: "We prepared a very detailed plan and discussed it with Director Bauer at the Ministry of Research. After a while, we got warning signals from Bonn. We had a new discussion with the Minister and the official opposing the concept, Gerd Eisenbeiss. The latter argued one had better start very, very small and create a Committee." That dealt the idea the fatal blow.

Thirdly, in 1987–88, the Bölkow consultancy firm prepared a study "to establish the cost of electricity production based on PV on a large scale, employing the technology available at that time" for the Ministry in Bonn. When they presented the results to the Minister, he was most satisfied and proposed holding a press conference. This was sabotaged by Siemens Solar, who attacked the presentation as "vision and wishful thinking without acknowledging the rigorous business analysis we did". Jülich then withdrew its support. The study was never published. "After the management at Siemens had changed, the new people in charge confirmed to the Bölkow consultancy that the results produced were ok and Siemens agreed with them."

In Brussels, we were not aware of the German saga on PV's future. We initiated and financed our own extensive studies on the same subject some years later. Our approach was from a different angle; we will return to this in a later section.

6.4 PV Ups and Downs in Japan

Japan began its solar energy drive with the famous Sun Shine Project, which was established after the oil price shock of 1974. The project was managed by the Ministry of International Trade and Industry (MITI), since 1980 in association with the New Energy and Industrial Technology Development Organization (NEDO). Since the early 1980s, PV became one of its strategic priorities. Until recently, Japan actually gave PV top priority among all the renewable energies, above wind, hydropower or bio-energy — I do not know any other country that had such a strong commitment.

Following the example of the US, the Sun Shine Project Committee had been keen to establish realistic cost targets since 1986. Their conclusion was that US$1/Watt to cover module manufacturing costs was feasible only if an annual production volume of 100 MW per year could be attained. Crystalline silicon and amorphous silicon alike were considered to be possible routes to achieve the goal. But the feeling was that "amorphous silicon was better suited for mass production".

The central driving force for PV promotion in Japan was Prof. Hamakawa from Osaka University. We benefitted from his keenness to develop international contacts; his wife is a famous concert pianist who often performs in Paris and other big cities around the world. Under his guidance, Japan set up the specialised Asian PV conferences. Over the years, I was often invited to Japan as a speaker; it gave me the opportunity to see all the different regions of the country where these conferences were held. Some years ago, a deceleration in the domestic PV market in Japan coincided with health problems Hamakawa was having. I personally

Figure 6. On the Great Wall, China, in 1986: Hamakawa (3rd from right); W. Palz (4th from right); Schock (2nd from right) and Pfisterer (1st from right) from Stuttgart, Germany.

Figure 7. From right: Ikki, Hamakawa and Palz.

think that there is a link between them: Hamakawa was less able to make sure that things developed as rapidly as he wanted!

As a scientist, Hamakawa became a key developer of amorphous silicon solar cells. Tremendous stability problems plagued this technology from the beginning: the hydrogen ions in the material are so small and mobile that it is almost impossible to keep them "quiet" in their positions in the crystal lattice. But his persistence eventually paid off and he succeeded in perfecting the technology. Today, amorphous silicon is produced in large volumes by the industry in Japan and elsewhere.

For several years, the PV sector in Japan kept rather quiet, until a New Sun Shine Project was created in 1993. In December 1994, the Japanese government issued, within its New Earth 21 framework, a fresh initiative on PV: "Basic Guidelines for New Energy Introduction". These guidelines called for a PV capacity of 4,820 MW to be installed in Japan by 2010. PV for export was not even considered a part of this target. Up to 80 percent of all this capacity was planned with building integration in mind.

Until 1996, public funding provided a 50 percent subsidy on the cost of PV systems. This was subsequently reduced to 33 percent for private homes but kept at 50 percent for PV façades and roofs of public buildings. The low interest rates in Japan's capital markets provided an additional incentive. Right from the beginning, the programme was quite successful: between 1995 and 1997 the domestic market increased ten-fold to 37 MW.

In later years, the Japanese government progressively reduced the subsidy to zero. Not surprisingly, the PV market followed and shrank markedly.

Nevertheless, the overall result has been respectable: by the end of 2008, the accumulated PV capacity in Japan was 2,150 MW. It cannot be ruled out that by the end of 2010, the original target may eventually be reached. Japan is undeniably one of the world leaders as far as domestic capacity goes. Later in this book, we will explore some new governmental decisions in 2008 concerning the adoption of the "feed-in-tariff" system, which promises to boost the PV market.

Japan has traditionally had a strong PV industry. It has more than a dozen major manufacturers of PV cells and modules; most of them are heavyweights on the international markets. They specialise in all classical crystalline and thin-film technologies, except CdTe. Sharp has one of the longest-standing involvements with PV on the international markets; the company has operated on these markets for four decades. Until recently, Sharp has been aggressively extending its production capacity — up to 1000 MW per year. It was the world leader in crystalline silicon cells and modules; not everybody understood why they switched to thin-film cells at the very moment they had achieved that commanding position.

In 2008, the Japanese PV industry produced over 1200 MW; by 2012 this production may have increased three-fold to 3000–4000 MW.

6.5. UNESCO

The United Nations Educational, Scientific and Cultural Organization in Paris — better known for its World Heritage List — was once a principal centre for international networking on solar energy. We have already discussed how, in July 1973, the UNESCO Congress on the Sun in the Service of Mankind in Paris set the ball rolling for renewed interest in renewable energy in modern times.

The people in charge at UNESCO at that time were Dr. Glitsch, later followed by Dr. Lustig. Particularly good relations developed with Prof. Boris Boiko, who took over in 1976. I worked with them as a consultant; my office in the 7th Arrondissement in Paris was within walking distance of the UNESCO building. I will never forget how Boiko's charming wife invited us to a Ukrainian lunch at their home. Together with Werner Bloss, we decided to forget the recent tension between Germany and the Soviet Union — with many toasts to our new friendship.

In October 1976, UNESCO convened a small group of experts in Genoa, Italy, hosted by the Italian government, with the aim to review possible co-operation on solar energy. It, or rather we, recommended forming a European Council on Solar Energy under the aegis of UNESCO. It could have been the start of a bridge between East and West: the group had delegates from Moscow, Bulgaria and East Berlin, as well as three officials from the EU Commission and other experts from Western Europe.

Later, I wrote the book *Solar Electricity, An Economic Approach to Solar Energy*, which was published in seven languages under the auspices of UNESCO in 1976–78. As I did not have much writing experience, certainly not in English, I relied heavily on the help and advice of Lucian Crossby, UNESCO's scientific editor. They published some 35 science books a year. Bing, as he called himself, became quite interested in our book project and we became very good friends. He was much older than me and told me interesting stories about his life, such as his experience as a British soldier in Palestine. But as I moved from Paris to Brussels in early 1977, my direct contacts with UNESCO began to fade.

In the 1980s, a new manager for alternative technology at UNESCO came on board: the Russian Boris Berkovski, who had good contacts to his Soviet homeland. An American friend told me recently how he had shared a joke with the Soviet man. He had asked the Russian what he would do if tanks invaded his country from all sides. "Mobilise my troops!" Berkovski had declared. "No," the American had replied, "You should invest in the tank business!"

The Moscow Energy Club was created in the mid-1980s by academician Sheindlin with the support of Boris Berkovski from UNESCO. I was invited to join together with some important energy stakeholders of those days: Luis Crespo, the "godfather" of solar thermal technology in Spain, and a good friend, Hans Blix, the Director of the Atomic Energy Agency in Vienna, who later tried desperately to convince George Bush Jr. that Iraq had no weapons of mass destruction. I remember that when I gave an optimistic outlook for solar energy at one of our meetings in Moscow, he replied angrily that this was all nonsense. When we were taken to see Swan Lake at the Bolshoi, my neighbour in the theatre, a prominent Britain from the Club of Rome, complained that each time he came to Moscow, he could not escape his hosts taking him to Tchaikovsky's famous ballet.

In February 1993, I was invited to a UNESCO round table in Beijing on "Strategic Energy Issues in China". This meeting was supported by many of the different stakeholders in the Chinese government. From the 15 foreign experts who attended the meeting, I was appointed Co-chairman of a session on the "Energy System and its Future Prospects". I remember well how the delegate from EdF, Monsieur Christian S., took advantage of the opportunity given by this conference to lobby our Chinese hosts on atomic energy. The meeting took place at Tsinghua University in Beijing, and extensive proceedings were later published by Atomic Energy Press in Beijing.

In July of the year commemorating the 20th anniversary of the 1973 solar congress, UNESCO, in association with EUROSOLAR, organised a world solar summit that brought together leading experts from all over the world to Paris. In association with this congress, I organised a second Euroforum on Renewable Energies. At that time, our EU R&D programme included more than 750 contracted institutions that co-operated in 200 projects with an overall budget of €200 million. The Euroforum was an occasion to provide a cross-section of our activities to the general public. Proceedings were produced later featuring all the presentations, with Michel Rodot as the general editor.

On the initiative of UNESCO, a World Solar Commission was created in 1995 and a "World Solar Programme 1996–2005", which was endorsed in a UN General Assembly Resolution in October 1998 as a contribution to the overall sustainable development agenda. To support the implementation of the programme, meetings were held in 1998/99 by UNESCO in Bamako, Tbilisi, Quito, Harare, Tenerife, and Moscow. Unfortunately, these activities came to a halt when Boris Berkowski retired from UNESCO in the early 2000s. Since that time, UNESCO has largely withdrawn from the international solar energy scene.

6.6 The European Union

The EU research and technology development programme (RTD) of the European Commission, the executive arm of the European Union, is a real powerhouse today: the current five-year programme, which runs until 2013, has a budget amounting to more than €50 billion, and all technological areas you can think of are covered.

The programmes and all their many components are decided by the EU Council of Ministers. The European Parliament has codecision power, with the exception of nuclear and fusion energy, with a budget that currently amounts to €2,500 million for a five-year period. The programmes are centrally managed through multinational contracts by the Commission in Brussels; they are not subcontracted to the EU Member States, but delegates from the Member States sit on all the Consultative Committees, which have legal rights.

RTD in Brussels started in the early 1970s with a small programme on the environment. An energy programme headed by Albert Strub followed in 1975. I took charge of its Solar Programme in January 1977 and remained in that position of responsibility without interruption until 1996. I mentioned previously that

Figure 8. Director General Günter Schuster, the father of the EU Science and Research Programmes; W. Palz.

the first RTD programmes in Brussels were set up by Dr. Günter Schuster, who became the first Director General of Research. He conceived the five-year Framework Programmes (FP) that are the basis of the EU development organisation until today. He was the one who hired me as an EU Commission official in 1977.

The outline of the renewables programmes was prepared by the Commission services under my leadership, screened by our Consultative Committee and eventually decided upon, as mentioned previously. During my tenure, we had five successive programmes of approximately four years each; the budgets increased from just €50 million in the first term to 100 million for 1992–94.

For the period 1989–91 the funding was drastically reduced; we were lucky that it was not halted altogether after a new decrease in oil prices seemed to have given decision makers the impression that energy was no longer an issue. It should also be noted that the European Economic Community (EEC) — now called the European Union — received three additional Member States in the course of the programme: Greece in 1981 and Spain and Portugal in 1986. The number of members would increase to 15; today, the EU comprises 27 countries with over 500 million people.

At that time, the EU budget available to us for the solar programmes made up no more than 5–10 percent of the overall spending on solar R&D in Europe. However, it played a central catalytic role. Firstly, we connected the stakeholders in transnational programmes and projects; in the early years when the EU programmes began, we found that people in one particular country were well informed about what was happening in the US — then everyone's "point of reference" — but were ignorant of what European neighbours across their common border were doing. Secondly, we multiplied the weight of our own

Figure 9. The committee of the 1st EU PV Conference in Luxembourg in 1977. Sitting from left: Krebs, Buhs, John Goldsmith, Strub (Chairman), Magid, Durand, Fischer, Palz, and Reinhartz; standing from right: Boiko (2nd), Pizzini (3rd), Moe Forestieri (5th), Brandhorst (6th), Treble (9th), van Overstraeten (10th), and Rodot (11th).

budgets, as our contribution to a particular project would cover only a minor portion. Member countries and industry stakeholders were more than happy to provide the remaining funds from their own portfolios, as they felt it was a privilege for them to become part of a programme where competition for contracts was very strong. A case in point was Germany; its national solar energy budgets were arguably a good deal larger than ours.

An important aspect of creating a "European PV community" was the organisation of the European PV Solar Energy Conferences. As already mentioned, the first such conference was set up by the EU Commission in 1977. In the early days, deciding which cities should host these conferences on a rotating basis was political. As the city with important headquarters of EU institutions, Luxembourg was chosen for the first conference on September 27–30, 1977; my boss and good friend Albert Strub became its Chairman. It attracted 500 delegates from all over the world and its proceedings had 1400 pages.

The second conference was held in Berlin in April 1979, the obvious idea being to support this dangerously isolated city, even though the conference's means were modest. The Chairman was the Flemish Prof. Roger van Overstraeten — for him, also an opportunity to improve his German. The conference took place in the "Oyster", the new Congress Centre that had just been completed as a gift from the American people to Berliners.

In front of the Conference Centre, we had installed a 5-kW PV generator to drive five water fountains. When Commissioner Brunner came for the opening, it was raining slightly — but still light enough to operate the fountains! Shortly after our conference, the building partially collapsed and since then was no longer used for congresses.

After Germany had to come France. Conference Number 3, at the end of October 1980, was held in Cannes in the Palais des Festivals on the Croisette, and I myself was Chairman. By then, attendance had already swelled to 700 participants. A major event was a PV auction in the prestigious Palm Beach gardens. The auctioneer was Paul Maycock, our friend from the US Department of Energy (DOE); it was an American-style auction and Paul has a good strong voice for that. French cartoonist Reiser had graciously provided a few PV paintings, which were auctioned to the highest bidder — the winner was van Overstraeten. Then, with the money gained, we held an auction to find which PV company could deliver the lowest possible module price; the winner was Solarex, represented by its Chairman, my good friend Joseph Lindmeyer. The module price achieved there was €5 per Watt; the modules went to a PV pump for the poor.

For the fourth conference, we went to Stresa in Italy because it is close to the EU Research Centre Ispra. That was in May 1982; Werner Bloss was its Chairman.

The fifth conference was held in Greece, the origin of European culture — which had in the meantime joined the EU. F. Fittipaldi from the Board of Enel in Rome was the Chairman. The prestigious Opening Ceremony was held in the ancient Odeon of Herodes Atticus, right below the Acropolis. It was attended by Greek actress and politician Melina Mercouri (star of the film *Never on Sunday*), three

Figure 10. The "Oyster" in Berlin, venue for the 2nd EU PV Conference in 1979; the PV generator and the fountain were specially mounted for the event.

Figure 11. Committee of the 3rd EU PV Conference in Cannes, France. From right: Krebs, Wrixon, Grassi, Sirtl, Schnell, Paul Maycock, van Overstraeten, Palz (Chairman), Backus, Barnett and Treble; on the left: Barkats from Aérospatiale in Cannes.

more Greek ministers and VIPs. The sessions took place in an hotel in Kavouri. The organisers, myself included, learnt only later that the hotel had actually just been sold and the personnel had been planning a strike around the time of the conference, but luckily that potential catastrophe was avoided.

Figure 12. The 5th EU PV Conference in Athens: the panel at the Opening Session.

Figure 13. W. Palz with the Greek Minister Melina Mercouri.

The next conference was held in London, England. In 2010, we have the 25th in Valencia, Spain.

Obviously, there was an interface between our EU programmes, with the Implementing Agreements carried out by the International Energy Agency (IEA)

Figure 14. The 11th EU PV Conference in Montreux, Switzerland. Clockwise from top: Leopoldo Guimaraes (Chairman), Luque, Wrixon, Scheer, Pontenagel, Götzberger, Hill, Bloss, Palz (Conference Director), Strub, Treble, and Mrs. Hill.

Figure 15. The 12th EU PV Conference in Amsterdam, Conference Committee. The Chairman was Prof. Hill (2nd from left).

in Paris. They cover all the OECD countries: the Europeans, the US, Japan, and so on. The difference between the EU and the IEA programmes was that the EU had a relevant budget and the IEA did not. As a result, it was often the case that our EU contractors attended — financed by us — all the various IEA meetings — one

Figure 16. W. Palz, Conference Director, making a point at a Committee meeting. At left: van Overstraeten; at right: Hermann Scheer.

Figure 17. Meeting with friends. From right: Charlie Gay (a long-time leader in PV in the US, now the President of Applied Materials), Varadi, Dipesh Shah (then MD of BP Solar), Hill, Ossenbrink, and colleagues.

week in Japan, then in California, then in New Zealand, etc. We from the Commission were obviously not happy with this situation — at least not in those days when I had a say.

My strategy was to have our own co-operation with the US programmes, which were leading the world at that time. To that end, I had an agreement with

Paul Maycock, then the man in charge of solar energy at the US DOE in Washington DC, to organise a joint EC/DOE workshop on "Medium-size PV Power Plants" in Sophia Antipolis, a nice spot on the French Riviera. The workshop was held in October 1980 with some 100 speakers from the US and Europe. The proceedings of the meeting were produced with the Office for Publications of the Commission and published by a commercial publisher in Holland.

Later on, I pushed for a direct political co-operation agreement between our EU programme on solar energy and the American one. The DOE was more than reluctant to have an agreement, as they were happy with the IEA, which had been created after the first oil shock and which, for them, was the right network with the rest of the world. But it happened that I got some support from the responsible councillor at the US Embassy in Brussels. An official diplomatic agreement was established between the DOE and us. It entered into force on December 17, 1982; I was "designated to co-ordinate the EEC's participation in the now-agreed Co-operation". But, in practice, it did not help very much: Bob San Martin, the Deputy Assistant Secretary for Renewable Energy, whom I knew anyway from previous meetings, invited me for lunch in a nice restaurant at the L'Enfant Plaza in Washington. And that was it. Why should they have sought additional co-operation when they were so happy with their IEA?

From the start, photovoltaics were a major component of the EU energy programmes. The aim was to lay the groundwork for large-scale deployment of solar energy in Europe's future energy system. That was a time when the PV global market was less than 1 MW a year; today it is 5000 times larger!

We did not want to miss out anything that could have been important for rolling out a renewable energy, and in particular a PV, strategy, so we prepared and published the results of the corresponding actions, including:

- A strategic assessment of the technologies and their market potential.
- A resource assessment of the solar potential in Europe. In co-operation with Europe's major meteorological offices in a 10-year effort, the Commission prepared the first Solar Radiation Atlas for horizontal and inclined surfaces. We did the same for Africa from satellite data.
- A systematic effort with specialised institutes — like the Fraunhofer Institute in Karlsruhe, Germany — to quantify the social and environmental costs of energy production and usage. After the conventional energy industries saw the conclusions, which did not look promising for them, they immediately embarked on a counter study to prove that oil, gas, coal and uranium were not all that bad.
- A clearing-house activity on misinformation and misunderstandings, such as the energy pay-back time for PV modules [see Part I, Sec. 5.7] and the capacity credit of PV power.
- An observatory of industrial development and the activities of the public research establishments in Europe. In the early 1990s, Europe had the highest density of cell and module manufacturing. Fourteen large or small companies

were involved in crystalline silicon cells and the whole range of thin-film cells from amorphous cells and the newly invented micro-crystalline silicon to CdTe, CdSe and CIS.

At the EU Commission, we put our development priorities into (i) solar cells, and (ii) system technology, full-fledged PV generators that produce electricity when needed. A major priority was crystalline silicon technology. Our R&D effort produced a continuous stream of improvements. Highlights in the early 1990s were the MONOCHESS project for mono-crystalline cells and MULTICHESS for the multi-crystalline variation of silicon solar cells. There was no technological leapfrogging! I would like to disagree here with some PV experts like the prominent and highly qualified Prof. Martin Green from Australia, who keeps talking about 1st generation, 2nd generation, 3rd generation silicon cells.. For me, it has just been a case of incremental improvements. When I returned from yet another PV conference, my superiors at the Commission kept asking me if there was finally a breakthrough. I had to explain to them that this was not the problem — the silicon of the day was already well suited for mass production. In the field of thin-film technology, our programmes focused first on CdTe in the framework of EUROCAD. But our enthusiasm was severely tested when the key player, Battelle in Frankfurt, closed down. Much work was done on CIS, namely, in the framework of the transnational network EUROCIS.

From the outset, our EU programmes placed a particular emphasis on system technology. This was all the more necessary as other national PV programmes, the American one excepted, had put it on the sidelines. Our programme was actually very sophisticated and complex.

Figure 18. The EU PV Pilot Plant at Cork, Ireland, for milk farming, 1983.

Our first priority was PV for building integration. We initiated and supported financially the construction of some of the first PV solar houses in Europe; today, in 2010, there are hundred of thousands of them in Europe. Another key component was the European PV Pilot Programme: 16 different PV systems were built across Europe, covering as far as possible all of the region's climate zones. We put great emphasis on joint technological development: all 16 contracting groups were regularly brought together for meetings that I chaired personally. All groups shared their knowledge and worked together to create a common European base for PV system technology.

All projects, with just a few exceptions in Italy that were late by a few months, were commissioned on time in 1983. The overall power installed was 1 MW, the largest plant being built on the island of Pellworm in Germany. Some of these systems are still operating today.

One of our key advisers on PV systems was the late Mathew Imamura. I got to know him in the late 1970s when he chaired the solar activities at the US company Martin Marietta. He was born on the island of Guam and spent WWII in Japan before eventually becoming a US citizen. After he had finished a job as manager of the 500-kW PV plant that the US had installed in co-operation with Saudi Arabia near Riyadh, I convinced him to come to Europe. He was extremely dedicated to PV and a very hard worker.

The detailed report on the EU PV systems development work is given in an annexe (M. S. Imamura, *PV System Technology Development in the European Community*). The paper was published in 1988 and gives not only an overview of the design work but also reports on the experience of the first five years. The paper is part of the proceedings of the Euroforum New Energies Conference, which we

Figure 19. The PV Pilot Plant at Pellworm, Germany, 1983; the largest PV plant in Europe at that time.

Figure 20. The solemn inauguration of the PV Pilot Plant extension at Pellworm, 1992.

Figure 21. VIPs at the Pellworm inauguration: German State Secretary Neumann (2nd from left), Eckehard Schmid (2nd from right) from AEG, who built the plant, and W. Palz (3rd from right).

organised in Saarbruecken, Germany, in October 1988. It provided a cross-section of all our activities and achievements in renewable energy R&D. But we also had invited presentations from Moscow, California and Japan. The conference Chairman was Oskar Lafontaine, a well-known German politician, and the Chief Editor of the proceedings was Commissioner Karl-Heinz Narjes, Vice President of

Figure 22. M. Imamura (right) with Luque and Palz.

the EU Commission. A much more detailed analysis of the Commission's PV systems development work in the 1980s is provided in *Photovoltaic System Technology — A European Handbook* by M. S. Imamura, P. Helm and W. Palz. This 550-page book was published in 1992 under copyright of the EU Commission by H. S. Stephens, Bedford, UK (ISBN 0-9510271-9-0).

Also in the early 1990s, when we felt the large-scale implementation of solar energy and all the renewable energies was approaching in Europe, we embarked, through my EU development programme, on a new initiative to prepare instruments for integrating these energies into the various levels of society, to prepare for a new energy paradigm. The corresponding programme was called APAS 1994, a French acronym for "preparatory and support actions". Funding of €25 million for contractual contributions through the Commission had been made specially available by the European Parliament. Hundreds of stakeholders from all over Europe participated in a collaborative effort.

This massive programme was divided into five sectors. Here, we mention some of the highlight projects:

- *Urban planning*: Towards zero-emission urban development
- *Decentralised PV electricity generation*: Study of large-scale manufacturing of PV. Co-operation with developing countries
- *Eurenet*: European Regions Network for Renewable Energies
- *Financial resources*: Assessment of international financial resources and opportunities for the development and use of RE

Figure 23. Meeting of the European Solar Council on October 21, 1994, at Sophia Antipolis, near Nice, France.

- *Prodesal*: Towards the large-scale development of decentralised water desalination

In 1993, I set up the European Solar Council, within the framework of the APAS programme. It was an informal platform on renewable energy, also called "The Club de Paris on Renewable Energy", as its first meeting in 1993 was held in Paris — in the Louvre as it happened. It did not handle any budgets and had no legal structure of any kind. It was set up in accordance with an EU Council Decision putting in place the EU R&D Programme 1994–98, which demanded specifically that "actions will be arranged by means of networks, several of which will be linked in a 'Major network for the development of RE'. It will include among others thematic sub-networks, major European electric utilities, leading architects and building engineers, specialised research centres, pilot towns, regions and islands". The Solar Council actually linked those EU networks (which will be introduced in more detail in a later chapter), including:

- EUREC, the EU Research Centres Agency on RE
- READ, the network of architects and building engineers
- CERE, the association of municipalities and towns
- EURE, the network of electric utilities

On the scientific and industrial fronts, we strengthened our PV co-operation with the US, Japan and Asia in general by giving an additional push to our PV conferences. I arranged with Denis Flood, then from NASA Lewis, to hold the 1st World PV Conference with the US/IEEE and the Asian PV Conference organisers PVSEC in 1994 in Hawaii, with Denis in the Chair.

The 2nd World PV Conference followed in Vienna in 1998; the Chairman was Jürgen Schmid from FhG/ISET. At this conference, at my request, we also created for the first time a World Award for Merits in PV. The award winner was Hermann Scheer, MdB. The 3rd World Conference took place in Japan and the World Award went to my friend Prof. Hamakawa, who I would call the father of Japanese PV.

The fifth conference in this series is being held in September 2010 in Valencia, Spain. Like our European PV conferences, the World Conferences attract thousands of delegates from PV science and industry; the associated exhibition in Valencia receives 50 000 visitors — more than the America's Cup that was held in Valencia in 2008.

We entrusted EUROSOLAR in Bonn with organising the first European Solar Prize in 1994. Since that time, the Prize is given regularly to reward the commitment of European stakeholders in solar energy.

A milestone in the political promotion of PV in the EU came in 1993 when Hermann Scheer, President of EUROSOLAR, proposed a 100 000 Roof Programme for the EU. Two years later, he proposed a 100 000 Roof Programme for Germany in the German Parliament; the conservative majority in the Bundestag turned it down. Subsequently, Scheer's initiative inspired other politicians to come forward with programmes of "mass dissemination of PV" on their own. It became part of the EU Parliament's "Mombaur Report" in 1996, Clinton's 1 Million Roof Programme in the US, and the EU Parliament's "Rothe Report" — but they were never implemented. Scheer's proposal began to be realised only on January 1, 1999, following a decision by the Bundestag after he had introduced the proposal again and it had received support from Finance Minister Oskar Lafontaine.

It is true to say, however, that in the early 1990s — while early pioneers like Germany or Japan had already put in place supporting legislation for renewable energies — the EU had not yet done so. The European Parliament called for several Resolutions on the need for more support for renewables. In its Resolution of January 1993, the EU Parliament made a farsighted statement that stands out as one of the defining moments in the development of solar energy: "The EU Parliament considers that a swift transition to energy systems based on renewables, coupled with the necessary moves towards rational use of energy, is the only means of halting the Greenhouse Effect." This Parliamentary Resolution was followed, as mentioned earlier, by the "Mombaur Report" in 1996 demanding a "100 000 PV Roof Programme" and the "Rothe Report" in 1997 demanding a "1 Million PV Roof Programme". (In the spring of 1997, the US was actually putting together "The President's Million Roofs Solar Power Initiative".)

But the European Commission did not react. And without the Commission, no new legislation in the European Union can be put forward. It was only in 1997 that we were able to change that. At that time, the EU Energy Commissioner was the Greek politician Papoutsis, and we were able to establish a direct contact via my Greek colleague at the Commission, Arthouros Zervos. Eventually, in the autumn of 1997, under the responsibility of the Energy Commissioner, the EU Commission published a White Paper on renewable energy, which was favourably received by the European Council of Ministers and the Parliament. The White Paper included some specific targets that I had worked out and published at the Renewable Energy Forum in Amsterdam in May that same year: doubling the share of renewables by 2010, achieving 3000 MW (3 GW) of PV power by 2010 (the European PV market that year stood at 30 MW) and creating 157 000 new jobs. I will come back to the philosophy behind these PV figures in Part II of this book, but at this stage it should be mentioned that by 2010 our PV targets had been comfortably exceeded.

With the 1997 White Paper on renewable energy, the floodgates for EU legislation on solar energy and all other renewables were opened, and the EU was finally in the driving seat:

- In 2001, the first Directive (EU Directives are legally binding for the 27 Member States and its 500 million inhabitants) made it mandatory to generate 21 percent of electricity from renewable sources by 2010; that goal has largely been met.
- A mandatory Directive from 2007/08, worked out under the German and French EU presidency and published in April 2009 calls for a target of 20% renewable energy penetration in the European Union by 2020. In this framework the EU member countries commit themselves to achieving a share of 40% electricity from renewable sources by that time — a very ambitious target indeed. As for the share of PV as part of the conglomerate of technologies for that electricity generation, the Directive demands that each one of the 27 EU member countries submits its own target for approval by the end of 2010.

6.7 The G8

The G8 (France, Germany, Italy, Japan, the UK, the US, Canada and Russia) once represented a grouping of the world's most influential states. But that has changed in the last 5 or 10 years: with the emerging economies of China, India, Brazil, and a few others, the G8 are no longer the only important players on the global stage.

In any case, at the G8 Summit in Okinawa in 2000, it was decided that a task-force would be created to put forward recommendations on "sound ways to better encourage the use of renewables in developing countries". VIPs from governments and industry around the world, the EU, the IEA, etc. joined this task-force under the leadership of Corrado Clini from the Italian government and Mark Mood-Stuart from Shell. I also made a contribution as a member of the EU Commission's delegation.

An extensive report was produced and published in time for the following G8 Summit in Genoa, Italy, in 2001. It recognised in much detail all the benefits of renewable energies for the global economy. For PV, it projected an installed global capacity of 120 GW by 2020, increasing to 650 GW by 2030. The cost projection for PV was not unreasonable either: between US$1000/kW and US$2000/kW, depending on the need for additional investment to reinforce existing grid lines.

In its Genoa communiqué, the G8 came out with encouraging phrases like: "We recognise the importance of renewable energy for sustainable development, diversification of energy supply, and preservation of the environment." The reality was different: the newly elected George W. Bush disliked the whole enterprise, and the report ended up as waste paper in the archives.

6.8 *The Energy Empire Fights Back*

The established energy business is not something small. In 2007, the European electricity market alone generated total revenues of over €300 billion. It is understandable that this conventional energy establishment will, initially at least, not sit idly by as a new generation of solar enthusiasts stands up and tells them that their business is bad and has to go. It takes time for an old system to understand that a new era is coming and that it has no choice but to adapt.

At first, the conventional utilities disregarded the defenders of solar energy as "quantité negligeable". But some tried to ensure that they would not even get a chance. In the early 1980s, the French government reduced the national budget for PV to zero — after the Reagan administration in the US had just about done the same. At the time, a director at a national French utility declared that wind energy was like an HIV infection for the grid and that he would immediately retire should a turbine be connected. But things change: today that utility is a leading player in the wind and PV businesses.

When Hermann Scheer MP succeeded in getting the first "feed-in tariff" for wind, small hydro, and a few other renewables through the German Bundestag in 1990, he took the established system by surprise. Political friends of the conventional utilities had not expected that in one of the last parliamentary sessions much would happen — and they had already departed for their summer vacations. When the electric utilities discovered that the new laws had given them the key role of promoting wind energy, they reacted furiously: they went to the highest courts in Germany, claiming that the law was illegal since subsidies are incompatible with international competition regulations. But the courts found that these were not subsidies, as the utilities had to pay for the tariff, rather than any public sources. The German utilities subsequently went to the highest European Court — only to lose again.

The conservative energy establishment also had its friends in the media and in administrations. Franz Alt, a well-known German TV journalist with a programme on nuclear plants, new energies and the environment, which was very popular on state TV in those days, had to continually defend his case against his own management.

In July 1990, a courageous Greek expert working at DG XVII — at that time the code name for the Energy Directorate of the EU Commission — had projected the first GW of wind power ever for European grids by 1996. Komninos Diamantaras published his analysis in "Information on Demonstration Projects" in September 1990. His superior, the Director Fabrizio C.D., found the projection monstrous and removed Diamantaras from his post. But the projections were adopted a year later by the European Wind Energy Association (EWEA) in "A Plan for Action — Wind Energy in Europe", which received harsh criticism from the nuclear lobby. Today, we are fast approaching 100 GW of wind power in Europe's electricity networks.

At that time, I had a prominent position at the European Commission in Brussels as the Head of the Division for the Renewable Energy Development in Europe. My group was located in the same building and under the same hierarchy as the European Directors for Nuclear Energy and for Nuclear Fusion Development.

My solar energy agenda was considered too "noisy", and my DG, an Italian professor of biology, decided that I had to become the European "Testing Chief". Many Members of the European Parliament and influential friends in Brussels tried to intervene on my behalf, but my superiors refused to revise their decision. Eventually, the German government discreetly intervened and I was moved back to the "solar job" with the personal apologies of the Commissioner, another Italian politician.

After that, throughout the early 1990s, there were incremental efforts to restrict my freedom to do the job. Directors Herbert A. followed by Ezio A. were instrumental in that respect. It ended with a little "scandal" around 1995, when my superiors, with the explicit approval of the responsible Commissioner, downgraded one third of our PV proposals that we had — strictly following the rules — retained for funding; the dispute concerned a budget of over €30 million. The European Parliament began an investigation and German television reported it in a major programme.

The EU Parliament would later remove the whole Commission from office because of other mistakes that same Commissioner had made. And it was a particularly competent Commission: it was they who had facilitated the introduction of the euro as Europe's common currency.

Part II

Solar Power for the World

1. Basics for a New Solar Age

1.1 *The Ethical Imperative of Photovoltaics*

Solar energy is not like any other energy! The Sun plays a central role in our existence: the Sun heats the globe to an acceptable temperature; otherwise, thanks to geothermal effects, it would be just a bit more than the −270°C prevailing in the Universe. And the Sun is at the origin of life. Photosynthesis, a highly complex quantum process that involves four photons hitting a molecule at the same time, is the source of life. And that is not even acknowledging the fact that to begin with there was no free oxygen at all in Earth's atmosphere: all the oxygen we breathe today has at some point been liberated through photosynthesis.

Life and photosynthesis are gifts from Nature. Not so photovoltaics: PV cells are very much man-made. The effect is actually less complex than that in photosynthesis, but to have invented it is one of mankind's major scientific achievements. And this particular science has not just provided us with a nice new theory but with something very practical: a noble way of harnessing the energy from the Sun — our "mother" — in harmony with our natural surroundings.

Solar PV is a unique semiconductor process that has no equivalent among all energies at our disposal. Its inherent characteristics are unique, too: it is clean, it generates no "greenhouse gases", it is available everywhere on Earth — the ultimate decentralised energy source — from microwatts to megawatts to gigawatts, and it will be available, for all practical purposes, forever! The same is true for the semiconductor materials that are needed for the production of solar cells, in particular the silicon that is non-toxic and almost everywhere around us; it is one of the most abundant materials in the Earth's crust.

Power for the World by W. Palz
Copyright © 2011 by Pan Stanford Publishing Pte. Ltd.
www.panstanford.com
978-981-4303-37-8

1.2 *Cost and Social Acceptance: Ingredients for a Viable Energy Strategy*

In adopting a strategy for PV implementation, it is wise to aim for a basis that is acceptable to all players in the overall global energy field. That ultimately concerns all of us, but it is also a global business worth over US$5 trillion a year, half of it involving power generation. It is better not to link this to ideology because there are so many different ones: in business terms, solar ideology does not carry much weight.

A fair playing field for trade-offs in energy markets and strategies is simply cost. It is the right common denominator for all kinds of energy. In view of the extraordinary budgets involved in the energy sector — see the figures previously mentioned — it must be a major concern for national and global strategists alike to ensure the most rational possible allocation of funds. There are so many different options among the fossil, nuclear and renewable energies that decision makers without a specialised background in energy can easily be led down the wrong path, which may ultimately entail the waste of tremendous financial resources.

Costs and prices are things everybody can understand. But costs must obviously be comparable and established with full transparency outside of vested interests — by eliminating overt or hidden subsidies or other distortions of taxpayers' money, by demonstrating all hidden extras in the clients' utility bills, and by doing away with any other form of intervention with public money.

Although hardliners in the energy business are perhaps yet to appreciate it, cost considerations do also have to be balanced by social considerations. Ultimately, all energy solutions must serve people; they must be socially viable. And given that decisions about investments in energy may have implications that extend over a period of 50 years or more, they must be sustainable not only for ourselves, but also for generations to come.

We have already underlined the following characteristics of PV, among others:

- it is clean in every respect
- it is available everywhere: there is no security of supply issue and no need for imports;
- it meets energy needs in any form, decentralised or otherwise;
- the lifetime of power plants easily exceeds 30 years;
- no waste disposal is needed;
- decommissioning of power plants is unproblematic.

In the following, we quickly review some essential characteristics of the other energy options considered for large-scale deployment now — and in the future in competition with PV. For all fossil and nuclear energies that come in large centralised units, one has to consider the following costs to be added to the net market prices as we see them today:

- subsidies using public money of several US$100 billion that interfere in the markets each year to manipulate the prices of conventional electricity or fuels;
- the cost of the grids, pipelines, and maritime routes for transport and distribution of electricity and the various fuels;
- the loss of electric energy within the grid, which typically exceeds 10 percent of the production leaving a power plant;
- the all-too-often non-transparent way that utility operators define the transmission and distribution costs to their clients;
- as far as transport by ships is concerned, the cost for society of a military presence to provide security in some dangerous waterways in the Gulf or Southeast Asia;
- the cost of diversification of supply: if, for instance, Western Europe cannot be confident enough that the natural gas from Russia passing through Ukraine will be reliable all the time, other sources that may be more expensive and cumbersome have to be kept "on the radar"; bringing gas after liquefaction by ship from Qatar provides the needed diversification, but it adds to the cost.

A separate problem is the consumption of coal, the worst of the fossil fuels as far as carbon dioxide emissions are concerned. On the other hand, it is relatively cheap and abundant: coal will still be available when the last droplets of oil and gas have long since been pumped out of the wells, and coal — "dirty coal" in particular — is plentiful in some of the most energy-hungry countries, such as the US, China and India. Hence, "clean coal" has recently been placed on the agendas of energy experts around the world in association with the term "carbon capture and storage", or CCS. Clean coal is certainly an interesting concept, provided that some of the following conditions and the costs they imply are considered:

- the efficiency of such power plants, which include the extraction of the carbon dioxide after combustion, is considerably lower than those without;
- such plants are a lot more expensive to build; one of the concepts being developed today is additional coal gasification instead of direct burning;
- the delays in building such plants are comparatively longer, a fact that adds to the financing costs;
- pipelines for bringing the CO_2 from the power plant to the sequestration site add to costs and raise issues of acceptance by the people living nearby;
- the sites for CO_2 sequestration are costly to develop and to maintain; there may be additional costs associated with insurance against accidental release of the gas.

Another hot item on the agendas of decision makers on energy matters around the world is atomic energy. I have some personal insights into this area as it was part of my syllabus when I was a student of physics in Karlsruhe, a well-known German centre for nuclear technology. My lecturer on atomic power plants was Prof. Karl Heinz Beckurts, who was later the boss of Siemens and tragically became a victim of a fatal attack by a home-grown left-wing terrorist group.

Before addressing the cost aspects of atomic power, one has to look at the social challenges. With much controversy surrounding it, nuclear power can only be considered when there is a broad consensus on the subject in society — no matter how low or how high the production costs of electricity may be. And as not everybody can be an expert in nuclear technology, people have the right to be well informed about all the pros and cons before they make up their mind. For instance, transparency has ensured that people have been made aware that living close to a nuclear power plant is not without risks of leukaemia and other cancers.

Another aspect that lies somewhat outside cost considerations is the interface with nuclear weapons. The fact that this may be a particular headache is exemplified by the ongoing discussions with Iran on the subject. Good luck to all those who are considering nuclear power in Abu Dhabi, Kuwait and Africa.

In addition to the net building and operation costs of a "sustainable" atomic power plant — putting aside that for some people this is a contradiction in itself — there are further considerations that may be summarised in the following:

- All construction costs should be borne by the owners, without exception. Financing is also an issue: it is current practice that the industry enjoys preferential rates of interest, borne by public funds.
- The security of nuclear power facilities is quite another matter compared to a plant that uses fossil fuel. Confidence in France was shaken in late 2009 when the government had to intervene to warn builders of the latest generation nuclear plant in Normandy that the security system on the drawing board was inadequate. In this case, considerable additional costs resulted from the need for hasty modifications.
- Plants need to be fully insured against accidents, including the severe case of an explosion. Insurance coverage must be borne by the owners, not by public resources.
- Plant owners must bear the cost of proper nuclear waste transport and disposal.
- The cost of decommissioning a plant at the end of its life must be borne by the owners; such costs can be as high as the costs of construction.
- Construction time over many years and typically also delays result in costs that do not exist for PV and other renewable energies.

1.3 *PV as Part of a Holistic Approach Towards Renewable Energy Implementation and Energy Conservation*

Could solar PV alone meet the planet's total electricity demands?

According to the US Energy Information Administration, the overall installed PV power capacity worldwide stands at 4,500 GW in 2010. To illustrate this abstract figure, consider that the US has installed 1000 GW in total, China has 800 GW, and Germany has 120 GW. Consider, too, that 1 GW stands for 1 million kW, and 1–10 kW is the range of power capacity that a household in an industrialised country typically demands. By 2025, the US Administration projects that worldwide power capacity could hit 6000 GW. By that time, China will have surpassed the US as the country with the highest electric capacity installed.

If we wanted — just in theory! — to meet all our electricity demands worldwide with PV, we would need roughly five times that capacity of 6000 GW projected for 2025: 1 kW of PV capacity generates on average some five times less electric energy over time than the conventional electric power plants of today. This means that we could meet our worldwide electricity needs with PV generators of 30 000 GW in total deployed across the globe. All in all, these hundreds of millions of PV generators would require an area just half the size of France.

Hence, finding enough space would not be a problem. But storage would be a challenge. In an energy system utilising solar PV exclusively, the need for storage to bridge periods of deficient radiation would be extraordinarily large. With the mobilisation of the best hydrogen, electrochemical battery, pumped hydro, and compressed storage technologies, among others, it may eventually be possible — though not likely by 2025 but later.

Thus, the answer to our initial question is "yes". Technically speaking, it would be possible for PV to monopolise the global electricity system — but for cost reasons, such a proposal would simply be nonsense.

The solution will need to combine PV power with other forms of solar-derived power and the other renewables. Ultimately, a comprehensive hybrid system in which solar power, wind power, bio-power, hydropower and others complement one another will not only be the cleanest but also the cheapest solution. The time will come when the renewable energies together will beat all other conventional forms of energy in terms of cost effectiveness. PV is smart and attractive; it is the orchid among energies — but you also want to see other flowers in your garden, not just orchids. Ecologists call it the monoculture issue. These other abundant flowers should be the renewable energies, not the classical non-renewable ones: you would not want your backyard full of cooling towers and chimneys from coal or oil plants, or the domes of nuclear reactors.

The first alternative renewable energy that comes to mind is biomass; in many parts of the world, it is the dominant renewable energy in use today. Indeed, in quantitative terms, it is always going to be important. In the organic world in which we live and of which we are ourselves a part, the cycle of life ultimately creates, in enormous quantities, material such as agricultural and forestry residues, sewage and other liquid effluents from farming; the most appropriate way of making use of them is by extracting the energy content by means of burning or biological digestion.

There are also important additional markets like short-rotation forestry and other energy farming, liquid biofuels such as ethanol from sugar, starch and vegetable oils, and biodiesel from oil seeds. Biomaterial is used as wood and other cellulose material for burning — that is as old as man's discovery of fire — and in more modern forms as wood chips and pellets. Methane is produced from landfill waste or using dedicated biogas systems where effluents are digested sometimes in combination with whole energy plants like maize.

Wind power has to be included in our analysis as well. Wind power has developed tremendously over the last 15 years or so: within a very short time it is accounting for 4 percent of the world's total installed power capacity; by late 2010, close to 200 GW of wind power capacity will be in operation. In Europe at least, it is true that the way wind turbines are installed today needs improving: some are not working at all and others are real eyesores. This is the result of a rapid growth in wind turbine business between 2005 and 2008 and too much easy money for the installers. But these setbacks are not serious. There are essentially two issues standing in the way of even more extensive deployments of wind turbines: firstly, the lack of sufficient grid capacity for electricity transport between wind-rich areas and the main consumption centres, such as in China; secondly, the challenges of developing the market for offshore wind farms, such as in the North Sea.

Hydropower is a heavyweight among the renewable energies as well; it is going to maintain its strong position for the foreseeable future. China, which dominates the world market in this sector, has a particularly strong industry, in particular for the mini-hydro segment (plants of up to 10–50 MW power capacity).

Among all renewable energies, hydropower and bio-power systems can be operated continuously, while solar and wind resources are intermittent. By operating several of them in combination, one can ensure a 24-hour continuous supply. This has already been demonstrated in Germany, where plants in different parts of the country connected via the national grid were run in "harmony"; they were managed in such a way that permanent supply was available as needed at any time.

How much solar PV power capacity will make it into the world market in the long run can only be a matter of speculation at this stage. Twenty percent of global electricity generation, corresponding to an installed power capacity of 8000 GW distributed across every continent within 50 or 60 years is not an unreasonable projection. In practice, what we have achieved at best today in 2010 is a 2 percent electricity contribution in Germany coming from 10 GW of PV, i.e., 10 percent of the national power capacity installed.

Another solar power option different from PV should also be mentioned in this section. Plants harnessing solar thermal power (usually called concentrating solar power, or CSP) employ the concentration of solar radiation via mirrors to produce temperatures of 300°C to 500°C, depending on the type of plant, in a circulating fluid. The fluid drives a turbine, often coupled with a natural gas-driven power generation system.

There are more than half a dozen different types of systems, the dominant ones being a tower system surrounded by many reflecting mirrors that concentrate the Sun's rays on a receiver on top of the tower, and another with curved reflecting mirrors in the form of troughs.

The first solar tower system in Europe and the world was Eurhelios. It was built at the end of the 1970s in Adrano, south of Mount Etna, in Sicily, by an international consortium under contract with the EU Commission. I was the project manager. The 1-MW plant was eventually commissioned and fed electricity into the Italian national grid. Later, the Commission discontinued its activities in this sector, as the experience gained was not conclusive. PV was considered more promising and became the preferred option for further development.

Other thermal power plants of this type were built in the US, Spain, France, and other countries. Solar Energy Generating Systems (SEGS) in California, which uses parabolic mirrors, is the largest solar power plant in the world and was built around the time of President Carter in the late 1980s. But international interest in both options — tower systems and the trough-mirror plants — vanished very rapidly.

Over the last few years though, CSP systems have been attracting renewed interest. The Germans and others are international leaders. By the end of 2009,

Figure 24. The 1-MW Eurhelios, the world's first solar thermal power plant, at Adrano, under the Etna volcano, in Sicily, Italy, 1979/80.

Figure 25. The field of heliostat mirrors and the receiver tower at Eurhelios.

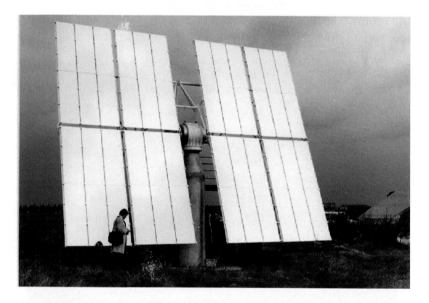

Figure 26. A heliostat at Eurhelios and W. Palz, project leader, 1979.

80 MW of new CSP systems were running in southern Spain; there is still much interest in the dry, sunny areas of the south-western US.

The Germans have also put CSP on their agenda with proposals for a large-scale implementation in the Sahara Desert with transmission of the generated solar electricity to Germany by cable. The so-called Desertec project would cost some €400 billion.

Figure 27. W. Palz at a 50-MW solar thermal power plant with "mirror trough" concentrators at Puertollano, Spain, October 2009.

CSP for electricity generation from the Sun cannot, in general, compete with PV. Unlike PV, CSP technology is strictly limited to deployment in "clear-sky" regions, and it only comes in multi-MW units. PV can also be built in large area units and in hot climates, but it is not restricted to this type of use; and it does not have the drawback of needing a cooling source as CSP does to get the thermodynamic process running. Supporters of CSP usually stress the possibility for continuous operation via a built-in thermal storage of the heat energy collected by the optical concentrators. PV does not offer this possibility, as radiation is transformed directly into electricity without a thermal process.

As previously explained at length, the best way to differentiate between competing technologies is cost. The figures publicly available today provide sufficient evidence that PV is cheaper to install and operate than CSP. The addition of costly thermal storage puts CSP at a further financial disadvantage. Millions of PV systems of all sizes and for all climates — from the South Pole to the hottest spots on Earth, and even in space — have been installed to date, compared to only a handful of CSP plants. Today, the total capacity of PV installed worldwide is over 25 GW — more than an order of magnitude more than that of all CSP systems combined. The confidence in the operational, financial, commercial and social aspects of PV really cannot be compared.

To conclude this section, a word about energy conservation. As the saying goes, there is no cheaper energy than that which is not produced nor consumed. Rational use of energy and energy conservation are extremely broad and diverse areas of interest, encompassing both the supply and the demand sides. Energy conservation deserves the highest priority, but it cannot be the only energy story: one has to first talk about energy generation before there is something to speak about conserving. Going into any detail on this subject is beyond the scope of this book.

1.4 *What about the Power Plants on the Road?*

1.4.1 Car drivers and their power plants

We have already seen that — with the notable exception of PV — 99 percent of global electricity generation involves the intermediate step of "mechanical power" conversion. In virtually all cases, this is achieved via a turbine — except for a few diesel generators.

The automobiles on the world's roads are all driven via mechanical power as well. And there are many of them: more than a billion cars and trucks, each year joined by more than 50 million new ones. As long as we have driver's licences and vehicles, we are all power plant operators!

All cars are driven by "piston engines"; they do not employ turbines like in electric power plants (they would have to be micro-turbines — and that turns out to be less practical). The combined power capacity of all automobiles adds up to a tremendous figure: more than 60 000 GW of "piston engine" power. The mechanical power of combustion engines in cars on the roads today is 10 times the power of the global electricity generation capacity in terms of "turbine" power!

Only when one considers the energy produced, does one find a better equilibrium. A typical large turbine in an electric power plant runs some 6000 hours a year while a car drives in general no more than 200 hours during that period. The energy generated by electric power plants globally is somewhat higher than that of all the world's cars in a year.

Today's car engines are technological marvels. They represent an extraordinary maturity of development. They are extremely reliable, too, and it is for this very reason that anybody can drive a car without having to constantly rely on a mechanic.

When applying the criterion of cost as the benchmark for energy trade-offs as we introduced previously, we again find an impressive advantage for the car engine: its cost per kW installed is some 20 times lower than that of the turbine. It is only when it comes to operational lifetime that the turbine comes out on top; it can easily run 30 times longer over its useful life than a car engine.

In conclusion, it makes a lot of sense to also consider car engines for stationary applications of electricity generation, provided one can reduce the running time and increase lifetimes accordingly. This can obviously be achieved via the addition of an electrochemical battery. Ultimately, this means a hybrid system, a stationary version of a hybrid car engine, of the type used in a Toyota Prius.

In comparison with a traditional heating system in a building, the hybrid system can provide full autonomy, as the same fuel serves to meet all the electricity demands of the building.

The efficiency would be much higher than that of an equivalent system in a hybrid car. In a car, the heat generated in the combustion engine goes into the air as waste energy; in buildings it could be used to meet heating demands.

This type of stationary system is currently not yet in use anywhere. One reason may be that it has only been a few years since the emergence of the hybrid car. We will discuss this issue further in a later chapter on future strategies.

1.4.2 Mobilising PV for transport

Operating PV on vehicles is a much greater challenge than integrating it into buildings or having it firmly installed on the ground. Stand-alone systems of this type cannot rely on the electric grid for back-up power and must include a battery if they require bridging times where there is insufficient radiation. It is true that space satellites also belong to this category; but there the power needs are relatively small and are not for propulsion.

Early on, the challenging task of PV for vehicle propulsion was taken up with enormous enthusiasm. And the people involved had fun — very much so in fact! Passenger boats that cruise on lakes have been built and a few of them are in operation here and there. Even ultra-light aeroplanes, relying exclusively on PV to drive their propellers, have been built and flown — raising enormous public attention in the media. "Solar Challenger" was the first solar aircraft; it

Figure 28. A Swiss solar PV boat.

Figure 29. The world's first solar electric car, the 1912 Baker Electric Brougham, probably with selenium solar cells.

was built in the US by Howard Hughes and Paul MacCready. In 2012, Bertrand Piccard plans to fly around the world in his aircraft "Solar Impulse". (Piccard was part of the first team to fly around the world in a balloon in 1999.) The plane's four electric motors will be solely powered by PV. A prototype of the aircraft already exists and got a first roll out at Zurich airport in Switzerland in 2009.

Perhaps the most impressive development of them all was PV for automobiles. One of the first light trucks fully covered with solar cells is from 1907. The photo in Fig. 29 shows one from 1912. Switzerland, a pioneer in the field, became involved in solar cars in the early 1980s. The starting point was the involvement of the Swiss in a European programme, COST, to build a "hybrid car" with Volkswagen. Since 1985, solar car races have been held in the Alps — the "Tour de Sol" — and since 1987, solar cars have raced 3000 kilometres across Australia from north to south in the World Solar Challenge. The US Sunraycer was the winner of the first race. GM developed the car with NASA software for US$15 million. Fifty-three cars from countries such as Japan, the US and Switzerland took part in the third event in 1993, which was won by Honda. Dick Swanson, a Becquerel Prize Winner who had provided the solar cells for the Honda car, personally attended the festivities in Adelaide, Australia.

Representing the Swiss colours at the World Solar Challenge, the "Spirit of Biel" (Biel/Bienne is a city in central Switzerland) was co-sponsored by Nicolas Hayek, better known for his Swatch watches. Hayek understood that the world would have a problem if everybody had a car with a combustion engine. Hayek talked to Renault about it and eventually announced his solar electric "Swatchmobile" in 1990; a year later, he had a joint venture with Volkswagen to develop it in Biel. The design followed one that the glider manufacturer Bucher in Switzerland had developed in 1985.

What remains today of Hayek's forays into automobiles is not an electric car, but the "Smart" car built by Daimler-Benz, the result of another joint venture between Swatch and the car industry.

These days, electric cars are back in the public spotlight. Toyota's Prius hybrid, already in its third generation, is available upon request with a PV roof. But this does not help with propulsion — it is more of an ecological "goodwill flag".

2. Driving Forces

2.1 *Aspirations of the People*

For many years, opinion polls conducted on the acceptance of solar energy have demonstrated overwhelming support and virtually no opposition. This is surprising, since at the beginning of the appearance of PV on the market, people did not have a precise idea about what solar energy really meant. Even the elites and normal politicians had never heard the term "photovoltaics" — well, it's true it is a bit of a tongue-twister.

People were even ready to pay a little extra to get green energy instead of the conventional ones: this was the idea of the green tariffs. In those days — in the 1990s — my own position was against the idea of higher prices for cleaner energy: it could not be right that people buying "clean" energy had to pay more than those who bought the polluting energies. In fact, this is the background behind the "feed-in tariffs" (FITs) that triggered a revolution in PV deployment; we will come back to this and other support schemes later. FITs actually ensure that people who

Figure 30. Ecological people's club, Tokyo.

Figure 31. Ecological women's club, Kathmandu, Nepal.

buy polluting energies pay more per kWh than those who buy clean solar energy. The extra financial support that is needed to trim the solar cost down to a competitive level is borne by everyone, not only those who commit themselves to a noble endeavour.

In the past it was normal practice for the subject of electric power generation to be kept out of public discussion — a deficiency in the democratic system that periodically caused conflicts in society. This was the case for atomic power plants in particular, but also for coal and for large hydroelectric dams. Even the deployment of wind turbines suffers to some extent from the NIMBY, or Not In My Backyard, effect. To dampen such opposition, promoters of projects often offered some financial benefits on top of local job opportunities and tax benefits.

PV has not suffered the same problem: it is not only widely accepted, but people have demanded it — and right in their backyard. This is an entirely new situation that we have hardly ever seen before when it has come to finding sites for power plants. Opposition has only been raised when some developers have planned to install mega-size PV plants on, for instance, good agricultural land or in areas of natural beauty that are of a high value for leisure. But there are enough other sites available — in Germany the first such plants were installed on abandoned military sites — but above all in particular in the built environment.

2.2 Preserving Nature and Alleviating Climate Change

I remember well that even in the late 1990s, there was not yet general acceptance that man's activities were causing climate change. In the Enquête Commission on

Energy of the German parliament in Berlin in 2000–02, of which I was a member, we spent a great deal of time refuting claims that changes in the global climate were simply due to the Sun's activity cycles.

When the Intergovernmental Panel on Climate Change (IPCC) declared that there was indeed a "global change" in Earth's climate due to man-made greenhouse gas (GHG) emissions, there was a general political consensus to combat those emissions — namely, those due to the operation of conventional fossil fuel burning power plants. In 1997, the Kyoto Protocol was adopted, fixing emission reduction targets by 2012. It was ratified by 187 states, but crucially not by the two nations with the highest emissions: the US and China. Although the EU — the "top of the class" — is about to achieve the commitments it made, globally the CO_2 content in the atmosphere never stopped growing (except in 2009 when the financial crisis brought down GHG emissions by 2 percent).

In December 2009, negotiations took place in Copenhagen in an attempt to find a treaty that could replace the Kyoto agreement. Though the presence of the world's most powerful heads of states and governments was an encouraging expression of the awareness of the gravity of climate change, the talks in Denmark ultimately made little headway. With national interests taking precedence, it seems that our planet is destined to continue on its collision course, and the consequences of inaction are certain to affect everyone — in particular those in developing countries.

Solar energy and the renewable energies in general are the only options — next to energy conservation — to solve the climate crisis, and this conclusion enjoys wide consensus among energy experts around the globe. There are certainly those defenders of "clean coal" and atomic energy, but both remain the subject of social and political controversy that has escaped the renewable energies. Solar energy and PV in particular do not emit GHG. This is also true of wind, hydro, and other renewables; even the use of biomass is neutral in terms of CO_2 emissions. We will see later in this book that they could be exploited extensively, rapidly, and at reasonable cost.

Carbon dioxide is not the only problem associated with conventional power generation that does not apply to solar energy. Coal combustion is a particularly severe polluter. Emitted compounds such as sulphur reacting with water molecules produce acid rain, and Europe had to make a comprehensive effort to equip its plants with emissions cleaning technologies to mitigate the problem. China, too, is enforcing strong measures to promote "blue skies" in its cities: in Beijing, heating with coal furnaces is being progressively eliminated, with a very real effect on air quality.

There are also pollutants that cannot be seen nor smelled. The radioactive effluents from nuclear waste reprocessing plants at Sellafield in England and at La Hague in France can even be detected at Murmansk in the Arctic Sea. In the early 1990s, our EU programme funded the first wave power plant at Inverness on the north coast of Scotland. At one stage the whole power block fell into the sea and could not be recovered: it was officially declared radioactively contaminated by the waters of that sea.

Switching to solar energy kills two birds with one stone (that's only a saying: solar energy kills nobody!) as it does not release harmful and noxious emissions or contribute to climate change.

2.3 *Peak Oil*

Logic dictates that the quantity of fossil and uranium resources in the ground must be limited; if they were somehow replenished, we would certainly know about it. Serious insiders in the oil industry claim that "peak oil" — the point in time when half of the recoverable global petroleum resources have been consumed — is now upon us. This is based on predictions by the geoscientist M. King Hubbert, who had correctly predicted that peak oil for domestic reserves in the US would occur between 1965 and 1970.

Colin Campbell is just one of the many oil experts who also defend the oil peak theory. In the US, many reports have been written on the subject. In Europe, the Energy Watch Group, under the auspices of EUROSOLAR, is a prominent analyst of peak oil. It is associated with the Bölkow Foundation with the protagonists Jörg Schindler, Werner Zittel, Harry Lehmann, and others, with political support from Hans-Josef Fell MP.

However, the theory also has many opponents. Listening to some representatives of the oil industry or of the US Geological Survey, you could easily get the impression that there must be an unending supply of oil and natural gas. As far

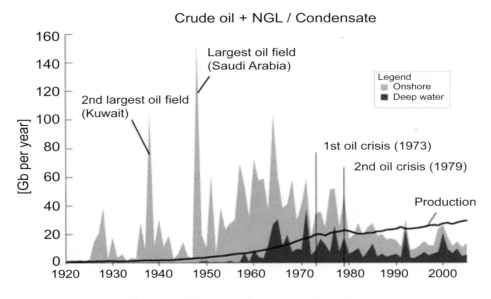

Figure 32. Oil consumption and new discoveries.

as the IEA in Paris is concerned, we have recently seen a shift in position. Once an advocate of almost unlimited fossil resources, more recently it has joined the swelling tide alarmed by the prospect that all of the world's hydrocarbon resources might be exhausted 50 years from now — oil a bit earlier; natural gas a little later. Only coal seems to be in sufficient quantities to last well into the next century — but that is the most odious climate killer of them all.

Once again, solar energy is the obvious alternative, a way to free ourselves from the worries of peak oil and the risks of damaging market speculation, as we saw in 2008. While the fossil resources represent a static deposit of energy capital, solar energy is absorbed permanently by our planet in huge quantities: the energy received from the Sun was around 10 000 times the average of 15 000 GW of global consumption in 2008. And that supply is, for all intents and purposes, unlimited: there will be no "peak solar energy" for billions of years to come.

2.4 *Energy Security of Supply*

Large blocs of countries, like many individual nations, rely to a large extent on energy imports. The EU currently imports half of its total energy; Germany imports 62 percent. Japan is one of the most vulnerable countries, importing 83 percent of its energy. China and the US rely heavily on their own coal resources, which makes them look a bit better in terms of energy independence — but worse as GHG emitters. The US still imports 31 percent of its total energy. China relies on coal for 71 percent of its consumption but has to import 40 percent of its petroleum. The upward trend of fossil energy imports looks unstoppable: politically and economically, the world's major economies are becoming increasingly vulnerable.

At the same time, it is common knowledge, at least since the oil price shock of 1973, that the hydrocarbon resources in the ground are concentrated in the Middle East, which is not a very politically stable region. The risk of military conflict is inherent in the region and the costly deployment of armed forces is a necessity. As for European gas supply from Asia, three different pipelines are in use or under construction, but ultimately it will be hard to escape Russian control — as some politicians would prefer.

This situation is clearly not sustainable: the risks of possible disruptions to supply are just too high. A responsible policy for a sustainable and secure energy supply must be sought from domestic resources. Unlike fossil energies, renewables are available everywhere, and they are the only viable solution from that point of view.

In the autumn of 2009, Germany made some very enthusiastic proposals for the deployment of huge solar plants in the Sahara Desert (the Desertec project) that would supply electricity to the country via long-distance cables. But on the crucial point of security of supply, such plans are not realistic. Priority should be given to the deployment of solar plants in North Africa to meet its own needs, rather than for export; Germany has adequate solar resources of its own.

Such conclusions do not mean one should not balance solar or wind exploitation between different regions via transport by electric cables. One example is China, which has ample renewable resources in the west of the country far away from the consumption centres on its eastern coast. Others are the US, with its high solar irradiation in the southwest and strong winds in the Midwest, or Germany, with the huge wind resources of the North Sea. Connecting favourable production sites with the major demand centres makes good sense, but the exploitation of renewable resources at only the most favourable sites should not be the exclusive approach. In particular, at the micro scale, solar energy is a possibility just about anywhere. Building-integrated PV is indeed a ubiquitous solution, ensuring security of supply on the individual scale.

I have had a PV generator on my house in Brussels for many years. It was one of the first in the city, built before the Belgian government introduced any subsidy schemes. My solar generator also has a large storage battery. In September 2009 when half of the city was without electricity for several hours, visitors to my house wondered how we had electricity while nobody else did. This is security of supply for the people.

3. The Role of Stakeholders in Society

3.1 Governments and Administrations

Solar PV is very much a newcomer on the world's energy markets: before 2000 it was largely unknown by the general public. Mass production is the key to reducing costs and gaining competitiveness on liberalised markets: historically, PV was an expensive technology, partly because, for a long time, its key market was extremely small and restricted to applications in space where the main criterion is weight rather than cost. Intervention by governments has been indispensable in providing the necessary impetus to get out of the vicious circle of "no market — no mass production — no market". Measures to push technologies considered by a number of authorities have taken various forms: carbon trading, green certificates, quota systems, and renewable portfolio standards (RPS), to name just a few.

We have already seen that the financial support that was first tried, in particular in Germany and Japan, was for "PV roof programmes". The first such programme was initiated in Bonn by Walter Sandtner on behalf of the German Research Ministry; it financed 2,200 PV roofs, beginning in 1989. From 1994, a similar programme was conducted by the Japanese. A new German 100 000 PV Roof Programme ran from 2000 until 2003; it was actually a 300-MW PV programme that included a "zero interest rate credit" and a 12.5 percent subsidy. The purpose of the FIT that was decided later was to strengthen the 100 000 PV Roof Programme and

Figure 33. "Solar conspiracy".

enhance its profitability. Initially, only projects that were already part of the 100 000 PV Roof Programme were eligible for further support by the FIT.

The 300-MW cap was removed by law — despite the opposition of Jürgen Trittin, Minister of the Environment at the time — following a "preliminary law" (*Vorschaltgesetz* in German) initiated by Hermann Scheer in the Bundestag. This happened in autumn 2003 to avoid any gap in the funding for PV; the Renewable Energy Sources Act (EEG in German) became law only in 2004. To compensate for the disappearance of funding via the 100 000 Roof Programme, which had been completed in the meantime, the FIT was increased by some 20 percent to 59 euro cents. It was this measure that initiated the boom in the German PV market that began in early 2004. Stocks in German PV companies soared by up to 2000 percent in value.

This exciting turn of events proved that the FIT system was the only PV support policy that really worked. FITs were introduced by the German government after a vote in the Bundestag as early as 1990 and had first proven successful to promote wind power. Through our EU Programme on Renewable Energies, we had previously provided contractual support to EUROSOLAR, chaired by Hermann Scheer, to formulate the details of this scheme. I remember complaining about this new "German" phrase *feed-in*, which sounded strange to me. The "feed-in law" was renamed the Renewable Energy Sources Act (*Erneuerbare-Energien-Gesetz*, or EEG) in April 2000.

The EEG obligated electricity grid operators to provide free access to the grid for all PV and other forms of renewable electricity generators, without the need for contracts with plant operators; it did away with all the superfluous administrative hurdles. Not much of a role was left for administration in the

EEG — at least not in Germany — though it may be different in some "Latin" countries.

An important financial improvement of the "feed-in" system over the PV Roof Programmes is that rather than operate on the investment cost at the front end, it provides a commitment for a period of 20 years (25 years in Spain and other countries) to regularly pay the PV electricity supplier a fixed price per kilowatt hour. In summary, the difference between the two is as follows:

- The EEG provides a purchase guarantee of all the electricity produced over 20 years or more.
- In the German version, as is also the case in most other countries, the tariff height is such that the owner is paid the total cost and a profit on top.
- The owner must in principle (in practice there is additional financial support from public sources and banks) find a bank loan for the full cost of the investment. In the PV Roof Programmes, it was only for part of it.
- Owners must ensure the quality and correct certification of installations. If there are malfunctions or quality issues, the owner is not paid. Although PV modules do not degrade easily and are easy to maintain, it is important to obtain proper certification by approved bodies — the International Electrotechnical Commission (IEC) in Geneva and the PV Global Approval Programme (GAP) (chaired by Peter Varadi and of which I am a member), Germany's TÜVs (Technical Inspection Associations), and so on. For instance, PV manufacturers tend to label a module's power higher than the actual output. Manufacturing imperfections may also occur.

The EEG for PV in Germany has enjoyed quite extraordinary success. With the PV market being global in the full sense of the word, companies from all over the world came to Germany to install their modules — until 2007–08 when an important domestic PV industry had established itself. By 2010, about one-third of all PV worldwide has been installed in Germany alone.

Obviously, this situation will not continue. Over 50 countries have now adopted national "feed-in" financing systems, though PV markets in the US, Japan and China have not taken off as rapidly as expected. It was only towards the end of 2009 that things also began improving in these countries. We will review the latest developments in a later chapter of this book.

The EEG has to deal with a fundamental problem: market prices are fixed by governments in an otherwise "fully liberalised" market (we have a World Trade Organization, after all). Fortunately, PV competitiveness is making great strides in important market sectors. It will only be a few years before support of the kind provided by the EEG will lose its key importance. The point where we reach "grid parity" is fast approaching — when PV electricity becomes cheaper than that provided by the grid. This is not going to happen everywhere at the same time — as it depends on local solar resources and conventional grid tariffs — but where it happens, the role of governments and administrations will be diminished one step further: the only regulation needed will then be "net metering" policy. In the US, net metering is widely applied today, ensuring that excess PV electricity that you produce is bought at the same price as the purchase price you pay to grid operators for conventional electricity.

However, before grid parity arrives, it may well be appropriate to introduce more competition into the PV markets controlled by the EEG. In particular, the excessive profits we saw in 2008 in Spain, and more recently in France, Germany, the Czech Republic, and other nations, indicate that there is a need for some fine-tuning of the EEG. One way may be the "declining clock auction" discussed in the US, which means that the right to sell power to the grid goes to the lowest bidder. This would give up an essential tenet of the EEG — the right for everybody to sell PV energy at a fixed tariff — but it could accelerate market introduction without the challenges the EEG is currently facing in the 2010s.

Ultimately, public financial support will become obsolete. Subsidies worth hundreds of billions of US dollars are currently given to the conventional energy providers around the globe. PV has the advantage that it will not require any such subsidies — quite apart from the enormous environmental and social benefits it provides. This is indeed a good thing: our experience with the emergence of the world of informatics has shown that prices come down most effectively after public support programmes are withdrawn.

3.2 *Industry and Finance*

No PV implementation is possible without a competitive and profitable PV industry. It is not the politicians and research scientists who produce the modules and install the hardware. Also, reducing cost through economies of scale is more a matter of industrial development than of political implementation. As we have already stressed several times, the decrease in the cost of producing PV in recent years has been due to increased global mass production. Technological progress has, in fact, not played the determining role.

Since the initial modest appearance of PV in Europe, I was aware that we have to actively promote the PV industry if we want it to succeed. In the early 1980s, I met with Giovanni Simoni, at that time an influential industrialist and politician in Rome. We agreed that a European PV industry association was needed; the European Photovoltaic Industry Association (EPIA) was created, and Giovanni became its first President. The actual organisational work of the EPIA was undertaken by Joachim Benemann from Germany, who developed it into a professional organisation. For many years, the late John Bonda was its Secretary General, turning it into an "institution" and the European voice for PV. He was a very good friend of mine, always standing firmly by my side in my political struggle for PV in Brussels. When he passed away, I gave the funeral address at the Abbey of La Cambre, here in Brussels. At my suggestion, the EPIA created the John Bonda

Figure 34. W. Palz with John Bonda, then Secretary General of EPIA.

Award for PV, but after some years the new managers discontinued the prize, the same way they discontinued the festive evening events that the EPIA had organised during our PV Solar Energy Conferences. Today, the EPIA still exists as a rich and influential lobby group in Brussels.

It is important to note that, generally speaking, the interests of the supporters of PV in the world's markets and those of the PV industry are not identical. This can already be seen from the fact that the markets will boom when the PV prices offered are low. On the other hand, industry tends to keep local prices artificially high to maximise profits. We are currently in the midst of a revolutionary PV development in the markets with strong national support: national regulations like those in Germany and France have resulted in high profit margins, as production costs have decreased much faster than the yearly regression originally planned for the PV FITs. Only the competition from global markets — in this case, coming from the Chinese PV industry — has ensured that profit margins do not all end up in the pockets of the national PV industry, but that it helps to strengthen the local market even more, to the benefit of consumers.

The overall goal must be to promote PV in society, in industrial countries and developing nations alike; there is a moral imperative to do so. The objective should not be to maximise the development and profits of the PV industry. Even the creation of jobs is not our primary aim; not all jobs in the solar energy sector are "good" jobs that deserve support. We have learnt from other sectors that job creation alone is not a sufficient criterion — the production of arms and landmines also creates also jobs. This is the kind of debate we are having right now in Belgium.

The overarching goal for PV is to benefit society. The development of the PV sector and the creation of employment in that sector represent an essential means of achieving this; they are not the goals as such. In recent years, with the PV markets and the PV industry, we have seen the same kind of greed as in other sectors such as information technology: intense market speculation in the stock of new PV companies led to a spectacular bubble in 2008, with all the effects of consolidation that followed.

Dozens of people became PV millionaires; a few made it to the billionaire level. Some have suggested that the fifth richest man in China made all his money with PV — at least, that was before the recent stock market crash. It has also been said that the eighth largest personal fortune in France was amassed by the man who first promoted the wind energy and PV business. The CEOs of the two largest PV companies in Germany get salaries in excess of US$1 million. However, one of them is losing money: in the first nine months of 2009, Q-Cells accumulated losses of €958 million.

You might say that this is just business as usual. Why should the PV business be different from any other? Well, PV is, and will remain, a bit special; this is why:

- If PV entrepreneurs are riding the ethical wave "for a better world", should they not be showing more social responsibility?
- The profits of the PV industry are currently gained through special levies; no matter what form they make take, they all reflect the extra that people have to pay because it is solar.
- PV businesses in some countries have benefitted from extremely favourable loans on the investments in their manufacturing plants. Why else should the German PV industry be so nicely concentrated in and around Saxony?
- PV businesses exporting into countries with favourable legislation have the problem that they have to rely on the financial incentives offered by a foreign country and its citizens. (Of the four points, this argument is obviously the weakest since we are, fortunately, moving towards a globalised society.)

Financing PV projects has also become routine, as with any business with high up-front costs, such as the construction sector. In countries like Germany, banks have learnt to provide special support for PV projects with favourable conditions.

A problem remains for the poor and the elderly, who, by and large, are excluded from PV development, as they are not entitled to obtain the necessary bank loans. This is not really an acceptable situation.

3.3 *PV Costs and Benefits for Society; A Special Role for the Grid Operators*

The current booms in the PV markets in Germany and around the world remain unthinkable without the support of the FIT system. But the burden on society of this kind of subsidy worries some people: it can run into the billions of euros.

Just do the maths for Spain in 2008, with its roughly 3 GW of installed PV power. On average, each Watt of the 3 GW installed produces 1.5 kWh over a whole year — 4.5 billion kWh in total. With a kilowatt-hour purchased at 45 euro-cents by the grid, the total cost is in excess of €2 billion. By law, the grid is committed to paying this amount for 25 years. The result: the PV power installed in just one year in Spain entails a long-term commitment of over €50 billion. The Spanish government suspended its FIT in September 2008 and dramatically scaled back PV subsidies. Spain's was a "one year's stand": little in 2007, a spike of 3 GW in 2008, and almost nothing in 2009.

What happened in Spain in 2008 was repeated in France in 2009. That year, France was offering the highest FIT anywhere in the world. Applications for 4.5 GW of new PV installations in the country were made — with two-thirds of that wattage in the last two months of 2009 as a dramatic decrease in the generous tariffs was expected the following year. The French government calculated that together those applications would require a €50 billion commitment over the 20-year guarantee period; in January 2010, applications in the last two months were subsequently rejected. In the end, rather than the original 4.5 GW, only 1.5 GW of new PV installations were approved in France.

Germany, the champion of FIT-supported PV markets, does the calculation differently. In its latest report in June 2009 on renewable energy, the Environment Ministry (BMU) balances its renewable energy expenses in 2008 against the benefits generated thus:

- While all the renewable energy operators — PV, wind, bio-gas, etc. — were paid a total of €9 billion, this amount is halved when one factors in the cost of the conventional electricity that was replaced.
- The extra cost was distributed equally across all electricity consumers, except for some 500 "energy intensive" enterprises. Each consumer had to pay an extra 1.1 euro-cents in 2008; for the average household, that equated to €3.10 per month. In Germany in 2008, 5 percent of the cost per kilowatt-hour of 21.6 euro-cents (33 US-cents) accounted for the FIT/EEG contribution.
- The additional expense is compensated several times over by the benefits gained:
  o external savings of €2.9 billion from the conventional "polluting" electricity that would have otherwise been employed;
  o the saving of €2.7 billion of imported conventional fuels;
  o the saving of €5 billion due to the "merit-order effect", whereby the FIT law reduces the average cost of electricity by affecting the wholesale price. Under the effect, plants with the lowest costs are used first to meet demand. If the FIT tariff is lower than the price from the least expensive conventional plants, then the average cost of electricity decreases.

One may also highlight the money saved on social benefits as a result of job creation. The CO_2 benefits of green electricity are already partly accounted for under "external costs".

The Federal Association for Solar Economy stresses the fact that the direct and indirect tax incomes from the solar industry and its employees exceed by €3 billion all the expenses raised by the FIT for solar PV electricity.

In Germany in 2007, wind electricity benefitted from 45 percent of the overall FIT support, while 23 percent went to PV. However, over the year, PV contributed only a tenth of the electricity that wind plants fed into the grid. Hence, it is fair to say that considering the overall support afforded to renewable energy, PV is the greater beneficiary.

By its very nature, the FIT system benefits only grid-connected and not off-grid systems; grid operators have the central role to play. In Germany, the financial support for PV is collected, as already mentioned, by the grid operators from their customers. This is not the case in Spain, where grid operators are remunerated through taxpayers' money for the PV electricity they provide. This shows up on the national balance sheet, which is not the case for Germany's EEG.

4. A New Energy Paradigm

4.1 *Centralised or Decentralised PV*

We have already mentioned that in the 1980s Dahlberg and Bölkow in Germany proposed deploying huge PV surfaces in the multiple-GW range in the Sahara; the solar electricity produced would be supplied to German households via hydrogen pipelines or electric transmission.

In 2009, a similar project called Desertec was initiated by important stakeholders from German industry, electric utilities, banks and insurance companies, and even Greenpeace. The idea is to generate electricity from the Sun and the wind in the Sahara and the Middle East for distribution to consumers in Europe, and in Germany in particular, via a multitude of new grid lines crossing the Mediterranean Sea. With a budget of €400 billion, which has yet to be found, the Desertec Industrial Initiative aims to meet 15 percent of Europe's electricity needs by 2050. The promoters' preferred technology is not PV but solar thermal power employing mirrors on a massive scale.

This gigantic project has raised the eyebrows of people concerned by a new kind of international energy dominance. Obviously, it can only make sense when the additional irradiation in the Sahara desert with respect to the solar energy available in Europe can justify the extra costs of transmission and distribution. And this is doubtful: even without taking into account the high transmission cost, 20 percent of the electricity produced in the desert will be lost on the way to Europe. In addition, the technology being promoted, thermal power, is currently more expensive than PV, and there is no particular reason why this should not also be so in future. The solar radiation in the Sahara is just 50 percent higher than that at good sites in Central Europe; in winter, it is only half of that in summer.

The way forward is a very different one, and we will elaborate upon this in the rest of this book.

Scepticism about the idea of transmitting solar electricity on a huge scale all the way around the Mediterranean Sea does not mean that large power transmission via electrical networks is unworkable in general terms; this is certainly not the case.

At the end of 2009, nine northern European countries signed an agreement in Brussels to develop a network of offshore wind power plants in the waters of the North Sea. Generally, wind electricity generated at sea is more expensive than that from wind farms on land, but the attractiveness of offshore wind power is such that it is bound to secure part of the overall renewable energy market of the future — this has to do with questions of improving security of supply and easing intermittency problems. For wind power, the situation differs somewhat from solar PV: the "natural" PV market is building integration; wind parks, on the other hand, are connected to the grid and are typically in the MW range. Consequently, a network of electric cables under the North Sea may be worth developing.

Coming back to the proposition of PV deployment in huge plants on the GW scale, one should mention PV power from generators installed *outside* the Earth's atmosphere. In space, the Sun's intensity is 40 percent higher than on the Earth's surface. In geostationary orbit, 36 000 kilometres above the equator (37,500 kilometres above us at higher latitudes), the Sun's radiation can be captured all year round (except for one hour when, for astronomical reasons, a PV panel is briefly in Earth's shadow). Above the atmosphere, one can harvest more than five times the solar power of the sunniest places on the ground.

Deploying many square kilometres of PV in space would be a major enterprise, but it would certainly not be impossible — leaving aside the question of cost. The bigger technical, political and social problem is rather the electric transmission to the surface, which would have to be done by powerful microwaves — and that may well be an insurmountable challenge. In fact, the concept had already been proposed in the 1970s by Peter Glaser in the US. At that time, there was hardly a solar meeting in the US where Glaser did not come forward with this project. By 2007, NASA and the US Department of Energy had spent over three decades and US$80 million studying solar power generation in space — without much to show for it.

Since 2009, something of a revival of these ideas has taken place — and there is not just one proposal, but several. A Californian start-up company called Solaren has made an agreement with the electric utility PG&E to provide it with solar electricity from space as soon as 2016, though to date they lack the financing. Also in the US, an association called the Moon Society proposed a similar project at an International Space Development Conference in Orlando. In Europe, a Swiss company called Space Energy has proposed putting a PV plant into orbit with an area of 4 km^2 via 40 satellites. The budget being discussed is US$200 billion — a little less than what would be needed for Desertec.

The Japanese are weighing in, too. Mitsubishi Electric and 15 other companies have formed a research group to develop the technology for a space project within four years. Like the Europeans, they envision 4 km^2 of PV panels in orbit. The objective is to produce 1 GW of power and the whole project cost is estimated at US$21 billion. In many respects, this Japanese project looks a bit more realistic — if one can talk of realism at all — than the other mastodons. But realisation is still some way off: they aim to launch a small experimental satellite by 2015 and hope to have an operational station ready only by 2030. Some Japanese consultancies have let it be known that the estimated price tag of the project is 100 times too high.

4.2 What Role can Conventional Power Utilities Play?

Obviously, the utilities and in particular the grid operators are of fundamental importance for our supply: essentially, all electricity is provided to consumers through electricity networks. One of the reasons is that today's electricity is generated in large power plants. Even in the case of PV development, not much is possible today without electricity grids. However, given that the Sun's radiation is widely available, in future we will not need the electricity distribution network to the same extent as the conventional system.

In a previous section, we considered the aggressiveness of certain quarters of the conventional utilities towards renewable energies and their supporters. But this is not the whole story. In 1990, we initiated, through our EU Commission programme, a working group called European Utilities for Renewable Energies (EURE). Fourteen large European utilities participated in, or made contributions to, this informal group, which was chaired by RWE from Germany. In 1993, a poll was conducted on our behalf by Powergen in the UK to gauge the views of the utilities towards renewable energy; forty-six utilities, representing a total conventional power capacity of 300 GW, responded. The position on renewables as a whole was one of "quiet" optimism. The utilities declared that they took them seriously — and there was considerable support for PV. In the period 1983–93, 29 electric utilities in Germany, Italy, Greece, Austria and Spain operated some 478 of their own PV plants with a total installed capacity of 2.4 MW.

Up to the present, European utilities' attitude towards PV has continued to be one of quiet optimism — with the emphasis on "quiet". In the booming German markets, renewable energy generation has remained on the sidelines. But things are improving. The big utilities in France, Italy, Spain and other countries are gradually realising that there is a potential for money to be made in renewables.

In the US, many state governments are demanding increased production of energy from renewable energy sources through "renewable portfolio standards". This is one of the reasons why a growing number of utilities — rather than independent power producers — have launched multi-million dollar initiatives to operate their own PV plants.

The regulation does not mean that the plants must in all cases be big multi-MW blocks. An example is Duke Energy, one of the largest power companies in the US, which, following the approval of the North Carolina Utilities Commission, plans to install, operate and own up to 400 separate PV arrays on homes, office buildings, warehouses, shopping malls, and industrial plants. The power generated from the various arrays will range from 2.5 kW to 1 MW.

Currently, the US has 77 MW of utility-driven PV projects operating. That constitutes only 5 percent of the total PV capacity installed in the country, but over the next few years US utilities have announced 4.8 GW of large "utility-scale" projects — something of an explosion. By 2020, US utilities are expected to have added 21.5 GW of solar PV. Large PV blocks are planned for installation in the dry areas of the South West. The company NextLight has announced PV blocks in California, Nevada and Arizona of up to 230 MW each.

4.3 Communities and Regions Mastering Their Own Energy Supply

Over the last few decades, there has been a developing awareness that, in addition to recognising the importance of the interests and responsibilities of nations as a whole, individual citizens' interests and responsibilities require political support at a local level.

The stage was set with Local Agenda 21, which was adopted as part of the UN's Agenda 21 programme by 180 nations at the Earth Summit in Rio de Janeiro in 1992. The Local Agenda 21 is an introduction to sustainable development at the local level. A survey in 2002 showed that 6,400 local governments in 110 countries had become involved in drafting sustainability plans locally.

From our own renewable energy development programme at the EU Commission, we founded the European-wide network Communities of Europe for Renewable Energies (CERE) in 1991. Its purpose was to promote the development and implementation of renewable energy with particular regard to the interests of municipalities, regional authorities, and local energy suppliers. Master plans for integration and networks of excellence for municipalities and regions were promoted, and energy packages for municipalities and regions for the creation of new supply structures using renewable energy were worked out.

Contractual support was given for urban planning maximising the use of renewable energies in the framework of the URBAN PLANNING network. One of its objectives was to investigate the urban microclimate and environmental factors such as access, wind movement, daylight availability, pollution, noise, social activities, modes of transport and waste disposal. For all the climatic regions in Europe, other projects concerned the establishment of new guidelines on energy rating of buildings, urban electricity networks, and the integration of renewable energy components in buildings and common storage.

On May 21, 1992, at a meeting on the island of Corfu, Greece, opened by a Vice President of the EU Parliament and a Vice Prime Minister of Greece, the project REBUILD was initiated. The aim of REBUILD was the promotion of renewable energy sources in buildings and historical centres of high cultural value in European cities. The project was co-funded by the EU Commission's Structural Fund. One of the efforts was in the municipality of Venice in Italy — it is almost impossible to install PV on the highly protected buildings of the old city.

The overarching goal of all these projects was to develop a strategy for "zero-emission urban developments".

EURENET was our European Regions Network for Renewable Energy. Mid- and long-term strategies were developed, including ecological and economic objectives, calling on the experience already available in some regions and at European institutes and associations. Tools for regional planners and decision makers were developed.

By the end of the first decade into the new millennium, we are seeing a good deal of development of this important sector. Thousands of cities, large and small, have committed themselves to zero CO_2-emission strategies for the future. With the liberalised energy markets of today, many have decided to take more direct control of their energy supply. In Germany, many municipalities are buying back the local electricity utilities they sold to private investors long ago — regional utilities such as RWE.

One thing is certain: there will be no community engaged in developing a sustainable future with zero-emissions targets that will not be employing plenty of solar PV on its buildings.

4.4 *The Autonomous Energy House: Solar Architecture and the Building Industry*

"Solar architecture is not about fashion, but rather about survival," declared one of the world's top architects, Lord Norman Foster, chairing a conference on solar architecture that was held under the framework of our EU Programme on Renewable Energies in Florence, Italy, in May 1993. And Lord Foster is not just anybody. Among many other prestigious projects, he is responsible for a number of iconic buildings in Germany and China. For the Summer Olympics in Beijing in 2008, he created the capital's new airport, a jewel of modern architecture.

In the 1990s, Lord Foster remodelled the historic Reichstag, the German Parliament building in Berlin, by adding a large glass dome. The carbon emissions of that particular feature are actually zero — achieved through passive solar gains, natural air ventilation and renewable energy applications, among them a PV generator. We provided a small financial incentive from our EU solar programme to achieve these environmentally friendly objectives — as we did for Renzo Piano's redevelopment of Potsdamer Platz in Berlin. For this latter project, the large

Figure 35. Meeting of the READ group at Renzo Piano's Building Workshop near Genoa, Italy. Anti-clockwise from left: Thomas Herzog, Norbert Kaiser, Stefan Behling, Lord Norman Foster, Lord Richard Rogers, Renzo Piano, a colleague, W. Palz, and A. Zervos.

PV panel that had been envisioned was cancelled at the very end of the planning process — but that is another story.

In 1991, we initiated READ — Renewable Energies in Architecture and Design — within the framework of our EU renewable energy development programme. Leading European architects joined forces: on April 22, 1994, Lord Norman Foster, Lord Richard Rogers, Thomas Herzog, and two EU Commission representatives met with Renzo Piano in his workshop in Genoa, Italy. The group agreed to work together on the concept of a "solar city" following the principles of sustainable energy and social integration. The idea was to demonstrate that the solar aspect of buildings is not new but that it has always been the most important factor for architects and planners throughout the centuries. The European Charter for Solar Energy in Architecture and Urban Planning was eventually developed on behalf of this group; it was edited by Herzog and published in 2007 by Prestel Verlag, Munich.

Since the days of the READ initiatives on solar architecture, tremendous progress has been made in terms of regulation and practical implementation. Germany is leading the way in the integration of solar PV: by 2008, half a million PV arrays had already been installed on German buildings. By the end of 2009, 80 percent of all PV installations in Germany went into the building sector. In addition, hundreds of thousands of German citizens have invested in PV: in virtually all towns and cities across the country, PV Citizens' Funds have been set up, contributing to the financing of new PV systems on public buildings of all kinds.

In Switzerland and elsewhere, "Solar+" buildings have become popular. The purpose is to generate, in the course of a year, even more electricity from a building's

PV generator than the building itself consumes. Over the last few years, the Swiss Solar Prize, organised by Gallus Cadonau's Solar Agency, has been awarding a whole portfolio of such buildings that combine PV innovation with architectural excellence.

PV building integration is far from absent in the US — quite the contrary — but rather like in Spain, the overall trend for large-scale PV deployment there is towards ground-mounted systems in the MW range. As far as China is concerned, it was reported in September 2009 that the Ministry of Finance and the Ministry of Housing and Urban-Rural Development plan to allocate €130 million for building-integrated PV. In the near future, two important new players in building-integrated PV are set to emerge: France and Japan. In their national FIT regulations for PV, they have recently established special tariffs for building integration that are particularly generous.

For many years, the EU has paid particular attention to the building sector. In terms of European energy demand, it is a key sector, representing almost half of the continent's consumption. At the end of 2009, after having imposed an energy demand certificate for all buildings, a new milestone in the EU's regulation of energy standards was reached. In future, all 27 EU member countries will have to fix minimum standards for the overall energy efficiency of buildings. By 2020, all new buildings will have to conform to the highest standards; solar panels will become mandatory. The EU Commission estimates that the new regulations will result in a 6 percent decrease in Europe's overall energy consumption.

I do not want to conclude this section without a brief mention of my own personal thoughts regarding the energy supply of individual homes. Most of us use

Figure 36. W. Palz's residence in Brussels. The 1-kW PV plant was the first to be installed in Europe's capital in 2000, without a subsidy. The array also has some coloured cells from Japan. The support structure is untreated timber.

our own boilers to heat our buildings and operate the engines in our own automobiles; the only energy we buy from a network is electricity. Why not produce our own electricity as well? The first step is PV power installed on the house, but complete autonomy is certainly possible — and at no extra cost.

I mentioned previously that my own house in Brussels draws most of its electricity from a battery that is charged by a PV array integrated on the façade. Heating is provided by a gas boiler connected to the city's natural gas network.

The following is a novel concept for autonomous buildings — private houses and public or commercial buildings alike. It is a further development of the system employed in my own home.

- Electricity is provided by a central battery that is fed by different renewable sources of electricity. New types of electric batteries, which are being developed for the electric and hybrid cars of the future, are becoming attractive for this application as well.
- The potential to collect solar energy in winter is not what it is in summer; for wind energy, it is often the other way around. Hence, PV arrays and solar heating panels on roofs or façades are combined with small wind turbines in the kW range. In late 2009 in China, I became acquainted with a new design of wind turbine: a vertical axis Darrieus type with magnetic levitation, which is extremely quiet in operation and runs at the very low wind speeds that are typical in urban areas. It is true that small wind turbines produce power at a higher cost than larger ones (physics dictates that turbines in the 2–3-MW range typically generate electricity at the lowest cost), but for building integration large ones are obviously unsuitable.
- The boiler is replaced by a cogeneration system that uses the same fuel to produce not just heat but also electricity. The engine used for this purpose is based on a car motor, which is extremely reliable and at least 10 times cheaper per kilowatt than the very large conventional power plants on which we rely today. In a car, the heat produced during driving is wasted; in a building, it can be reused efficiently. Initially, the engine can be fuelled by natural gas; eventually, biogas or liquid bio-fuels will be available in sufficient quantities for this application. The engine does not run permanently but only intermittently to feed the battery — similarly to what happens in a hybrid car.

Not a single system of this kind has ever been built. I wonder why!

5. Power for the People

5.1 *Starting a Global Strategy: 10 Watts per Head*

Currently, in 2010, the overall electric power capacity installed worldwide amounts to 4,500 GW. This corresponds to an average of 650 W per head.

Figure 37. W. Palz's office at the EU Commission in Brussels in the early 1990s.

The calculation takes into account all 6,900 million people living on Earth, no matter whether they are children or adults, working or retired; it also includes the 1,500–1,700 million "energy poor" who have no access at all to electricity. If we exclude the energy poor from the calculation, we get some 850 W per head on average for today's grid-connected people. Only part of this is used by people in their homes. Most of the electricity is consumed in other ways: by industry, commerce, railways, street lighting, desalination in arid countries, and so on.

Twenty years ago, PV's contribution to global power capacity was essentially zero. At the time, I considered how the world would look if we had an average of just 10 W per head of PV power — a significant proposition, as it represented about 1 percent of the total power in place. The Swiss, who were then among the most advanced in PV implementation in Europe, had a fraction of 1 W per head of PV installed and were discussing a goal of 10 W per head.

For people in an "energy poor" country, 10 W per head makes a world of a difference: for a family of five, 50 W provides a minimal degree of lighting as well as power for a radio or telephone. At the village community level, it translates to solar power for survival.

5.2 PV for the People in the Industrialised World

When preparing the EU "White Paper on Renewable Energies" that was adopted by the Commission in 1997, A. Zervos and I adopted a target of 3.7 GW of PV implementation by 2010. This figure was equivalent to 10 W per head, as the EU population stood at 370 million people that year. Ultimately, other Commission services reduced the target to 3 GW. Logically, however, it translates to 5 GW to reflect the realities of today: in the meantime new member countries joined the EU and in the current target year of 2010 we do indeed have 500 million European citizens.

To my surprise, the optimistic target set 13 years ago has actually been more than comfortably achieved. By the end of 2010, some 20 GW of PV will be operating in the EU; at the end of 2009, we had already achieved well over 14 GW, the lion's share of over 8 GW being in Germany alone.

Obviously, targets are not self-fulfilling. The EU had no promotional instruments for PV market implementation: the EU Directive on electricity from renewable sources in 2001 defined targets again but not the instruments to achieve them. In fact, the targets in the 1997 EU White Paper would never have been realised without Germany's EEG and the FIT for promoting clean electricity. By 2004, H. Scheer and H.-J. Fell, both members of the German parliament, had succeeded in getting this law adopted in Germany, and seeing its unbelievable success in the marketplace, most EU countries followed suit, anxious not to be left behind.

On average, the 500 million citizens of the EU are currently benefitting from some 30 W of PV per head — three times the very optimistic projections of little more than a decade ago. But shares of the PV pie within Europe are highly uneven: in early 2010, Germany has a total PV capacity of 100 W per head on average, which roughly corresponds to a commercial panel of 1 m^2 in size. And that is for each individual German citizen, whether young or old, not just for each home.

By comparison, the capacity of conventional electrical power in Germany stands at 1000 W per head; hence, PV has now reached 10 percent of that capacity. Even though the share of electrical energy this represents is only around 2 percent — solar radiation feeding PV power plants being a precious, but rare, good — the result is impressive. It gives us an idea of what is possible on a wider scale around the world: Germany is a highly industrialised country, while the intensity of its solar radiation is at the level of that in Alaska.

It would be wrong to think that German elites are all solar fanatics. When I was sitting as an expert on the "Energy Commission" of the German Parliament in Berlin from 2000 to 2002, renewable energy had a hard time getting the recognition it deserves. Even the German solar industry in those days was far from being proactive in promoting PV. Germans are solar enthusiasts, but to get the policies in place, the leadership of a few "solar hardliners" was absolutely essential; we owe a lot to them!

Second in the league of PV market development is Japan: on average its 128 million people can count on more than 20 W per head. Accordingly, the Japanese are roughly on the same level as the average EU citizen. This is currently also the case for California, which has a population of 36 million and an installed capacity of over 1 GW. For the time being, however, the rest of the US remains well below the 10 W per head mark.

5.3 *PV for the People in the Solar Belt*

Supplying the world's poorest people with clean, decentralised solar PV was part of the "Power for the World" concept that I presented in my capacity as the EU Programme Head for Renewable Energies for the first time in public in 1990 and at numerous conferences and publications since.

The proposal called for a global action plan to supply electricity to the millions of people who have no access to it today — a highly distributed and decentralised electricity system with a capacity to meet basic minimum needs. Such a scheme is the only realistic approach that may be implemented in a foreseeable time frame of 20 to 25 years; setting up a conventional electric grid to cope with the problem is often too costly and time-consuming.

If the world's poorest, about 25 percent of the global population, were to have 10 W per head, or 15 GW in total, it would represent less than half of one percent of the power capacity we, the other 75 percent, currently consume. PV is the ideal technical solution for supplying that capacity: it utilises an omnipresent resource, it is modular and available on large or small scales, and it is very reliable and has no moving parts.

The goal of the original "Power for the World" concept was for 1 billion people to be electrified within 20 years; it was deemed unrealistic, even in the most optimistic scenario, to cater to all of the 1.7 billion people who currently go without electricity. Today, the number of energy poor remains unchanged, and the concept is more relevant than ever.

In the following, we quote from the original proposal, which remains wholly valid today, 20 years after it was written:

The infrastructure building process that is employed in conventional development schemes is inappropriate to supply vital and basic needs to the poor. As a first step, building power stations with distribution networks is unrealistic leapfrogging, exceeding available financial resources and taking an excessive amount of time. In contrast, villagers need electricity quickly even if it is on a small scale. A small but significant step now is better than a big step in 50 years.

The fact is that PV, even at 50 cents per kilowatt-hour, is the cheapest source of electricity in large parts of the world's rural areas today. Eventually, there will be enough evidence to show that PV is the answer for village electrification. However, at the current pace of implementation, the villagers of the "Third World" will never reap the benefits of technical progress that could be accessible to them; the dangers of the poverty explosion in the "Third World" are increasing.

It is expected that a minimum degree of personal comfort in poor rural areas in the "Third World" through lighting, communication, TV, etc. would have a dramatic effect on fighting the exploding population, social unrest, and migration to cities and emigration to other countries.

Among the essential needs that electricity may serve, there is first of all the minimum for survival — at least 20 million people die each year because of contaminated water and a lack of basic health care. Electricity is also urgently needed for personal comfort, in education, and in cottage industries. Just a few Watts of reliable electricity can bring about a dramatic change in standard of living and social progress.

The "Power for the World" concept aims at making the maximum possible impact on the lives of the world's poor, in no more than one generation, by employing minimal and mostly existing financial resources. The programme intends to give an additional push to the moral imperative to help the world's poor; it is in the interests of the "First World" not to let the demographic situation degrade until it explodes.

At an estimated cost of US$6 per Watt of installed PV power, the total cost of the programme will be US$60 billion spread over 20 years, or US$3 billion per year. This corresponds to just 3 percent of present aid budgets made available by donor nations; it is less than 1 percent of the global military budget.

The programme can be achieved not by requesting additional billions of dollars in subsidies but by better utilising some of the existing ones and by adopting a global approach and a comprehensive organisation from the top. A well-defined and structured programme will greatly increase efficiency, minimise cost and accelerate implementation.

A unique global approach is possible in this case thanks to the modularity of PV: no more than a few dozen system types are required with only minor adaptations for the villages in the "Solar Belt". This leads the way for mass production and economies of scale for PV.

The supply options for village PV are manifold, such as:

- *water disinfection and supply;*
- *health care, dental care (in towns orthopaedic treatments);*
- *lighting: in homes, on streets, for sports, leisure;*
- *refrigeration for foodstuff;*
- *communication;*
- *electric tools.*

In such a joint venture between the rich, industrialised countries and the poor, developing nations, everybody will benefit — not only the rural poor who tend to be mostly overlooked in current development schemes but also the developed countries, which will more rapidly get access to low cost solar electricity for their own needs and markets.

There are some important criteria that must be followed in any case for successful implementation:

- *This electrification programme must not affect valuable existing traditions and customs.*
- *The programme must enrich social life in the villages, particularly in the evenings.*
- *The programme must involve young people, stimulating agro- and cottage industries and job creation.*

- *The programme must not disturb any existing PV industry and emerging solar energy markets in the "Third World". These must be included, not eliminated.*

As for financial contributions from local communities, experience has shown that some contribution is necessary to motivate villagers to operate and maintain the installed devices.

To deploy 10 GW of PV within 20 years, one needs a solar cell production of 500 MW per year; this would trigger mass production of PV.

In conclusion: Supporting the world's poorest people with modest electrification is not only a noble task but also mutually beneficial for all. In the "Power for the World" programme, there will only be winners.

On March 29, 1993, the Parliamentary Assembly of the EU and ACP (Africa/Caribbean/Pacific) countries approved a Resolution requesting an intensification of solar energy development.

The programme was actually never implemented but it did have an effect. As a follow-up, we investigated the mass production issue: at what cost could PV modules be produced in the manufacturing plants envisaged in "Power for the World" with a solar cell production of 500 MW per year, when compared to those produced by the 5-MW plants that were in operation in Europe in those days?.

In an investigation code-named "Music FM", the best available European experts in PV science and technology from industry and academia worked together under our EU Commission programme to address the following question: What would be the manufacturing cost of PV modules produced in 500-MW plants with the technology available in the early 1990s? This endeavour had two lines of action, one for crystalline silicon cells and the modules made with them, and the other for thin-film cells and modules.

The result was that thanks to mass production and economies of scale — not taking into account the expected technological improvements — silicon modules could eventually be produced at €1 per Watt, and thin-film modules even for 20 percent less. The detailed results were presented at our 1994 European PV conference (PVSEC) and can be found in the accompanying proceedings.

The implementation of PV since 1994 has not been as we proposed in "Power for the World". Instead of massive deployment in the developing countries of the Solar Belt, PV markets have taken off in the industrialised world. We have considered the reasons for this in previous sections.

We can be happy that PV is now widespread in the markets of developed countries, but it is unacceptable that the energy poor in the Solar Belt have been left out of this development: less than 1 percent of the 7000 MW of PV that was newly installed in 2009 in global markets went to the poor rural in the developing world.

I call this the *one percent scandal*.

6. Power for the Poor

6.1 *Getting Involved*

As already mentioned, the presence of solar PV in developing countries is very small. It may sound impressive that several million "solar home systems (SHS)" (50-W PV systems to provide lighting and power small appliances), tens of thousands of PV water pumps and other community systems were set up in the developing world over the last two decades, but this is very little compared with the global needs of the "energy poor".

The World Bank in Washington has an excellent record of past efforts, and there have been a great many private initiatives, like that of the Solar Electric Light Company; special reports on both are included in original contributions in this book. A lot has also been done by the UN and UNDP, the Global Environment Facility (GEF), the EU and its member countries (including organisations from Germany such as the GTZ and KfW), Japan, and many others.

To get a flavour of the enthusiasms and frustrations experienced over many years of activity by supporters of supplying PV to the rural poor, allow me to share with you a quick glance through my own curriculum vitae in this field:

- Around 1972, I was approached by Paris-based Yves Houssin, who organised school TV in the villages of the French-speaking nations of West Africa for the French government and the national television broadcaster ORTF. The programmes were broadcast by satellite. He wanted to do away with the dozens of jeeps that were employed to transport over long distances the lead-acid

Figure 38. The world's first PV water pump, Corsica, France, 1973.

batteries used for charging — and replace the need for charging with PV. I advised him in several meetings and wrote a report for him. He was one of the first to get very rich from PV.

• Around 1973, I assisted Dominique Campana — now Director of International Affairs at French Environment and Energy Management Agency ADEME in Paris — with her thesis on the development of the world's first PV water pump in Corsica. One follow-up project was the deployment of such pumps in West Africa by the Guinard company and Father Verspieren, a white Catholic priest.

• In the mid-1970s, I became friends with Thierry Gaudin, Chief Engineer of Mines and Director of the Technology Department of the French Industry Ministry in Paris. I informed him about PV and he taught me innovation policy. He was writing a book called Listening to the Silences about innovation and institutions that concerned the role corporations have played in Europe over the centuries in withholding technical knowhow from the general public.

Under contract with the West African Development Bank (WADB), we visited Togo, Upper Volta (now Burkina Faso) and Niger in 1977 and had comprehensive discussions with their governments. In the report we wrote for WADB, we proposed the installation of hybrid systems employing PV and small wind turbines for local villages following a prototype that had just been set up in the village of Campagna in the French Pyrenees.

To finance the project, we proposed a system of leasing to the villagers: the essential part was to be financed by a new para-tax added on to the

Figure 39. T. Gaudin from the French Industry Ministry inspecting PV for communications in Niger, 1977.

Figure 40. A French Catholic priest, Father Bernard Verspieren, who installed hundreds of PV pumps in West Africa; he arranged to make the boreholes and do the installations in one day.

Figure 41. Visit to Ashgabat, then part of the Soviet Union, in the 1980s.

tariffs of all grid-connected utility clients in the cities, as well as a small part by the villagers. (We even calculated that in Burkina Faso's capital, Ouagadougou, one could collect 168 million F CFA and in Niger's capital Niamey, 198 million F CFA over a year.) If this sounds familiar, it is because the idea was a predecessor of the German EEG, which was reinvented 25 years later.

- A few years later, I visited several countries in West Africa again as part of a UN delegation. From that trip I learnt about the Peul people, who seem to be very influential in all West African countries. But what I remember most was that the government people complained in private that almost every week another investigating donor delegation would visit — talk and more talk, instead of action! I certainly did agree with those officials.

- Years later, I was with another UN delegation visiting some villages in the "Second World". We went by jeep to the villages around Ashgabat, then in the Soviet Union, now in Turkmenistan on the border with Iran. The head of our delegation was a UN official called D. Lovejoy; what a wonderful name, I thought.

- In March 1979, on behalf of the EU Commission, I organised the governmental conference "The Sun in the Service of Development" in Varese, Italy. It was chaired by my boss, Günter Schuster, then the Commission's Director General for Research, Science and Education. It was attended by official governmental delegations from 80 nations from all over the world, several UN organisations, and international banks and institutes. The conference was opened by, among others, Sir George Porter, a Nobel Prize winner and Director of the Royal Institution in London. Various regional seminars led up to the full conference; I was active in preparing and chairing them and even wrote out the bank cheques for the participants on behalf of the Commission. The seminars were the following:

  o September 1978 in Nairobi, Kenya, for East Africa, with official representatives from Kenya, Ethiopia, Lesotho, Mauritius, Rwanda and Zaire (Congo). It was opened by the Kenyan Minister for Water Development.

Figure 42. West African Workshop in Bamako, Mali, in 1978. From right: B. McNelis (4th); W. Palz, chairman (9th); Pierre Lequeux (8th), EU official for development aid. On the left: Father Verspieren.

Figure 43. Latin American Workshop in Caracas in 1978 to prepare for the EU conference in Varese, Italy, "The Sun in the Service of Development". Lower row: W. Bloss (4th from right); W. Palz, chairman (2nd from right); behind him, M. Phéline from Paris.

- September 1978 in Bamako, Mali, with representatives from Mali, Mauretania, Niger, Chad and Togo. Three ministers from Mali and Upper Volta (Burkina Faso) participated.
- October 1978 in Amman, Jordan, for the Arab World, with representatives from Jordan, Syria, Sudan, Saudi Arabia, Kuwait, and the Arab Physical Society. It was also attended by the Crown Prince of Jordan and three of his ministers, three institution presidents, 10 director generals, and six ambassadors.
- October 1978 in Caracas, Venezuela, for Latin America, with representatives from Venezuela, Argentina, Bolivia, Brazil, Chile, Mexico, Peru, the Latin American Solar Energy Society, and Latin American Energy Organization (OLADE).
- October 1978 in New Delhi, India, for Asia, with representatives from India, Pakistan, Sri Lanka, Singapore, Bangladesh, Indonesia, and Thailand. It was opened by the Secretary for Science and Technology from the Indian government.

The full conference stressed the need for co-operation in various areas, such as research, transfer of technology, education and capacity building, as well as environmental and social aspects. As far as PV was concerned, it was interesting to note that some participants expressed a reluctance to depend on imports from industrial countries for their supply. It is a typical argument: this is our Sun.

Figure 44. Asian Workshop in New Delhi in 1978: W. Palz, chairman (centre); with Peter Dunn, an expert.

- In 1986, the EU Commission started the Regional Solar Programme (PRS) in West Africa. That year we had invited my colleague Hans Smida, then the Director for Infrastructure Development in West Africa at the EU Commission's aid programme, to attend our PV SEC conference, which was being held in Seville, Spain. He came up with the idea of developing PV water pumping in his area of influence: in the rural areas of the Sahel, where only 25 percent have access to clean drinking water. No sooner said than done: I helped him to get a new PRS for approval through the committee of the EU member countries — which is never a minor task. A convention was established with the CILSS, an organisation fighting the drought in West Africa.

  In total, 626 PV water pumps were installed in nine West African countries. In addition, 660 community systems were installed in the villages for lighting and cold storage, totalling 1,257 kW. The contribution to the cost from the EU's budget was €34 million, of which €24 million was for the PV generators. The villagers themselves bore the operation and maintenance costs, so that the water supplied from the PRS was not completely free. Almost 1 million people benefitted from the programme.

  After its completion in 1996, the programme and all the systems were thoroughly evaluated. Various investigations were undertaken and in a meeting in Bamako in 1999, the governments of the countries concerned and the EU decided to go ahead with a follow-up programme, PRS II. As an official in the EU Commission's Aid Directorate in 2000, I was involved in approving this follow-up programme. The budget over the period 2001–2007 was €73 million: 210 water pumps and 280 PV community systems from PRS I were refurbished and 465 PV pumps were newly installed.

- In 1991, we organised a conference in Zimbabwe in conjunction with EUROSOLAR and sponsored by Germany's Friedrich Ebert Foundation. Held

Figure 45. PV in a non-electrified village in Nepal.

Figure 46. Visit to Pulimarang, Nepal. Sitting from right: W. Palz, Petra Schweitzer, and (5th from right) Neville Williams, a great promoter of PV in South Asia.

just a year before the Earth Summit in Rio in 1992, the conference was attended by 160 experts from 22 countries who approved the solemn Harare Declaration on Solar Energy, which called for a "solar revolution" and the establishment of the specialised agency IRENA.

Figure 47.   Where cultures meet: a marriage in Kathmandu, Nepal.

- Years later, Petra Schweitzer from the University of Magdeburg in Germany took me to Nepal several times. It was here that for the first time I became aware of what it means to live in a non-electrified village, where everything is plunged into darkness after sunset. I recommend it to anybody who wants to understand the hardships that the poor endure.
- In 2001, I attended a seminar on sustainable energy in Santo Domingo in the Dominican Republic. It was chaired by my colleague Garcia Fragio, in charge of infrastructure development in the ACP countries on behalf of the EU. In particular, it was supported by the Pacific states of Tuvalu, Kiribati, Fiji and Tonga.

  An important input was a report of the ACP-EU Joint Parliamentary Assembly meeting on Renewable Energy Sources in March 2001. One of the conference conclusions was "the recognition that renewable energy and energy efficiency have a key role to play in sustainable development and poverty alleviation, particularly for island states where high energy import prices have a major impact on macro-economics".

- In 2000–2001, I was a member of an internal inter-service group of the EU Commission to promote "Digital Opportunities for All" and to "Bridge the Digital Divide". It was a time of enormous enthusiasm for the emerging Internet markets. One reason for the exclusion of the poor from the Internet revolution was that they simply had no electricity.

  After having left the Commission as an official, in 2004 I agreed to embark on a new regional programme — with Fernando Cardesa, then the EU Director for co-operation affairs with Latin America — to deploy PV in

Figure 48. Visiting Latin America in 2006–07 to promote the EU PV project Euro-Solar to connect the poor to the world via the Internet.

Figure 49. A prototype of a Euro-Solar system at ITER in Tenerife, Spain, in 2005.

some of the poorest countries in Central and South America. The Euro-Solar programme was born. The purpose was to install small 1-kW PV generators together with small wind turbines and storage as community systems in some 600 villages in Guatemala, Honduras, Nicaragua, El Salvador, Ecuador, Peru, Bolivia and Paraguay. In 2005, I went with some of the

Figure 50. A remote village in the Andes: exploring possible sites for Euro-Solar systems.

Figure 51. Kids in Peru — they deserve our help.

administrators in charge of the programme to discuss the details of preparations with government officials and the villagers in Peru, Bolivia, Guatemala and El Salvador.

In 2005, the consultative committee of the EU Member States failed to pass the programme proposal; even Germany declined. It took the personal intervention

Figure 52. Exploration mission 2006 in Latin America. Sitting on the left: Gambini, EU Commission official, head of the Euro-Solar programme when it was initiated; Cendagorta, expert from Tenerife (sitting 2nd from left); W. Palz (standing 2nd from left).

of the EU Commissioner in charge of External Relations, Benita Ferrero-Waldner, to clear the hurdle. I was most impressed by her heartfelt support for PV; we owe her a lot. Eventually, the Financing Agreement between the EU Commission and the eight beneficiary countries was signed in Brussels on December 21, 2006.

Systems, which are connected to the Internet via satellite and installed in village schools, open after school hours, providing Internet access to villagers. The systems also provide lighting and water disinfection services. The contribution from the EU budget was €30 million; villagers and governments of the countries concerned provided the remaining funds, in particular to cover the cost of satellite connections. Expertise for the development and follow-up of the programme — all systems are permanently controlled via the Internet — is provided by ITER, the Institute for Renewable Energy in Tenerife, Spain.

6.2 PV Power for the Poor in the Developing Countries

Despite all this effervescence to promote solar energy in the Third World, the balance is a great disappointment. Why has PV deployment worked in the industrialised world but far less so in developing countries? For a moment, let us compare the essential characteristics of the support scenarios in both cases.

In Germany, a leader in PV implementation, the regulations encapsulated by the EEG, the Renewable Energy Sources Act, work as follows:

- There is a virtual absence of bureaucracy.
- There is no intervention of public funds.
- The investment funds, typically upwards of €5000, are provided by banks, both public and private.
- There is an obligation for grid operators to buy the kilowatt-hour produced by the PV plant operators, who in turn get the expense back by imposing it evenly over all their regular customers.
- The PV investors have zero expenses of their own; they even make a profit.
- The macro-economic benefits for the country are highly positive.

Compare a typical developing country, where you cannot count on any of these elements:

- PV systems in villages are not grid connected; it is not obvious how to meter the kilowatt-hour produced that ultimately has to be paid to amortise the investment.
- Grid operators do not have enough regular clients to get them involved in financing a massive programme to the benefit of rural people.
- The country does not have the necessary economic and macro-economic resources of its own for such a programme.

The problem is not the money, however. Donor nations provide each year on average some US$100 billion in subsidies; this figure does not even include bank loans from such sources as the World Bank, the European Investment Bank (EIB) in Luxemburg, the KfW in Frankfurt, and many others. The scandal is that virtually none of this aid money goes to PV deployment. Administrators should take note that we are no longer talking about a "Mickey Mouse" sector, but one that represents US$30 billion a year. Information is one of the first needs to be tackled.

The problem is that for aid money to reach the villagers in need, it has to pass not one but two hurdles: bureaucracy in the aid administrations and corruption in local governments and administrations. My own experience from several years in the EU's aid administration is that it is very well organised and structured, it is generally efficient, and people are highly committed. But it may take one critical position being filled by a bureaucrat of pure blood and you have a problem. When it comes to responsibility for public money, the executioner is never far away. A bureaucrat who repeatedly asks for justification reports for every last penny is always right and nobody can stop him — even if it kills the project. Mobbing inside administration services can deteriorate the problem further.

So, how does one move forward? From my own experience and from all we have learnt together over the years, I would propose the following:

- The involvement of public administrations should be kept to a minimum.
- As in Germany at the present, the driving force must be profit; all stakeholders at all levels must make money.
- As in Germany, the money invested in PV should come from banks. Microcredit is receiving a great deal of attention and would be a suitable instrument. As proposed by my friend Michael Eckhart from Washington, one could also develop a centralised "solar bank" with many local branches: it would be specialised in solar systems, ensuring that the earmarked money does not ultimately go to other branches, as might happen in "normal" banks.
- PV systems may be best implemented by adopting some kind of "solar roof" programme, as promoted in Germany and Japan in the 1990s. Banks should pay 100 percent of the PV investment cost plus a little extra. Module prices should be fixed centrally the same way the FITs are now fixed centrally in the larger industrialised countries. Tendering should be avoided as it promotes bureaucracy. The banks should require certification of PV systems to avoid fraud.
- An FIT system, even if adapted to the new conditions, may be more complex to implement. In any case, a bank has to provide the investment loan. The bank should ask for progressive reimbursement of the loan and accept for this a certificate for the kilowatt-hour produced. Such a certificate requires an additional step for metering, either by local utilities or even via the Internet.
- One would not have to wait long to find interested parties to participate in implementation; the message that there is a profit to be made would spread quickly. Preference must be given to the local civil society.
- The banks must be provided with the necessary funds from donors' aid budgets. To make this process viable, one may want a centralised banking system — why not a "solar bank"? — with a network of banks all the way down to regional centres for microfinance.
- Donor countries should not have to provide new funds for PV deployment in villages. They would benefit from the external effects previously mentioned and would have a well-defined programme for sustainable development never seen before.

One might complain that the scheme proposed here gives the central role to the banks, not to administrations or utilities — a legitimate enough concern given that it was the big international banks that are most to blame for almost ruining the world economy in 2008–09. This is another argument for creating a specialised bank: a "solar bank" would have difficulties doing business through rotten subprime certificates; having a dedicated bank would make possible the strict transparency needed for a major investment initiative of this kind.

6.3 *Power for the Poor in the Industrialised Countries*

Globally, some 100 million people in industrialised countries are classified as poor. In Europe, 17 percent of the population is at risk of falling below the poverty line. Among European children, 9.4 percent grow up in families where the parents are unemployed. In Germany, 1 million people have to queue for food handouts from charities. One may even be tempted to consider the fate of these people to be less favourable than that of the people of the developing world, as it is often associated with social exclusion; but in Africa and other parts of the Third World, the poor, women in particular, are additionally affected by civil war.

There is no reason why the poor in the industrialised world could not also benefit from the deployment of PV; it may help some get out of the poverty trap. We have already seen that in Germany and other countries the FIT support for PV does not necessarily involve investors' own money. The money is provided by banks that have as security the value of the PV plant that is financed together with the FIT law and its long-term income guarantees. This could work perfectly for the poor as well; it would not involve extra expenses for anybody. Only the banks would have to learn to work with the poor directly.

7. Power for Peace

What is the cost of a Stealth Bomber? You know, the one that can supposedly make itself invisible to radar? Ten billion US-dollars? Twenty billion? And who is the enemy? Extraterrestrial invaders? The poor farmer turned Al-Qaeda fighter? His

Figure 53. A village in Afghanistan electrified with PV by local women under the guidance and sponsorship of the Barefoot Institute, India. Picture credit: Barefoot Institute.

guns and explosives are made in the First World; Afghanistan has no arms production.

There is an increasing awareness that our young men and women in uniform can be put to better use than on the battlefield — by deploying them to stricken areas in the Third World to provide sorely needed development aid. This would have two important effects. Firstly, it would help local people improve their technical knowledge and develop a positive perception of Western values. Secondly, it would allow the young generations of the West to gain a deep appreciation for the cultures of the developing world — not from the vantage point of a military vehicle, but from within its villages. What ever happened to the wonderful American idea of the Peace Corps?

Of course, PV is no miracle solution that can heal all the world's ills. However, there is little doubt that PV can provide new hope for the planet's poorest. Poverty is the breeding ground for conflict and social unrest; alleviating that poverty through technological solutions can foster peace and stability.

The world's largest PV conference took place in Hamburg, Germany, in September 2009. In its closing session, Bunker Roy from the Indian Barefoot Institute gave a solemn lecture about their initiative to inform local women about PV so that they can take the initiative to install it in their villages. One example he gave was a village in Afghanistan that had PV lighting installed in this way. The 500 delegates from 70 nations in the room rose from their seats and gave Bunker a lengthy, heartfelt standing ovation.

It is more than symbolic that the first large PV plants in Germany were built on land that the Soviet army abandoned in 1990 when leaving East Germany after the "hot phase" of the Cold War.

One day, when the world's economies have gained independence from the oil and natural gas resources of the Middle East, the military presence that the West maintains at such high cost in that region for reasons of energy security will become obsolete.

Part III

PV Today and Forever

1. Solar Power 2009–10: A Wealth of Achievements

1.1 *The Global PV Markets 2009–10*

The detailed market developments described in the previous chapters can be summarised as follows:

- In 1978, global PV shipments for terrestrial applications reached 1 MW with a turnover of US$30 million.
- In 1988, the global market was 33 MW with a turnover of US$300 million.
- By 1998, the global market had reached 155 MW.
- By 2008, it had reached 5000 MW, or 5 GW.

It is interesting to note that in the decade from 1978–88 markets increased 35 times, and in the decade 1998–2008, they increased 32 times, but on a much higher level. During the decade in between, they increased only by a factor of five.

In 2009 global PV solar cell production reached 10,660 MW, exceeding 10 GW for the first time. Almost half of it was manufactured in China and Taiwan. The lion's share of 8,265 MW consisted of crystalline silicon solar cells; CdTe cells exceeded 1 GW for the first time. The ten largest solar cell producers in 2009 in order of volume were First Solar (CdTe cells), Suntech, Sharp, Q-Cells, Yingli Green Energy, JA Solar, Kyocera, Trina Solar, SunPower, and Gintech. Only 70% of global production in 2009 was actually sold and installed: depending on sources, between 6.9 and 7.2 GW. This is an increase of some 15% over the previous year for the global market. More than half, 3.8 GW, was installed in Germany, with the remainder going to the following countries (in order of market share): Italy, Japan, USA, Czech, Belgium, France, South Korea, China, and Spain. (Source: RTS Corp., Tokyo, in *PV Activities in Japan and Global PV Highlights*, Issues of April and May 2010.)

Power for the World by W. Palz
Copyright © 2011 by Pan Stanford Publishing Pte. Ltd.
www.panstanford.com
978-981-4303-37-8

Figure 54. 42 EU PV conference, Valencia, Spain, 2008. From left: McNelis, Palz, Wrixon, Mike Eckhart (ACORE Chairman) and Jodie Roussel.

Figure 55. Valencia conference, 2008. From left: Daniel Lincot (Chairman), Hélène Pelosse (French Ecology and Energy Ministry), Gallus Cadonau (Director Swiss Solar Agency), Palz and Peter Helm (conference director).

In fact, the details of the market vary by more than 10 percent between the various market analysts and observers. The reason is that the PV plants are highly decentralised and it is difficult to keep track of all of them: an estimate for 2009 puts the number of newly installed PV plants at more than 300 000. By comparison, the statistics for wind turbines, for instance, are a lot more precise, as those come in megawatt units and tend to be assembled in wind farms; some 15 000 wind turbines were installed in 2009.

Again, more than half of the new PV power capacity in 2009 was installed in Germany. Leaving alone the fact that 38% of Germany's newly installed capacity in 2009 was added in just one month, December! Clearly, however, it is not really sustainable that 50 percent of the global market is concentrated in a country where only 1.2 percent of the world's population lives. In terms of plant size, the trend in Germany is towards larger plants — 9 kW on average in early 2009 compared to 20 kW towards year's end. However, of the total PV capacity of almost 10 GW in Germany, some 90 percent is installed in plant sizes of less than 1 MW. Globally, according to some sources, 25 percent of all modules installed in plants are more than 1 MW unit size.

Global turnover also remained stable in 2009 at some US$30 billion. For the first time ever, production costs of less than US$1/Watt were achieved for thin-film CdTe modules, and of only slightly more than US$1/Watt for crystalline silicon modules. Correspondingly, strong competition in the global markets produced module price decreases of up to 50 percent over those of the previous year — and they are not expected to come back up again in the future: PV module prices ranged between US$1.2–2/Watt in 2009. For complete systems, excluding storage, prices came down to €3/Watt, even for small systems.

1.2 Political, Financial, and Industrial Environment

For the time being, the size of the global PV markets is unthinkable without the feed-in-tariff (FIT) system, called EEG in Germany. This political support remains indispensable for the healthy development of the PV markets; if it failed, the markets would be in danger of collapse. The security of political support is still essential for the security of investments and the attractiveness of the PV markets.

The market in Germany is currently the most attractive for two reasons: the profit margins for investors are especially high and the bureaucratic hassle is low. But the high margins are not sustainable in the longer run. The new German government that was put in place in September 2009 turned out to be no smaller supporter of the EEG than the previous one; this is good news. In the past, German legislators have demonstrated that support for renewable energy is not always unconditional: over the last few years, the financial support for biofuels and biogas has indeed been reduced and a whole new industrial sector in Germany has been on hold.

There are signs that besides Germany many other countries are about to follow new policies that support PV, too, and it can be expected that the global markets will broaden accordingly in future and global political support will continue to grow.

The steep fall in PV module prices on global markets in 2009 surprised many. Legislators were not prepared to react immediately, and as a result, the financial environment of 2009 was altered markedly. The 2008–09 financial crisis

Figure 56. Böer Award Ceremony, 2009, Delaware, USA. From left: Prof. Klaus Thiessen, Berlin, Karl Wolfgang Böer and W. Palz.

was, in fact, not the reason for this new situation; it only has enhanced it. In late 2009, there was a global overcapacity in module manufacturing of between 50 and 100 percent — depending on which analyst you talk to. Competition became very tough and profit margins in the cell and module industry came down sharply, even dipping into negative territory. Accordingly, there was a reduced appetite for further investments in module manufacturing; the orders for equipment fell by more than 70 percent in 2009.

Another loser in this new environment was the silicon supply industry. A few years ago, when the global PV market began to boom, "electronic grade" or even "solar grade" silicon was in short supply: the price for a kilogram of silicon feedstock reached US$500 on the spot market — some 20 times the production cost. This has now also become a buyer's market, and prices have fallen accordingly, leaving no further room for speculators. The beneficiaries were the sellers of, and investors in, full systems, and in particular, the installers of building-integrated PV, the promoters of large "in-field" plants of the MW size, and last but not least, the buyers and investors.

It is important to note that PV plant investments in Germany and other countries that adopted the FIT system do not only rely on favourable tariffs but also on a whole range of supporting measures such as low-interest banking rates or tax credits. Investments that were already profitable with the FIT alone are even more so through these extra benefits.

No wonder that profits reached even greater highs when the FIT as defined by law remained high while module prices, which represent more than 50 percent of the overall plant costs, hit rock bottom. In Germany, small investors who for tax reasons were used to investing in the shipping industry, switched to the

PV sector in 2009. Profits of well over 10 percent on the investment were not unusual that year. A rule of thumb in Germany says that the net system price must not exceed 10 times the yearly income through the FIT to achieve a profit between 5 and 8 percent. That was much better in 2009, a year that stood out for its unprecedented profits in PV.

However, this spike was never going to last for long. German cuts in solar subsidies in 2010, to as little as 25 percent of the 2009 tariffs, will rein in the excessive profits. In general, however, the trend is positive: investors will likely maintain their interest in PV as they have become acquainted with it and interest rates remain attractive; external costs for society will come down further; and industrial competition remains very strong. And the winner is: photovoltaics!

Another major event in 2009 was the emergence of thin-film modules in global markets. First Solar, a US company and producer of CdTe modules, finally dethroned Sharp of Japan and Q-Cells of Germany, the long-time global market leaders. But it would be dangerous to jump to the conclusion that the era of crystalline silicon is coming to an end — it certainly is not. For the foreseeable future, crystalline silicon cells will stand side-by-side with the thin-film cells, even if the latter will never again be relegated to niche markets; in 2009, thin-film cells earned their credentials to participate in mainstream global markets.

Besides the rise of First Solar, another driving force for the steep drop of module prices in 2009 was the Chinese PV industry — which is still focusing on crystalline silicon. Even after consolidation at the beginning of the year, hundreds of Chinese PV companies have remained in the game. Mainland China together with Taiwan sold 44 percent of all PV in the global markets in 2009; the Chinese mainland alone had one third of the world market. And with China's insignificant domestic market, this increased competition in export markets significantly. Germany's PV module producers were ultimately glad to have retained half of the country's domestic market for themselves in 2009, while losing 25 percent to Chinese imports.

The typical refrain is that Chinese industry has the cost advantage owing to its low labour wages. I do not think that this applies to the PV business. I visited the leading Chinese firm Suntech in Wuxi near Shanghai at a time when it was still a small start-up company. Their production line was already running fully automatically, using new equipment purchased in Italy; the human resources required were minimal. Indeed, module manufacturing is now automated around the world, be it in Japan, the US, Germany, Malaysia or China; what may differ is the management skills of the owners.

1.3 *The Technology Boom Goes On*

There exists an almost interminable list of research and development opportunities for PV semiconductors. Dozens of composite materials are still waiting to be explored. But it is grossly misleading to say that we first need technological development before PV can truly make it into the big energy markets. That argument is regularly used by opponents of solar power as ammunition to denigrate PV; it is totally wrong!

In practice, we currently have three main technological routes for PV semiconductors:

- thin-films of CdTe, the CIS family (copper indium di-sulphide, copper indium gallium sulphide, copper indium gallium selenide, and others), the family of silicon micro-morphs (a very thin double layer of amorphous silicon and nano-crystalline silicon), etc.;
- crystalline or semi-crystalline silicon cells;
- the GaAs family (components from the third and fifth column of the periodic table of elements).

The last category is expensive and only of interest in combination with optical concentration — and for applications on satellites. Up to 42 percent efficiency has been achieved and there is a realistic potential for more — through more research and development.

For crystalline silicon cells, the highest efficiency theoretically possible — around 25 percent — has almost been achieved already. The main interest here remains improvements in the technology of industrial production — more a case of development rather than research.

The highest potential for technological improvement concerns thin-film solar cells. Theoretical efficiencies are well over 20 percent for these materials; today's

Figure 57. ISFOC PV concentrator testing institute at Puertollano, Spain, 2008: Prof. Luque (left) with Prof. Götzberger.

industrial production achieves efficiencies barely over 10 percent. Next to CdTe, the CIS family remains the most promising contender for further development.

2. Outlook

2.1 *On the Threshold of Commercial Viability*

As already mentioned, the price individual customers like you and me currently have to pay to the electricity utilities is anything up to 35 US-cents per kilowatt-hour — including all taxes and without subsidies. The price is trending upwards. Conversely, the price of PV systems has dropped dramatically in recent years, and as a result, the kilowatt-hour cost of PV generation has followed suit; the prospect is for prices to head even further downwards and the trend seems unstoppable. Accordingly, PV's appeal for the public is widening.

The crossover point — called "grid parity" — happens when the kilowatt-hour price of PV electricity generated on the consumer's own house becomes equal to the price of conventional electricity offered by the grid. This is about to occur first at places most likely to fulfil two conditions at the same time: (i) where the kilowatt-hour price from the grid is highest, and (ii) where solar irradiation is most plentiful. Candidates include Italy in Europe as well as Hawaii and other US states. Some American analysts have asserted that PV generation and grid electricity have already experienced price convergence in certain market segments in California and New Jersey; by 2012 it is expected that as many as 10 states will have surpassed grid parity.

In 2010, we have already reached grid parity in the most favourable regions of Europe as well. However, grid parity is not enough to get people switching automatically to PV electricity generation. To make PV attractive for purely financial reasons, the price of PV electricity must not only be equal to that from the grid but also cheaper. Until this happens, institutional support remains important — at least for a while — even once grid parity is achieved.

PV support in the US is, generally speaking, somewhat unique, as it focuses on tax credits and net metering — customers only pay the difference between the electricity bought from or sold to the grid; such schemes do not provide net profits like FITs. This situation changes once grid parity is achieved: from that point in time investors in the US earn a net profit, too. Similarly, countries like Germany can abandon the FIT system once grid parity is achieved. From then on, the special interest rates and tax advantages are attractive enough for investors to gain an appreciable profit — provided that net metering becomes standard practice, as it is in the US today.

In conclusion, it can be anticipated that in the global PV markets grid parity is at our doorstep. Shortly afterwards, the floodgates will open for a new and sizable PV market that operates purely on commercial grounds. All the signals point to this happening before the end of the present decade. But it will require that political decision makers follow PV market developments closely, armed with the best expert knowledge, to avoid over-subsidising at certain times, like in 2009, or a too-hasty reduction of support at others.

2.2 Outlook Towards 2020

Since the turn of the millennium, the overall trend for renewable energies in the electric power sector has been encouraging. In the EU, next to PV and wind power, only gas power plants have increased their capacity; the net capacity of coal, oil and nuclear power has decreased. The trend in the US is similar.

Extrapolating these trends lends credence to the further vigorous growth of solar power and the other renewables in the foreseeable future — even though China, Germany and others are pursuing a strong investment programme in new coal power plants, and France, the US, Korea, Abu Dhabi and others are pushing hard for the nuclear option. At the same time, as long as the global PV markets continue to rely heavily on political support, any predictions about future developments must remain speculative.

In 2009, the worldwide PV market reached 7 GW. However, as already mentioned, without Germany's market explosion at year end, there was a risk that for the first time in 30 years the market might not have shown any growth. Annual growth rates of 50 percent or more had not been unusual previously. Still, the global PV industry is preparing for large growth rates in the coming years; that at least will not be a problem.

If global markets increase by a mere 7 percent per year on average, by 2020 an additional 100 GW of PV capacity will be added to the 25 GW in place today. If trends continue, at least half of that new capacity will go to Europe — and there is every chance that Germany will maintain its lead. A better extrapolation from what has happened in the first decade of the present century might be to achieve at least double that — 200 GW of new PV capacity worldwide by 2020, with half of it in Europe. As far as Europe is concerned, recent progress justifies some optimism about the coming developments in France and to a lesser extent in Italy. An additional 100 GW of capacity in the EU would mean that more than 10 percent of overall power capacity would be solar; in terms of electricity production, that would constitute 2–3 percent.

China: The Solar Dragon

In energy terms, China is currently the number one global player. Its dynamism is characterised by some extraordinary growth rates: every year, China adds up to 100 GW — as much as Germany's total — to its power capacity; in 2008, China accounted for 85 percent of global coal demand; China recently surpassed the US as the world's biggest CO_2 emitter; and in 2009, China became the largest automobile market in the world, with over 14 million new cars.

Figure 58. China PV Conference, Beijing, 1986.

Figure 59. High-Tech Forum, Shenzhen, China, 2009: Qin Haiyan (Director Chinese Certification Office) and W. Palz.

China is aware that it has a "sustainability problem". Its leaders are realising the need to maximise energy conservation, introduce "clean coal" power plants and close inefficient coalmines, and increase nuclear capacity by building new atomic power reactors. Above all, China is now a key player in the global deployment of renewable energy: one in three PV modules on the global markets come from Chinese manufacturers, leaving the former industrial PV leaders Japan and Germany behind. Sixty percent of the world's solar water heaters are installed in China, and they are of the highest quality. In 2004, 60 million square metres were installed; more than twice as much — 130 million square metres — was installed in 2008.

By 2005, China had only installed a meagre 1.2 GW of wind power; just three years later, it had 10 times as much. In 2009, China installed the largest new wind capacity anywhere in the world: 14 GW in a single year.

Three-quarters of the renewable energy installed in China is hydropower and the country is the world leader. Small hydropower (SHP) accounts for one third of the total. In 2008, the amount of SHP installed increased by 8 percent to 51 GW. Today, 300 million people in China's rural areas get their power from SHP. This is part of the country's "New Rural Village Construction Plan". Local stakeholders take responsibility for construction, management and operation of the facilities. For 20 million people in China, SHP is used for cooking, saving precious firewood. The decentralised nature of these endeavours is highlighted by the fact that 500 micro-credit systems are operating in the country.

China's bio-energy programme currently equates to the consumption of 100 million tonnes of coal. Efficient wood stoves have been installed in 146 million homes, and 30 million households rely on modern forms of biogas. The potential for biogas is enormous: the country produces some 1,100 million tonnes of cattle manure a year, from which biogas may be extracted. Currently, 35 percent of China's 770 million tonnes of straw is used for energy generation. As bio-alcohols for transport are becoming increasingly important, China is experimenting with sweet sorghum — where it is the world leader — and with cassava and sugar cane.

If current growth rates continue, China could have a 45–50 percent share of renewable electricity by 2020.

Figure 60. Meeting at the reception hall of the Chinese government, Beijing, 2009. Ma Kai, Secretary General of the State Council of China, (centre); Dr. Wang Tao, Chairman Chinese National Committee of World Petroleum Council and former minister (right), and W. Palz.

By way of comparison, by the end of 2013 Spain is expected to have a total of 2.4 GW of solar thermal power capacity — the country is the leader in Europe in that technology. This will represent less than 10 percent of the PV installed at that time; as explained previously, solar thermal power is rather wishful thinking by some vested interests, while PV is widely held to be the answer when it comes to solar power deployment.

On global PV markets, China stands out as the most promising country: the development of solar energy has considerable public support in China, and the Chinese government is aware of its potential and has ambitious plans in the offing — and China's PV industry is already leading the world today.

Needless to say, the US can be expected to stay competitive in PV as well, revitalising their leadership role that was so strong in the 1980s. Similar shifts happened in the wind power sector: in the 1980s, the US was a pioneer in the world; at the turn of the millennium, Germany led the charge; today, the leaders are again the US together with China as well as Germany and Spain. In 2009, 32 GW of new wind power capacity was installed, compared with 6 GW for PV; some analysts have predicted a global wind power capacity of some 700 GW by 2017.

2.3 PV as Part of a 100 Percent Renewable Energy World

We have considered at length the unsustainable nature of the world's conventional power and energy system for reasons associated with global climate change, security of supply, and the ultimate limitations of all fossil and uranium resources. Even experts in the energy sector are conceding that when the oil and natural gas finally run out, the "clean coal" and "nuclear breeders" options are highly problematic — not only due to the excessive costs involved but also because people's commitment to atomic power or large-scale carbon dioxide storage is questionable. It has become increasingly obvious that solar energy provides the solution needed for a world requiring an abundant and sustainable energy supply.

A world that is 100 percent reliant on renewable energies is a novel idea and — although it is attracting more and more attention — it is revolutionary. To test how far that idea might also be realistic, various simulations are currently being undertaken; we are going to mention just a few of them. All must obviously be speculative, as a "solar world" is not something one could achieve tomorrow, but perhaps in a few decades from now. However, the stakes are high for the world's future, its economic outlook and new opportunities for a peaceful society. It is time to have an inspired look ahead.

In the early 1980s, the French Groupe de Bellevue became the first to assess the opportunities for a 100 percent renewable energy supply for a large country: France. Details have been given in a preceding chapter of this book. The report confirmed the viability of such a vision.

In the framework of our EU APAS programme mentioned earlier in this book, an extensive study entitled "Long-term Integration of Renewable Energy Sources into the European Energy System" was undertaken in the mid-1990s. Results were published in a book of the same name by Springer in January 1998 (ISBN 3-7908-1104-1). The project leader was Helmuth Groscurth from the renowned ZEW Centre for European Economic Research in Mannheim, Germany, in co-operation with a group of 25 experts from CIRED in Paris, France; Polytechnic University in Mons, Belgium; Roskilde University, Aalborg University and Novator Consulting, all in Denmark; and in Germany, FhG ISI in Karlsruhe and the Wuppertal Institute.

Of the two scenarios considered, the group calculated a "Sustainable Future Energy System" for the EU15 (at the time the EU had 15 members) with a time horizon of 2050. Living standards were to remain the same as they were at that time in northern Europe — among the highest in the world. The authors also stressed that such developments in Europe should not come at the expense of other regions of the world, especially the poor ones.

It is important to note that the purpose was not an attempt to predict future developments, but a means to check concepts for a sustainable future energy supply for Europe for their technical and economic feasibility and to provide a positive vision for decisions to be made in the future. The results of the study were the following:

- The overall energy demand in Europe (EU15) should be reduced by 62 percent from 4,500 W per capita in 1990 to 1,700 W per capita in 2050.
- Ninety-five percent of the energy demand should be supplied from renewable sources (even when considering only those technologies that were already developed at the time of the study).
- Bio-power should supply 500 W/capita; solar thermal collectors, 330 W/capita; wind power, 210 W/capita; PV, 153 W/capita; and hydropower, 140 W/capita; and so on.

The following figures give valuable insight into what implementing PV in practice meant in the study. Only building-integrated collectors were taken into account: on 2,770 km^2 an overall PV capacity of 405 GW would have to be installed. For the EU15, the authors calculated a total surface area of 5,300 km^2 on rooftops and 3,900 km^2 on façades. Thirty percent of these surfaces would be equipped with PV — the remainder with solar heat collectors.

In 2000, the German parliament in Berlin, the Bundestag, set up an official Enquête Commission on Sustainable Energy Supply. It ran a number of different supply scenarios for a "sustainable Germany" with a time horizon of 2050. I was on the Commission together with Harry Lehmann, presently a director at Germany's Federal Environment Agency (UBA), and other experts and members of parliament from all parties. An official report (14/9400) was delivered at the end of the Commission's term in 2002. Harry Lehmann and his team carried out

Figure 61. The German parliament's Enquête Commission of 2002: W. Palz (left of the "lady in green") and Harry Lehmann (standing behind both).

one of the detailed simulations: to investigate the viability of a Germany 100 percent powered by renewables by 2050. As the report documented, he proved that this option is indeed viable.

The Enquête Commission found that with PV arrays deployed just on the roofs and façades of all existing buildings in Germany, 38 percent of all electricity demand could be met; counting the whole of Europe, the potential would be 44 percent, and on the global level, 100 percent. The report does not consider ground-based PV installations; doing so, one may easily arrive at a 100 percent PV supply for the entire world.

In one of the other scenarios calculated by the Enquête Commission, the nuclear option was considered with a view to reaching the goal of an 80 percent reduction in greenhouse gas emissions by 2050. The conclusion was that up to 70 new atomic power plants would have to be built in the country — a big embarrassment for the conservative party members of the Bundestag, who presumably would have wanted the whole affair to disappear into the archives.

It is worth mentioning that Harry Lehmann and his team also tried simulations for Japan and other countries powered exclusively by renewable sources. The results were positive in all cases.

In its November 2009 cover story, the prestigious journal *Scientific American* presented the most aggressive "100 percent scenario" imaginable: not just conventional electricity but all energy needs, including those for transportation and heating, would be supplied globally by 2030 — only 20 years from now!

Figure 62. Washington, 2010: W. Palz with Carol Browner, Director for Climate and Energy at the White House, previously long-time Director of the US Environmental Protection Agency.

Mark Jacobson of Stanford University and Mark Delucchi of the University of California calculated a world in 2030 in which the maximum power consumption of 16,900 GW at any given moment could be met — after taking higher efficiencies into account — by solar plants (40 percent), by wind power (51 percent), and by hydropower (9 percent).

Three-quarters of the solar power would come from rooftop PV panels, and all the technologies involved would be those already available today. Only 1.3 percent of the global land area would be needed for installation. The cost of generating and transmitting electricity with renewables would be less than the projected cost using fossil and nuclear fuels.

The *Scientific American* paper completely discards energy derived from biomass; this is surprising given that it is currently the world's most important renewable resource. In future, it would be nonsensical not to make use of at least part of the five billion tonnes of animal and human waste that pile up every year on Earth by extracting its energy via biogas; or to exploit some of the more than six billion tonnes of new solid material that grows every year in the world's forests; or to employ the energy contained in agricultural residues. And why not utilise "energy crops" that convert solar radiation into biomaterials ten times better than conventional forests?

We should not forget that we ourselves are part of the biosphere, and sustainability means looking after the interface we all share with the living world. Only robots can rely on semiconductors alone!

Despite any shortcomings it might have, Jacobson and Delucchi's study convincingly illustrates that shifting the world to 100 percent renewable energy is

Figure 63. The bright solar future: a large PV plant in front of a coalmine in Saarland (W. Palz's home state), Germany.

perfectly possible — given the will to do so. The planet's energy needs will never be met entirely by solar power, but PV will be one of the brightest stars among an ensemble of clean energies that generate the electricity of our future Earth.

3. Conclusions

Our generation is a generation in transition. When my father Karl was born in 1892 in Germany, virtually no family in the world had electricity. By the time my son Carlo reaches my present age, it will be the year 2060, and I fully expect that the world will be supplied primarily by solar energy and its derivatives. By then, the age of oil will have come to an end — and we can only hope that it will not be a tumultuous final act.

Later generations will remember ours as the generation that bestowed upon humankind the comforts of electricity and electronic communication while exploiting and consuming the hydrocarbon resources accumulated over millennia in little more than 200 years. Nothing to be proud of.

Will we at the very least also be remembered as the generation who made way for the coming solar age?

PV SYSTEM TECHNOLOGY DEVELOPMENT IN THE EUROPEAN COMMUNITY (1988)

M.S.Imamura
WIP, Sylvensteinstr. 2
D-8000 Munich 70 Germany

ABSTRACT

A comprehensive PV System technology work is being carried out by the Commission of the European Communities. The R&D activities consist of improvement of the original PV pilot plants, Subsystem development, and new applications. As a group, the 16 pilot plants represent the largest array of stand-alone System in the world, and they contain a unique variety of System topology and hardware. These pilot plants, with continued improvement and modernization, are thus expected also to serve as test beds that will provide invaluable technology Information for a long time in the future. The Subsystem development is comprised of seven concerted action tasks resulting from common technical problems experienced by the pilot plants in the past four years. The new projects include PV-powered applications up to 80-kWp from PV houses to domestic appliances, development of trans-former-less inverter and de motor/pump, and a study of a combined small hydroelectric/PV power plant. This paper also describes key lessons learned from the pilot plants and identifies technology issues and further development needs. There are about 30 of these R&D projects as of mid 1988. The total EC budget for above projects is 5.7 million ECU, a third of the total amount allocated by CEC for all PV technology work.

1. INTRODUCTION

In the past decade, the Commission of the European Communities (CEC) has been instrumental in initiating, implementing, and coordinating the photovoltaic (PV) technology activities among its 12-member countries. The Commission has actively pursued PV technology and its utilization since 1975. Their PV activities are now implemented mainly by the R&D, Demonstration, and international development programmes. These are carried out by Directorate General (DG) XII, XVII, and VIII, respectively. The Commission provides direct financial support on a cost-sharing basis for R&D (up to 50%) and Demonstration (up to 40%) projects. As of mid 1988, DG XII has authorized 30 R&D projects, and DG XVII has 48 Demonstration projects, all in the EC-member countries. It should be noted that the Joint Research Centre (JRC), a part of DG XII, located in Ispra, Italy, has been actively engaged in monitoring of PV plants, testing modules, calibrating solar sensors, and coordinating the European Working Group on PV plant monitoring.

Photovoltaics also play an increasing role in the Commission's development aid agreements, in particular with Africa, the Pacific islands, China, and Middle East. CEC is thus committed to ensure that the European Community maintain its world Position on photovoltaic technology and meet its large responsibilities with respect to the Third World.

The DG XII R&D activities basically fall into two categories:1) solar cell materials and devices, and 2) System development. Specifically, this paper provides a summary of System development projects recently initiated and describes key lessons learned.

2. CURRENT SYSTEM R&D PROGRAMME.

The system-level R&D projects are comprised of three groups of activities: improvement of and further experimentation in the pilot plants, advanced Subsystem development, and new applications and developments. About 30 projects are now underway, and we expect a few more to be approved by late 1988. They all have different Start dates (late 1987 to early 1988) but most of them will be completed in 1989. The contribution by the Commission for these projects is 5.7 million ECU, which results in a total project spending of about 11.4 million ECU by the contractor. This is about one third of the total R&D budget for the entire PV technology development under DG XII.

WIP is responsible for the coordination and control of the System R&D Programmes for DG XII. A part of their task is to prepare a Consolidated handbook of PV technology in the European Community at the end of the current programme. This handbook will cover the significant accomplishments and technology contributions from all projects identified herein, including useful guidelines in design, analysis, and testing of PV Systems.

All currently approved R&D projects, their responsible organizations, and significant objectives of these projects are described below in terms of three groups of work. For more detailed Information on a given project, readers are encouraged to contact the head of the project as identified herein. Table I is a list of all projects authorized by DG XII.

2.1 Pilot Plant Improvement Projects

The most visible activity in the past has been the 16 "Pilot PV Plants". These plants were installed during the period from 1982-1984 at several sites in Europe. Ranging from 30 kW to 300 kW and totaling 1.1 MW, they were the first major prototypes intended largely for research purposes to encourage the development of

Table I. New PV System R&D Projects Authorized by DG XII as of Mid-1988

Project No.	Title	Contractor/Location	Head of Project
Pilot Plant Improvement (Plant-specific Actions): CEC Contribution: *1.9 M* ECU			
122	Pellworm (300 kW)	AEG, Wedel, FRG	J. Lowalt
123/	Kythnos (100 kW) and	PPC, Athens, GR	P. Pligoropoulos
124	Aghia Roumeli (50 kW)		
125	Kaw (35 kW)	Seri Renault, Bois d'Arcy, FR	B. Aubert
127	Fota (50 kW)	NMRC, Cork, IR	G. Wrixon
128	Terschelling (50 kW)	Ecofys, Utrecht, NL	E. ter Horst
132	Tremiti (65 kW)	Italenergie, Sulmona, IT	F. Fonzi
134	Giglio (45 kW)	ENEA, Rome, IT	S. Li Causi
135	Vulcano (80 kW!	ENEL, Milan, IT	A. Previ
136	Zambelli (70 kW)	Serv. Tech., Verona, IT	D. Braggion
126	Mont Bouquet (50 kW) and Nice (50 kW)	Photowatt, Rueil-Malmaison, FR	B. Monvert
130	Rondulinu (44 kW)	Univ. de Corse, Ajaccio, FR	G. Peri
Advanced Subsystem Development (Concerted Actions): CEC Contribution: *1.0 M* ECU			
137	Battery Control and Management	NMRC, Cork, IR	S. McCarthy
138	Power Conditioning and Control	Fraunhofer Inst. ISI Freiburg, FRG	J. Schmid
139	Computer Simulation and Modelling	NMRC, Cork, IR	G. Wrixon
140	PV Modules and Arrays	Conphoebus, Catania, IT	G. Beer
142	Data/Plant Performance Monitoring; Array Structures; Social Effects; Control and Coordination of Concerted Actions Projects	WIP, Munich, FRG	M. Imamura/P. Helm
New PV Applications and Development: CEC Contribution: *2.85 M* ECU			
143	10 kW PV Plant for New Component Experimentation	EAB, Berlin, FRG	T. Mierke
515	80 kW PV System for Power/Heat Degeneration Plant	Stadtwerke Saarbrücken, FRG	W. Leonhardt
119	Agricultural Applications (6 kW)	IER-CIEMAT, Madrid, ES	L. Delcado
121	ANCIPA Hydro-electric Plant (3.5 kW)	ENEL-Milan/ Conphoebus, Catania, IT	A. Previ/B. Morgana
115	Detection of Radioactive Anomalies	Merlin Gerin Provence, Senas, FR	J. Jaillon
116	Passenger Cars	VW AG, Wolfsburg, FRG	A. Konig
525	Transformerless Inverter, 30 kW 3-phase and 3 kW 1-phase	Fraunhofer Inst., Freiburg, FRG	J. Schmid
117	Domestic Appliances	Systemes AMI, Marseille, FR	T. Fogelman
141	Encapsulation of PV Cells	Equinoxe Production, Cambronne-les-Ribecourt, FR	J.P. Hanique
120	DC Motors/Pumps (1.5 and 5 kW)	Univ. of Florence, Florence, IT	A. Rosati
528	A Study of Small Hydroelectric/PV Plant	WIP, Munich, FRG	P. Helm/ M.Imamura

new concepts and processes and to stimulate the European industries. To a large extent, these original pilot plants served as the springboard for many national organizations and industries to create markets for many photovoltaic applications within the European Community and abroad. Much useful information has already been obtained, but, as expected with research projects, many problems also became apparent. The pilot plant improvement, activities, therefore, were undertaken to improve and optimize the technical performance and efficiency of the pilot plants.

Table I lists the 13 pilot plants that will be involved in these activities, along with the responsible organizations. The remaining three plants were not able to participate because of co-financing problems. It is significant to note that the improvement work on four pilot plants will be handled by organizations other than the original designer. These are PPC of Greece for both KYTHNOS and AGHIA ROUMELI, Université de Corse for RONDULINU, and Ecofys for TERSCHELLING.

All plants are currently operating except MONT BOUQUET in Southern France which was badly damaged by the brush fire, and NICE because of damaged battery cells. The improvement areas involve battery charge and discharge control, inverter, plant control, data collection equipment, and proper utilization of available array power. Table II summarizes the major tasks and unique or interesting experiments to be performed.

2.2 Advanced Subsystem Development Projects

The advanced subsystem applications and development is comprised of seven concerted action tasks resulting from common technical problems experienced by the pilot plants. These are identified as: 1) PV arrays, 2) array structures, 3) battery management, 4) power conditioning, 5) data collection equipment and plant monitoring, 6) computer simulation and modeling, and 7) social effects study. The Commission selected four contractors to direct these seven tasks. Table I lists these projects, the responsible organizations, and their total CEC budget. Unlike those of the other two project groups which are funded at a 50% level, the Commission finances these Concerted

Action projects completely. Therefore, the results of these projects will be available to everyone without restrictions. These projects will be done in a concerted manner with all pilot plant leaders and other selected European specialists participating.

These activities should lead to comprehensive guidelines for the PV systems, to be published by the four organizations listed in Table I. The initial results should be available in late 1988.

A common aim in tasks 1) through 5) is to produce guidelines useful to the operation of existing plants and to the future PV plants in each respective area.

One activity under the Conphoebus project is the investigation of simplified solar irradiance calibration method. A team of principal investigators from several test centers in Europe has been assembled to evaluate the possible effects of geographical sites and seasonal variations on the sensor calibration factor and validate both calibration and analysis techniques. They will also look into the question of Si sensor types to be used for solar insolation monitoring purposes. The ultimate goal of this activity is 1) to allow field (in-situ) calibration of both thermopile and Si sensors, and thereby reduce their recalibration cost, and 2) replace the more expensive conventional thermopile sensors with the cheaper more stable Si-based devices for continuous outdoor monitoring of solar irradiance on both horizontal and tilted surfaces.

Each of the seven concerted action tasks has formed a working group consisting generally of all project leaders identified herein plus selected experts from Europe. This group now totals about 76, including representatives from three major projects that were not co-financed by the Commission (MADRID 100-kW, SENEGAL 10-kW, and DELPHOS 300-kW). They were invited to participate in these concerted action activities because of their unique contributions.

2.3 New Applications and Development Projects

Several new 'projects started in early 1988 (see Table I). The subsystem development projects include encapsulation of PV cells, dc motor/pump, and transformerless inverter. System hardware applications are PV houses, special sensing systems for crop and forest protection, passenger cars

Table II. Key Tasks and Experiments in Pilot Plant Projects

PV Array:
- Replace Broken Modules and Repair Shorts in Array Cabling (TERSCHELLING)
- Improve Lightning Protection (KYTIINOS)

Evaluate Array Performance (Most Plants)
- Increase PV Array Size (TREMIT1)

Battery and Charge/Discharge Control Protection:
- Replace Bad Battery Cells (All Plants)
- Improve Charge Control Method and Discharge Termination Criteria (All Plants)

Develop/Test State of Charge Indicator or Calculation Method (FOTA, TERSQIELLING, ZAMBELLI, VULCANO, KAW)
- Parallel Two Batteries Into One (FOTA, PELLWORM) Increase Energy Storage (TREMIT1)

Control/Power Conditioning:
- Install New Load Management Software (ZAMBELLI, PELLWORM)

Improve Plant Control for Better Energy Utilization and System Operation (PELLWORM, ZAMBELLI, TREMITI, AGIA ROUMELI)
- Replace Computer (ZAMBELLI)
- Split Control and Monitoring Software (PELLWORM, TERSCHELLING, ZAMBELLI)
- Reactive Power Compensation (KYTHNOS, TERSCHELLING)
  Add capability for grid-connected mode (VULCANO)
  Improve Load Management (PELLWORM)

Data Collection System and Plant Monitoring:
- Remote Monitoring Using ARGOS Satellite Data Transmission (KAW and RONDULINU)

Remote Monitoring Via MODEM (FOTA, TERSCHELLING, PELLWORM, ZAMBELLI)
- Add Wh and Ah Meters (Almost All Plants)

Collect/Analyze Plant Data Near Real Time (All Plants)
- Install New Data Collection Equipnent (TERSCHELLING, ZAMBELLI, KAW)
- Re-calibrate Solar Sensors
- Implement improved performance analysis methods (All Plants)

Consumer/Loads:
- Increase Water Storage Capacity (TREMITI)
  Improve Water Desalination System (TREMITI)
- Replace ac Motors with dc Motors (TREMITI)

System Experimentation and Evaluation:
- Battery: Charge Control Techniques, Cycle Life, Effects of New and Old Cells in Series, Effects of Long-term Operation in Partial State of Charge, Means of Monitoring State of Charge

PV Array: Effects of Tilt Angle Adjustment on Annual Energy
- Methods of Increasing Plant Energy Efficiency
- Evaluate Long-term Performance and Dust Accumulation Effects on Solar Irradiance Sensors
- Develop Simpler But More Effective Ways of Monitoring and Reporting Plant Performance

3. LESSONS LEARNT

At present all 16 pilot plants except Mont Bouquet and Nice are operational. Mont Bouquet's array was badly burnt by a local brush fire; Nice will replace all batteries to make the system operational. The batteries in the pilot plants have now operated for 4-6 years. A few PV plants have replaced a number of battery cells that have short-circuited.

Table VI is a list of key lessons learned under subsystem, system, and project management categories. It lists what must be changed or avoided in the future, what were demonstrated to be correct, and the unknowns that need answers.

One of the best lessons learned is in the area of plant monitoring. First of all, none of the data acquisition system (comprised basically of a computer, signal interface and conditioning circuits, and data storage devices) was found to be 100% reliable and able to collect/store 100% of data requirement continuously for a long time. This has improved dramatically.

Table III. Key Features of New Projects

Projects		Features
SAARBRUCKEN Power/Heat Cogeneration Plant	0	Hybrid PV/Heat Generation; 20 4-kW PV/Inverters for 20 Houses
BERLIN 10-kW PV Plant for Experimentation	0	Heat Pump and Other Consumer Operation New Inverter/Converter Evaluation
Agricultural Applications	0	Milking, Milk Shaking, Refrigerator, Pumps 6-kW PV, Boost Converter, Transformerless Inverter
ANCIPA Hydroelectric Plant	0	Sliding Gate and Trash-rack Rake 3.5-kW PV, Inverters, Batteries
Passenger Cars	0	PV Integration, Fuel Saving, Comfort Battery Charging, Reliability Improvement
Detection of Radioactive Anomalies	0	PV Supply and Radiation Detector
Domestic Appliances	0	Wireless Portal Bells; TV Aerial Antenna Build-in PV Cells
Encapsulation of PV Cells	0	Synthetic Supports and Material Outdoor Testing
Ecological/Crop Protection Sensors	0	Self-acting Irrigation; Early Warning Sensor New Consumers (Washer, Refrigerator)
dc motors/Pump	0	Heat-Pipe Cooling; Low-friction Sealing and Bearing Special Permanent Magnet
Transformerless Inverter	0	3-kW and 30-kW, 3-phase
A Study of Small Hydroelectric/PV Plant	0	Feasibility-type, Up to 2-MW PV Small Existing Hydroelectric Plant, Northwest Spain

Table IV. Summary of Key Problems Encountered by the Pilot Plants

PV Array
- Power degradation in some modules
- Module glass cracking from mechanical pressure
- Lightning damages on modules
- Most modules had lower actual power and efficiency than advertized; use of optimistic module rating by system contractor resulted in lower array output at installation
- Intra-module cable breakage due to loose free-hanging cabling
- Module string shorts from metal grounding of bypass shunt diodes and blocking diodes
- Water in cable trays causing intermittent shorts
- Array structures generally overdesigned and expensive
- Array power lower than estimated initially

Batteries
- Cell cases have ruptured from internal pressure build-up; one cell with evidence of "flash" burns
- Cell failures from internal shorts and low capacity beginning to occur more frequently
- Batteries have seen in both under-charged and excessively overcharged conditions
- Lack of battery-level data:
- Available capacity (Ah) vs days of operation
- Discharge and charge characteristics (new and old)
- Cost of "PV" battery cells is expensive

Data Acquisition System, Sensors and System Controller
- Many computer and software problems after installation; skilled people not available on site quickly
- Critical spare parts not available
- Often impossible to reprogram on site: poor software documentation
- Many stoppages and failures in DAS, especially recording and printing devices from dust and temperature extremes
- Signal isolation devices are very costly and noise problems on measured data
- Solar sensors not re-calibrated for several years due to high cost and uncertainties in re-calibration needs
- thermopile-based sensors degrades with use; have slower response time than Si sensors

Power Conditioning
- Inverter failures occurred frequently
- Long delivery time on certain inverters
- Large frequency and voltage instability of the grid (esp. from a low-power Diesel substation) caused inverter shutdowns
- On a daily basis, a single inverter system operates at low efficiencies
- Bias power supply failures from line voltage surges and spikes
- Inadequate cooling inside the electronic racks
- Lightning strike damages to power conditioning elements
- Electrical loads often significantly different from original design assumptions

Operation and Maintenance/Plant Monitoring/Other Issues
- Most owners and operators did not acquire the capability to reduce recorded data and analyze plant-performance
- Lack of spare parts for critical items
- Plant analysis, maintenance, and ownership responsibilities often not clear
- Maintenance needs underestimated
- Lack of adequate training of operators
- Inadequate operation/maintenance documentation
- Local utility not willing to accept and/or pay. for PV energy; or utility pays very little

French cartoon that the famous cartoonist and PV pioneer, the late Jean-Marc Reiser, painted in 1982 as a gift for Wolfgang Palz. The "speech bubble" reads in English: And when there is no sunshine, everybody stays in bed.

Chapter 2

My Solar Age Started with Tchernobyl

Franz Alt

TV Journalist and Publisher, Baden Baden, Germany

Until the great accident of Tchernobyl, I was a supporter of the atomic energy. I trusted those experts that had told us their ideology of a "safe nuclear energy". As a TV journalist and member of a conservative Party, the CDU in Germany, I was blindly believing the energy experts. But the disaster of Tchernobyl was a wake up call and eventually I started to make up my mind seriously. Tchernobyl was my "Damascus event".

In the meantime I did learn a few things:

- The Sun sends us, every second, 10 000 to 15 000 more energy than man on Earth is currently consuming
- The wind streams contain 80 times more exploitable energy than we need
- 10 times more biomass is growing than our total energy needs
- Just the hydro energy alone contains as much energy as we consume in one single day
- The marine currents and the wave energy can give us at least 100 times the energy we need
- The geothermal energy offers a multiple of our needs. 99% of our globe is more than 1000°C hot.

Hence, also in practice, the direct solar energy or the indirect ones of wind, water, and biomass are more than sufficient to satisfy the energy hunger of almost seven billion people. The Working Group 'Solar Energy for Environment and Development' of the United Nations stated: "It is well established that the over-all

Power for the World by W. Palz
Copyright © 2011 by Pan Stanford Publishing Pte Ltd
www.panstanford.com
978-981-4303-37-8

potential of the Renewable Energies lies in the order of magnitude of 10 000 times the world's current energy demand".

The Sun is the motor of everything that happens on our planet. She is five billion years old and will continue to shine for 4.5 billion years more. Each year she sends us 350 million billion (350 000 000 000 000 000) kWh of radiation energy to our planet. All sources of energy on Earth we owe to the Sun: the wood of the forests and the vegetation in the fields, and also the deposits of coal, oil, and natural gas, that stored the Sun's energy over millions of years. Moreover, the Sun drives the water circuits in the seas and rivers.

The Sun is the source of all life. She heats our planet, gives us light, and provides the plants with energy for photosynthesis. This is the most important chemical reaction on Earth giving life to plants, animals and man. And even during night, the Moon and the planets reflect the Sun's rays to Earth. The nuclear reactor "Sun" differs from the nuclear power plants on Earth in that it is safe for accidents and nuclear radiation, needs no recycling of atomic waste, and provides all people with energy free of charge.

Just the solar radiation that reaches the Earth this day could meet our energy demand for the next 180 years. The Sun is our only inexhaustible source of energy. Buying fruits mostly in the super markets, warm meals in the canteen and milk packed in plastic bags, we tend to forget how all this comes about. We owe the cow and the fruit tree, the cereals and the electricity from the wall socket — just like our own life — to the Sun's energy. No Sun, no life. The Sun is the trade secret of our existence. The Sun's rays are *the* great gift of the Cosmos to us. The Sun was and is the one and only income of our planet. Since millennia we live from these gains. For all that, the Sun is tragically and terribly undervalued.

The astrophysicist Klaus Fuhrmann from Munich University has calculated the following negative scenario for the Sun:

- The global mean temperature today is 15.9°C
- If we had no Sun anymore, tomorrow, the globally tempreature would be −15°C
- Just after three days without our Sun, our temperature would have come down to −40°C, all schools would be closed
- After four days, we would have −80°C, no car engine could start
- After a week, the temperature would have come down to −173°C, all life would numb
- After four weeks or so, there would be no plants, no animals and no human left. Our Earth would be just a big cemetery. And there would be nobody left to complain about it

If we have at our latitudes in central Europe 0°C in Winter and 20°C in Summer, then this is the effect of the Sun. Our only need is to level out the difference between Winter and Summer. But as we don't heat in the Summer, we only have to make up for 7°C on the yearly average. And even the balance between

Summer and Winter can be achieved by calling on the assistance of the Sun — her sheer unlimited offer of choices. Only that little difference is at stake. The balance between 7 degrees.

In previous civilisations, people were more aware of all life's dependence on the Sun. That's why she became a basis of religions and cults. Sowing and growth, harvest and feasts were related to her light and her warmth. Only the age of enlightenment and the "modern" churches blurred the awareness of the Sun as the all-embracing donator of life. Jesus still saw it differently: he was talking about his father in Heaven who lets the Sun shine over the righteous and the unjust alike. The "Sermon on the Mount" is full of pictures of God close to nature. But the churches have replaced them by insipid and abstract images of God. In all holly scripts of mankind, the Sun is a divine symbol. God: The Sun behind the Sun, the very first energy, to which we owe everything.

But it is encouraging to see that in Germany in 2009 already 800 Christian churches of the big two denominations have installed solar installations at their buildings. They offer at last landing decks to the Holy Ghost. Religion becomes concrete and practical. Since the late 2008, the Vatican got its first large solar PV plant right next to St. Peter's — heavenly energy for the German Pope in Rome. The God of the creation is no ascetic, but an inspired artist of sincere cheerfulness who created with no more than Earth and Sun a total artwork. God or the Goddess forgot no smell and no sound, no colour and no form, and not the light and the warmth. All that I had to realise step-by-step, me, the son of a coal merchant — a representative of the old energy system.

Industrialisation has sidelined the great interrelations of nature. But modern natural science and cosmology give us a new insight of how much we are dependant on the Sun. The Solar Age is coming. If we succeed in the next few decades to employ just a fraction of the Sun's energy then the energy problem is solved once and forever without contamination of the environment. This inexhaustible and primary source can bring wealth and wellbeing for all people. With solar energy we can start a new era of solar culture as it never existed before.

The greatest danger for the future of mankind is our wrong energy policy. Instead of gaining environment friendly energy from the Sun, the wind, the water, and the biomass, from solar hydrogen and the heat of the Earth we consume in a record time the energy resources oil, natural gas, coal, and uranium. With oil, natural gas, and coal we heat via CO_2 emissions the planet and create a hothouse atmosphere hostile to life. And atomic power plants are extremely dangerous by their sheer existence. For a very long time after they will have stopped operating, radioactive waste keeps emitting radiation for over 100 000 years.

The year 2010 — the year after the unsuccessful Copenhagen Conference — is a year of decision for a fundamental change of our energy supply. The purpose is to supply the world in the future securely and environment friendly with domestic energy sources in a decentralised manner. Already by 2020 — following calculations of the German Renewable Energy Associations — half of the electricity could become generated from renewable resources. But this requires a new political framework.

The future park of power plants must be complemented, next to the speedy deployment of the Renewable Energies, by flexible power plants that are capable of harmonising the variations of electricity demand and supply respectively. Or do we want to go back for decades by allowing for longer operational life times of the atomic power plants and for newly built coal plants? Atomic power plants cement the old park of power plants by their inflexible operability. New power grids and energy stores — indispensable complements of the renewable, decentralised energy supply of tomorrow — are blocked by the inert atomic mastodons. This way, the urgently needed extension of the electric grids does not come in support of the decentralised energies of the future, but as nothing more than the transport of the current from the fossil-atomic Mega power plants. A forward-looking policy offers the chance, via a bonus for combined power plants, to promote investments in biogas storage, batteries, and electric cars and in this way, to prepare the crucial steps for a full supply with Renewable Energies. Just in 2008, by the deployment of the Renewable Energies in Germany, 30 000 jobs have been created in the country. Switching to a 100% Renewable Energy supply in Germany and in the European Union could entail, following calculations of the European Commission, five million new jobs in Europe and more than one million of them in Germany.

1. Solar Policy is Social Policy

The great philosopher of the Atomic Age, Günther Andres, once said: "Our biggest problem is that we don't imagine anymore what we are doing". As I had many political and personal reasons over the last 20 years to change my ideas, I know what Andres had in mind. Only when I started to concretely and practically imagine what atomic power plants and the atomic arms race can do, I was able to shift my thinking. A new awareness is the prerequisite for a new way of thinking and of doing things. As long as I was not aware that with 1 litre of gasoline I pollute 10 000 litres of air, I drove the car some 15 times more kilo meters now. As long as I did not realise that we burn simply with our current energy practice in one year as much oil, coal, and natural gas as was needed to grow in one million years, I saw no reason to think seriously about alternatives to the present energy policy. Before Tchernobyl, I was not aware what an atomic accident could mean — hence I was not against atomic power plants. Before the great peace movement in Germany in the early 80s I could not imagine that the atomic arms race could quickly lead to an atomic war — and so I was not against nuclear weapons.

Günther Ander's philosophy is valid not only in the negative sense but also in the positive. Only by gaining experience with alternatives we strengthen the positive forces of change. Only after we got the experience with alternative sources of energy and developed their implementation, they can get a real chance. A particularly stupid claim is to say: "I am not able to change anything". Those who say this should not wonder why indeed nothing changes. We got since 18 years on our house in Baden-Baden, Germany, two solar plants, a thermal and a photovoltaic one and I can assure everybody that during that time the Sun never sent us a bill — while the price of the old energies increased four-fold.

A precondition for all real action of change is the deep conviction: "I can if I really want so". We also have to imagine positively what is possible with alternative energies.

But I also learnt that the old energy economy defends itself massively, with much money and legal expertise against the Renewable Energies. As a TV journalist I had to take my state television station in Germany eight times to labour court to be able to enlighten also my audience of millions of people about my new findings. Journalism means enlightenment. And this I want to do until the end of my life. To enlighten many so that the "Sun of the Father" (Jesus) does not shine for nothing. No fight, no progress. Citizens, to the Sun, to freedom!

Franz Alt

Chapter 3

More Electricity for Less CO$_2$

Yves Bamberger

Electricité de France (EdF), Paris

The 20th century left us a complex and efficient electrical system, ranging from the plugs in our homes to nuclear power stations, from the pylons along our country roads to major dams. The system plays an essential role in our everyday lives and our societies. It can be described as a network through which transit most of our energy needs and supply. In the 27-member EU, electricity accounted for some 17% of final energy in 1990 (Fig. 1); by 2007, the share had risen to 21%. In China, during the same period the electricity share on the final energy consumption grew from a bit less than 6% to more than 16%, narrowing the gap with Brazil. In the next decades, as shown in Fig. 2, electricity will continue to represent an ever larger share of energy consumption, accounting for about 30% in 2030, while enabling a significant reduction in carbon emissions as energy efficient electricity solutions are substituted to those relying on fossil fuels and as electricity mixes worldwide continue to be decarbonised. The growing role of electricity will thus be a key source of energy eco-efficiency, greenhouse gas emissions reduction and will warranty the independence in terms of fossil fuels. In summary, it will play a central role in sustainable development in our countries and in emerging and developing ones where electrification rates remain very low and where unfortunately human beings are suffering from many other lacks.

1. Electric Eco-Efficient End-Uses

On the demand side, the increasing role of electricity will come through better electric appliances, through a more efficient use of them but mainly through the substitution of end uses using fossil fuel by electricity. Heat pumps replacing

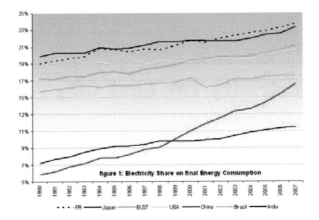

Figure 1. Electricity share on final energy consumption.

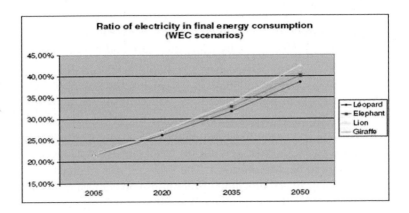

Figure 2. Ratio of electricity in final energy consumption (WEC scenarios).

boilers, plug-in hybrid vehicles transferring a part of oil consumption towards electricity, industrial end-uses like induction and again heat pumps for a better use of heat derived from the manufacturing lines, are examples of how to optimize the use of energy, taking in account the necessity to decrease our societies' CO_2 emissions. For instance, a fleet of one million plug-in hybrid vehicles in 2020 being recharged at night in France will allow to save every year one million ton of CO_2 and 0.5 Mtoe of oil. Of course, these developments have to be accompanied by other actions to reduce the global need of energy. The first lever in European countries is to better insulate homes and buildings.

Interestingly, these substitutions still make sense in terms of saved CO_2 emissions in all Western countries, and not only in France where the carbon intensity per kWh is very low (thanks to nuclear and hydro, 95% of EDF electricity generation is free of CO_2). Therefore, *all efforts* aimed at reducing the carbon dioxide emissions when generating electricity will emphasis the specific role of electricity to fight against global warming.

2. Achieving an Ever Lower-Carbon Electricity Mix with Nuclear and Renewables

Technologically speaking, we can expect major breakthroughs in terms of production costs and innovations on photovoltaic cells. Solar photovoltaic will be an integral part of the future generation mix with, every year, more efficient and less expensive cells. In addition to building-integrated panels, which have the great advantage to use already existing spaces located in the dynamic development sites, industrial, commercial areas with high density of population, photovoltaic plants will come on stream, raising questions of the land use competition. Wind power is relatively mature and generation costs have already fallen sharply. The main difficulty with a massive rollout of the technology relates to the challenge of operating and connecting to the network. Offshore wind power will also benefit from a reduction of generation costs thanks in part to the larger size of the turbines (up to 10 MW?) even if the cost of the offshore farms connection to the continental grid and the cost of turbines maintenance will remain high over the long term. Marine energy technologies are developing fast but successful demonstrations are still required to figure out what is the real potential of that source of energy.

Needless to say, the decisions of all players regarding investments in the future centralised generation facilities are based on economic considerations and try to take in account the instability of the regulations. As regards the renewable energies other than hydro, the development of the different technologies will depend, for the next decade, firstly on feed-in tariffs defined at a national level. What was observed in Germany, with the first EEG law in 2000, is that tariff was the first deciding factor for the development of renewables. Another example is the strong correlation in time between the incentive tariffs established by the French government and the fast growing development of photovoltaic. This raises immediately the question of lack of coherence in the incentive-based systems over neighbouring countries, especially within Europe, with a clear impact on the rapidity of growth in every country.

Another observation, that can be drawn from the fact that feed-in tariffs and their stability over time are the major key factors for the development of renewable sources, is that the organization of the electric sector, vertically integrated energy groups or totally independent producers, DSOs and suppliers, has no effective impact on the rate of development of renewable sources. As a matter of fact renewable energy sources have been developed in any kind of electricity sector organisation! This comment is perhaps politically incorrect for some economists or politicians but it is simply based on the analysis of the last 15-year history.

3. Networks: A Tool for Pooling Production and Integrating Renewable Energies

The "traditional" electric system — large generation facilities, transmission, distribution, customers — will evolve substantially as renewable energies gather momentum. This is particularly true of the distribution network which will become more of a "circulation" grid where all contributions from renewable and intermittent power sources are pooled. The truth is that the development of renewable energies only makes sense, notably in terms of safety and expansion, if they are combined together on the distribution network.

It implies to invest in the network to give him the "intelligence" to integrate the local generation (or storage capacity) without reducing the level of quality and safety which is required for the uses of electricity and long-awaited by the citizens-customers, at least in the developed countries and progressively everywhere. We are here coming to the "smart-grid" or "intelligrid", which will also enhance, coupled load management and load-shaving and, more globally, "flexibility" on the demand-side.

It is important to mention in this brief review that electricity storage will require (on all scales) to invest in research efforts. Depending on the costs entailed, this area could completely change the electric system. There are two key advantages being able to storage: first, it would facilitate the integration of renewable and intermittent power sources by guaranteeing supply, and secondly, it would help to smooth peak demand and allow the energy generated with store methods to be substituted for the high-carbon technologies relied on during consumption peaks.

4. Carbon-Free Electric Mix as an Opportunity to Develop New Industrial Facilities

Carbon content per kWh could in the very near future become a key criterion in choosing locations for industries or industrial facilities, as it affects companies' image and the lifecycle of finished products.

To conclude, we add that total CO_2 emissions generated by making photovoltaic panels, from manufacturing to dismantlement, depend in very large part on the carbon content of the electric mix in the country where the factory is located. An evaluation conducted by EDF R&D showed that CO_2 emissions resulting form the manufacture of the modules may change from a factor higher than 15: if we compare, photovoltaic panels produced in France with an electricity containing less than 50 g/MWh to panels produced somewhere with an electricity coming from a coal plant with 800 g/MWh, you see the difference! Some players in the photovoltaic market have already noted the "comparative advantage" of the French mix!

Yves Bamberger

- Chairman and CEO of EDF R&D

- Member of the French Academy of Technologies

- Member of the French Government Science and Technology Council

Yves Bamberger began his career in the R&D field, researching structural mechanics at the Central Laboratory of the "Ponts et Chaussées" (civil engineering). There, he carried out theoretical and experimental work on vibrations, stability and wave propagation and their applications.

He joined EDF in 1980 to set up the Numerical Mechanics and Models Department within the EDF Studies and Research Division. He was named Technical Director and Information Systems Manager at the EDF Generation and Transmission Division. He then became CIO of the common division shared by GDF and EDF.

In July 2002, he was appointed Executive Vice President, Head of EDF R&D (2000 employees in France).

Yves Bamberger has won an award from the Académie des Sciences (Prix Monthyon, 1988). He also chairs the French "Institut National de Recherche sur les Transports et leur Sécurité" (INRETS) and the "Laboratoire Central des Ponts et Chaussées" (LCPC).

Mr. Bamberger is a graduate of the French Ecole Polytechnique, the foremost engineering school in France, and a Chief Engineer of Ponts et Chaussées (civil engineering).

Chapter 4

Solar Power in Practice

Stefan Behling

Design Director and Senior Partner, Foster + Partners,
Professor for Advanced Building Construction,
Technology and Design, University of Stuttgart, Germany.

When Foster + Partners first introduced photovoltaic technology into one of its schemes back in 1975, both the technology and the reasoning behind its use was in its infancy. The Gomera Masterplan — a tourist development on the Canary Island of Gomera — aimed to encourage self-sustaining development based on a combination of indigenous construction techniques and renewable resources. Crucially, the island's constant sunshine and steady winds made it a natural test-bed for solar power.

The advantages and benefits of such a system were immediately clear to the practice and since this pioneering project, it has continued to develop and push the possibilities of the technology. In many ways the history and development of photovoltaic technology and how it can be incorporated into architecture can be read through a quartet of projects beginning in Duisburg, Germany, in the 1980s and which continues today with the construction of Masdar in Abu Dhabi, a new 6 million square meter sustainable city.

It is telling that the practice's first large-scale project to utilise photovoltaics should be in Germany. Much of Europe, and Germany in particular, is forward-thinking with regard to sustainable technology and has an enviable history of passing legislation at the highest level to provide incentives and tax breaks to those developers wishing to pursue a green agenda — after all, such designs and technologies are only as good as the political and social will to implement them.

Foster + Partners' work in Duisburg — one of my first projects at the practice — began with the Microelectronic Park in 1988. The design comprised of

Power for the World by W. Palz
Copyright © 2011 by Pan Stanford Publishing Pte Ltd
www.panstanford.com
978-981-4303-37-8

Figure 1. Masdar, Abu Dhabi. This pioneering development explores sustainable technologies and the planning principles of the traditional Arab walled city to create a desert community that will be carbon neutral and zero waste. Designed for a population of 90 000 people, it includes a new university and the headquarters for Abu Dhabi's Future Energy Company, together with special economic zones and an Innovation Centre. The city comprises two clearly defined quarters, the larger of which will be complete first. The surrounding land will accommodate wind and photovoltaic farms, enabling the community to be energy self-sufficient.

Figure 2. Illuminated at night the Reichstag's cupola becomes a brilliant beacon, visible across the city.

three integrated buildings for new-technology companies — the Business Promotion Centre, Telematic Centre and Microelectronic Centre — together with a linear park, all set within a dense residential district. It explored fresh approaches towards energy and ecology; and given the trend towards clean, quiet industries, demonstrated the potential to create attractive, mixed-use neighbourhoods.

Figure 3. The Reichstag's comprehensive environmental strategy relies extensively on daylight, solar energy and natural ventilation and makes use of renewable biofuel. The mirrored cone within the cupola reflects daylight into the chamber, while the moveable sunshade blocks solar gain and glare. Refined vegetable oil, when burned in a cogenerator to produce heat and power, is remarkably clean and efficient when compared to fossil fuels. The surplus energy produced, in the form of heat, drives an absorption cooling plant to produce chilled water for summer cooling. Excess hot and cold water can be stored in aquifiers, which are located respectively 300 and 60 metres below ground. All of these measures combine to make the Reichstag a highly efficient energy user. In fact its own requirements are so modest that it has become a net producer of energy, performing as a local power station in the new government quarter.

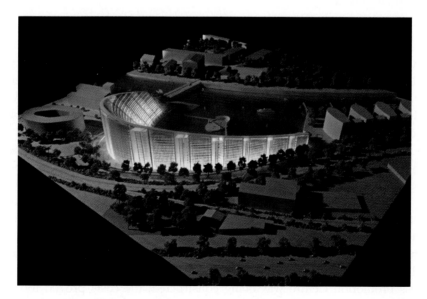

Figure 4. Presentation model of the Eurogate scheme, Duisburg. The crescent-shaped building rises from three to nineteen storeys and encloses a broad public esplanade. The building contains shops, cafes and entertainment facilities as well as offices, a hotel and conference centre. The south-facing façade contains a vast array of photovoltaic cells — the largest such array in Europe — which will allow the building to generate its own power in entirely sustainable way.

It was intended that the Business Promotion Centre's roof should be covered in photovoltaic cells which would allow the building to generate its own electricity and heat water to drive an absorption cooling plant. However, due to capital costs this proposal was discontinued, although it is capable of being retrofitted at a later date.

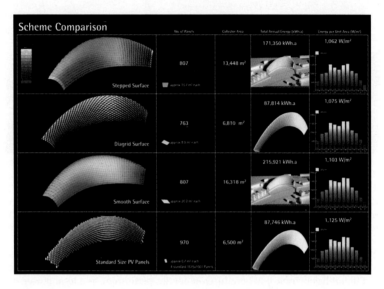

Figure 5. Scheme comparison examining the relative efficiency and power output of varying photovoltaic arrays at Eurogate.

The underlying themes of the project were reinforced by a masterplan for the physical and economic regeneration of the Inner Harbour — the final piece of which is Eurogate, a crescent-shaped building containing a hotel, conference centre, offices and shops, and which is located at the heart of the scheme. When completed, the Eurogate complex will have the largest solar array of photovoltaic cells in Europe. Designed to generate electricity for the building, as well as the wider site, the photovoltaics are integrated into the south facing façade, acting like a giant billboard and amply demonstrating how modern photovoltaic cells can be used to great aesthetic effect. By reducing the dependence of the harbour development on the national grid, the scheme takes full advantage of German legislation which provides incentives for the use of renewable energy sources.

This large-scale use of photovoltaics would have been unthinkable back in 1975, both from a cost and a technological perspective. Even in 1988 when the Business Promotion Centre was being constructed, the costs proved too great. However, from the mid-1990s onwards, a combination of enlightened legislation, together with lower production costs for photovoltaic cells has meant that the widespread use of photovoltaics in architecture is finally becoming achievable.

Figure 6.

Figure 7. Photovoltaic cells on the south-facing section of the Reichstag's roof power a moveable sunshade in the cupola. The shade is suspended on rails allowing it to track the path of the sun. In addition, the photovoltaic cells drive mechanical fans within the cone to help facilitate airflow.

Figure 8 and 9.

Figure 10.

While this is a major step forward, it is important to remember that the key to their effective use lies in a multi-faceted approach which combines the use of photo-voltaics with other passive and non-passive systems. The addition of some arbitrary solar panels to a building won't make it any more sustainable if all the other systems are inefficient.

This holistic approach was taken when redesigning the Reichstag between 1992 and 1999. The transformation of the Reichstag is rooted in four related issues: its significance as a democratic forum, an understanding of history, a commitment

Figure 11. City Hall, London includes an array of photovoltaic panels on its roof, which were proposed from the outset but not finally fitted until 2007. These now complement the many other passive systems deployed to reduce energy consumption. For most of the year the building requires very little active heating – the major potential energy demand comes from cooling. To assist passive or low-energy cooling, the building is naturally ventilated for most of the year, with openable windows in the office spaces; heat generated by people, computers and lights is recycled; and cold ground water is pumped up to supply chilled ceilings.

to public accessibility and a vigorous environmental agenda. At the heart of the scheme is a glazed cupola which while symbolic of rebirth, is also fundamental to the building's natural lighting and ventilation strategies. At its core is a "light sculptor", a cone that reflects horizon light down into the chamber. Photovoltaic panels on the roof are used to power the motor-driven sunscreen that tracks the path of the sun to control the amount of sunlight reflected into the chamber. Mechanical fans within the cone to help facilitate airflow are also driven by solar power. In addition, heat exchangers make it possible to recover and utilise a proportion of the heat from the extract air before it is expelled through the top of the cupola.

The use of photovoltaics is just one aspect of the building's sustainability strategy, by burning clean, renewable bio-fuel — refined vegetable oil ? in a cogenerator to produce electricity the building has seen a 94 percent reduction in carbon dioxide emissions. Surplus heat is stored as hot water in an aquifer deep below ground and can be pumped up to heat the building or to drive an absorption

Figure 12.

Figure 13. The roof of each building in Masdar will incorporate photovoltaic cells, supplying power directly to the city's grid. The use of shading and colonnades helps give respite from the sun and was influenced by traditional Arab settlement. Narrow streets with angled facades together with planting and water also help mitigate the fierce desert sun.

Figure 14. Masdar, which is similar in scale and density to Venice, is completely integrated — there are no separate zones for industry, education or culture.

cooling plant to produce chilled water. Significantly, the building's energy requirements are so modest that it produces more energy than it uses, allowing it to perform as a mini power station in the new government quarter: it is an object lesson in sustainability.

There are clear synergies between the Reichstag and London's City Hall, work on which began in 1998. City Hall advances themes explored in the Reichstag, expressing the transparency and accessibility of the democratic process and demonstrating the potential for a sustainable, virtually non-polluting public building. Designed using advanced computer-modelling techniques the building represents a radical rethink of architectural form. Its shape achieves optimum energy performance by minimising the surface area exposed to direct sunlight and maximising shading.

Its offices are naturally ventilated, photovoltaics provide power and the building's cooling systems utilise ground water pumped up via boreholes. While the use of solar panels was originally rejected due to cost, we ensured that the structure was capable of being fitted with a solar array at a later date, something which happened in 2007. The solar panels are raised up from the roof on a series of posts so that they act like a rainscreen, with the cavity beneath preventing any heat transmission.

To achieve a sustainable way of being we must work at all levels and scales and with both traditional and new technology. At the smallest scale this might involve the use of solar components and systems, such as the solar panels proposed for Gomera. The next scale up from the design of individual components is the design of buildings as integrated energy systems, like the Reichstag and from there the next jump is to neighbourhoods and cities.

The Masdar Masterplan represents sustainable thinking at a city scale. When completed it hopes to be the world's first carbon neutral, zero waste city. Mixed-use, low rise, high density, it is designed for a population of 47 500 people and will include a new university together with special economic zones for manufacturing and research. It will be entirely car-free with residents and workers negotiating the city using a combination of Personal Rapid Transport and Light Rail Transit systems. These will be powered from 100 percent renewable energy so will not generate any CO_2 or polluting emissions. To achieve this, a large-scale, concentrated solar power plant will be constructed on the outskirts of the city.

Masdar is the embodiment of Abu Dhabi's quest to create a new non-oil based industry for the country. In addition to MIST University — a collaboration with MIT — practical research and manufacturing will take place in the city. Fully equipped, air conditioned labs capable of carrying out the highest levels of scientific research will sit alongside high-tech workshops which will manufacture solar panels to enable a sustainable future for the region.

Collectively, these projects have become a paradigm for the practice, embodying a number of themes and concerns that are central to the search for sustainable solutions to life in the 21st century. As buildings are the largest single consumer of energy, it is only logical that we combine a source of power — solar — with the single highest demand for electricity. With increasing innovation, the possibilities for Building Integrated Photovoltaics are only limited by a designer's ambitions.

Stefan Behling

Stefan Behling received his Diploma from the University of Aachen in Germany and joined Foster + Partners in 1987. A specialist in ecology, sustainability and energy conservation, he has worked on a number of projects that have pioneered new techniques for energy management, including the new German Parliament at the Reichstag in Berlin and the Commerzbank headquarters in Frankfurt. He has also developed a vision for sustainable living and the wider use of renewable energy in the UK with the Department of Trade and Industry. At Foster + Partners he leads the research and development of new sustainable designs and the use of new materials and construction methods. He has been a professor at the University of Stuttgart since 1995, holding the Chair for Building Construction, Technology and Design. Since 2004 he has been a senior partner, a member of the executive board and a member of the design review board, which ensures design excellence, innovation and sustainability in the architecture of the practice.

Chapter 5

The Story of Developing Solar Glass Facades

Joachim Benemann

Former Industrial Leader in Germany, Berlin/Cologne

In April 1984, I got the job to take care for the solar business at Flachglas AG, a leading European glass manufacturer and the only supplier of parabolic mirrors for solar thermal power plants. More than 2.4 million square meters were supplied when in 1990 the oil price was low that solar generated electricity could not be competitive anymore. We discussed to shut down the solar activities.

Shortly before the final decision was taken by the Board, we received a paper, written by Strategies Unlimited, a US based consulting company focussed on the solar business. The paper forecasts an interesting future business in photovoltaics which was done at that time mostly in Japan and by international oil companies.

We very soon recognized that we were not able to compete big oil companies or governmental subsidized Japanese concerns with standard photovoltaic modules. But as a glass company we were deeply involved in the architectural business, we knew how to build glass facades and glass roofs. Therefore we developed a strategy to use photovoltaic in buildings.

The first step was to find a building to demonstrate the integration of photovoltaics into façade windows and very soon we found an quite interesting customer called STAWAG, the utility of the city of Aachen located just 100 km away from our headquarter. We presented our idea and received the order to install photovoltaic double glass façade elements into an existing office building which was foreseen for refurbishment.

We had to supply four different sizes of facades elements and because we had no specific manufacturing line we did it in one of the laboratories of the mother company. The general technology was based on the production of double glass noise reduction windows. A transparent and liquid resin was pumped between

Power for the World by W. Palz
Copyright © 2011 by Pan Stanford Publishing Pte Ltd
www.panstanford.com
978-981-4303-37-8

Figure 1. This building demonstrates the perfect integration of photoroltaic window elements.

two glass sheets and then hardened by ultraviolet light. Our little PV team embedded in addition solar cell strings between the glass sheets. A number of problems had to be solved: where should the connection box be installed and where the cables? Our engineers developed an ultra flat electrical connection system which replaced the standard boxes which usually was installed at the rear side of standard PV modules. We also started a close cooperation with one of the advanced German aluminium frame maker, called Gartner. This was necessary to find a technical solution to host the electrical cables of the PV façade elements within the aluminium standard frames. All this small but important problems were solved within a nine-month period and in early 1991 we installed the PV façade elements at the four floor staircase, of the office building. The job was done within a few days and the inauguration of the world's first photovoltaic facade took place either on the 8th or 9th of May 1991. The ceremony was very short. Only a few sandwiches were taken by the guests and the champagne were nearly untouched, because it was raining cats and dogs and therefore this event was forgotten very fast.

But the project itself was a great success. Hundreds of architects visited, within the next months that building and the utility had to hire a specific specialist to explain this new technology to interested visitors.

This success also convinced our board to invest in a specific production line which allowed us to manufacture various sizes of double glass PV façade elements and early 1993 the Prime Minister of the State of North-Rhine Westphalia and later President of Germany Johannes Rau inaugurated the plant at the city of Gelsenkirchen.

A large number of projects followed and were carried out in Germany and other countries around the globe. Three of the most remarkable were a 10 000 square meter sized roof at a conference building (1 MWp), a façade of the office building of the Ministry of Economics at Berlin and last not least the most spectacular PV roof of Europe's largest railway station also located at Berlin.

But we also had to solve problems such as the capacity loading of the production line. Sometimes we had to run the line in three shifts, 24 hours a day, sometimes we run the line just a few hours a day because of lack of orders.

So we decided to set up an additional production line for standard photovoltaic modules and to share the existing infrastructure as well as the personnel. To be competitive and to achieve a steady high quality we wanted to install a fully automated line. Our worldwide search to find a supplier failed and so we bought equipment from various companies. Some of the machines were still prototypes and some of them did not meet the acceptance tests and had to be re-designed which caused several months of delay. But last not least we started the production of standard sized photovoltaic modules in autumn 2002. It was the first fully automated line in Europe with an annual capacity of 15 MWp.

When we started our photovoltaic activities we had a long term vision: we should become a PV system supplier, we should deliver engineering for our BIPV customers as well as façade elements and also the electronic inverters. For the façade elements we needed a high quality cell supplier who was able to deliver a steady quality and quantity. Shell, the giant oil company, was one of the big players in the photovoltaic business and we could arrange a cooperation. Shell was setting up a new plant for solar cell production next door to our plant at Gelsenkirchen so that we could get the cells on demand.

In respect to the electronic inverters we had already for a long time contact to a small company called SMA which had developed very innovative machines in close cooperation with the University of Kassel, Germany. The company was owned by three young engineers. The company grows steadily but had some problems to achieve credits from banks because they were very restrictive on credits for the solar business. We offered to acquire 50% of the companies shares which was accepted and the banks became more flexible because of Flachglas AG financial reputation. This happened in 1995 and the cooperation was a benefit for both companies. SMA could grow and we had access to a high quality inverter supplier. The shares were sold back in 2002 and meanwhile SMA has become the world's leading company for PV electronics.

Within the last years the photovoltaic integration into buildings (BIPV) was done by several companies worldwide but it has not become a significant market segment. The interest on PV was focussed on mass production of standard modules for the extremely booming market while Building Integrated PV needs high qualified engineering and specific sales and marketing activities to convenience architects and investors. But the worldwide discussion on climate and the growing responsibility for the environment as well as the optical and technical attractiveness of solar facades will revive this technology.

Joachim Benemann

Chapter 6

Bringing the Oil Industry into the Picture

Karl Wolfgang Böer

University of Delaware, Newark DE 19736, USA

When I received from my parents books as a young teenager, they were usually about nature, science, and technology. One of them caught my attention "Du und die Natur", and it had a chapter about solar energy. It stayed with me for decades to come. It became clear to me that in order to harvest the sun's energy, one needs to understand much more about how sunlight can be changed into "something useful," beyond just heating. I had early in my teenage years my own chemistry and electronics laboratory, and I experimented a lot. There were also a few Se-photocells that gave me an idea. However, I learned that I needed a lot of these cells to even power one lamp. They had a mere 1% conversion efficiency and I could not figure out what was done with the rest of 99% of the energy.

Since I was a teenager, I always dreamt that I wanted to become a scientist to get a better understanding of nature and to help this understanding for the benefit of mankind. I wanted to learn as much as possible to gain this understanding, and shortly after the war I studied Physics at the Humboldt University of Berlin. After seven semesters, I received my diploma with *summa cum laude*. I was very fortunate that, by accident, Professor Frerichs left some CdS at the Physics Institute that I inherited. These were little yellow crystals, photoconductive, but not much more was known about them. I asked my teacher, Professor Robert Rompe, whether I could measure the conductivity of these crystals under various conditions. He agreed, and he also allowed me to have my own graduate students. Soon, their number increased, and we all took more and more measurements about the properties of the CdS. I based my Ph.D thesis on these crystals and a few years later my habilitation. It wasn't long before we were recognized

Power for the World by W. Palz
Copyright © 2011 by Pan Stanford Publishing Pte Ltd
www.panstanford.com
978-981-4303-37-8

internationally as a center for CdS research and I became a (full) Professor. At that time we had almost 50 publications with my group, and it so happened that my own Institute (the 4th Physics Institute of the Humboldt University) was created, and I became its director and Ordinarius (at the age of 35).

Then "the Berlin Wall", interrupted my way of life and work; and I immigrated to the USA, started again building a research team at the University of Delaware. We were soon publishing my 100th publication, all in CdS. Again an institute was created for me, the Institute for Energy Conversion (IEC). At this time the science had advanced, and 5% solar cells were invented, at about the same time in Si and CdS/Cu_xS technology. Again, little was understood about the actual mechanism in these cells, and I decided to become active, obviously with the CdS-based cell that was poly crystalline — but most of my more convincing results were obtained from single crystals. Therefore, in 1968 I asked my graduate student to investigate the behavior of this cell starting from a single crystal CdS. Three years later I learned enough to see that a CdS-based polycrystalline cell can be made inexpensively in a semi-automatic production line, and with 8% efficiency which we occasionally reached, one could power electrically from its roof an entire one-family home.

This fact was interesting enough for the Shell Oil Company to start negotiating with us. I had formed Solar Energy Systems, Inc. (SES) in the meantime. A year earlier EXXON had made a joint venture with a Si company, so it was obvious, that the other potential candidate needed support. Shell initiated this for applied research and for a pilot line at a tune of $3MM in 1973. Simultaneously the government sponsored the basic CdS work at IEC, and so did a group of several other industries and power utilities. In 1975, IEC had directly and indirectly

Figure 1.   The front of the Solar One house shortly after its inauguration in June of 1973.

together with other departments a total of 100 scientists, and SES had 80 scientists, all working in CdS-related research. It was a very powerful research team. Our work stimulated research in many institutes across the world. Altogether we learned more about this CdS-based solar cell than was known for any other solar cell at that time.

However, to make this research known to more than a handful of researchers, we carefully designed a showplace for the general public to demonstrate what can be done with solar cells. We created the Solar One House of the University of Delaware which showed that one can harvest electric power and heat from the same collector surface on the roof. The house was well conceived as a systems' house that included many active and passive solar features, had extensive electric and heat storage systems, and was monitored in any possible way. It was highly publicized worldwide and had great media coverage. In Delaware the house was visited by over 100 000 people in the first nine months. But soon we had to recognize that its design was a bit premature: It had a build-in solar collector surface that replaced plywood and shingles of the roof, and it was a truly hybrid system that generated electricity and heat from the same surface (taking heat away from the solar cells decreased their temperature and made them more efficient). It took three decades before such systems are now considered for the commercial market.

The following years, with two operating teams, I was free to spend my time on theoretical research about the operation and loss factors of the cell. With several ground breaking papers during this period, it also became increasingly clear

Figure 2. Dr. Karl Wolfgang Böer, Distinguished Professor of Physics and Solar Energy, emeritus at his 80th birthday, 2006, in the Senate's auditorium of the Humboldt University, Berlin, his old alma mater.

Figure 3. Karl W. Böer Solar Energy Award Medal of Honor (on the front, Karl W. Böer, on the back, the Solar One House), created by Charles Parks, a world renowned sculpture of Delaware.

to me, that, though we had essentially solved the degradation problem of the CdS/Cu_xS solar cell, we could not hope to surmount its efficiency limitation: it remained at 10% (reached at IEC). This limit is given by the lattice mismatch causing too high an interface recombination, hence a loss in power.

By this time two other thin-film solar cells, the α-Si and the CIS solar cell, had already exceeded 12% efficiencies and were on their way up, our research team had to recognize that we had to change the substance. I suggested to Shell to use CdTe instead of the CuS, but it was rejected, because "there is not enough Te in the world" I was told, to make it worthwhile for Shell. They decided to go the α-Si road. IEC, on the other hand, researched a wide variety of thin-film solar cells. Shell wanted to work in a joint venture with Motorola on thin film Si and left Delaware.

In the following years, after terminating my presidency of SES and my directorship of IEC, I became the elected President of the American Section of the International Solar Energy Society, and then changed this section into a professional society, with its own professional office, its publication office with a monthly magazine and two of its periodicals, and incorporated the society (ASES) in Florida.

During all of these years, I received many honors, including the highest awards of ASES and of ISES, and of the University of Delaware. I was elected a fellow of four professional societies AAAS, APS, ASES, and IEEE, and became inducted into the international Solar Hall of Fame. I finally was rewarded by the University of Delaware with the Solar Energy Medal of Merit that was minted in my name. The medal is to be given to an individual, who is selected by a committee

Figure 4. First Böer Award winner, Jimmy Carter, former President of the USA.

of eight leaders of the relevant professional societies. He is chosen out of a large number of candidates who are proposed by their peers. The award, with a medal and $50 000 attached to it, is given "for outstanding contribution to the advancement of the global use of solar energy with the intent to replace conventional fuels under conditions that enhance the quality of the environment and minimize visual and material pollution". The committee selected unanimously the former President of USA, Jimmy Carter, as the first recipient of the medal for his invaluable leadership to make solar energy worldwide known as a source of energy that has to be taken seriously and, under his leadership substantial financial support was funded for research and development of solar in the USA. In his acceptance speech in 1992 at the University of Delaware in front of more than 3000 students and distinguished guests, he stressed his continued high interest in solar energy and promised to continue his effort to promote solar energy wherever he can.

During the following years, the award was given biannually to many other international leaders in science, industry and politics from the US, Germany, Japan and Australia. Most recently Dr. Hermann Scheer, member of the German Bundestag, and the leading force behind the European utility industry to invest in solar energy, won the 2009 award. He is now the President of the World's Solar Energy Forum and has expanded his influence over most of the developed nations worldwide.

Very gratifying to me was my ability to help solve the puzzle that all of us had for three decades, why a thin layer of CdS improves the efficiency of CdTe and many other CIS-type solar cells significantly. I was able to publish my theory only a few months ago in the leading solar cell journals. To me, this seems to close the circle of all my life, investigating CdS.

Chapter 7

Factory for Sale, or the Long and Stony Way to Cheap Solar Energy: The Story of the Thin-Film CdTe Solar Cells; First Solar and Others — A Semi-Autobiography

Dieter Bonnet

PV Specialist, Frankfurt, Germany

After four decades of more or less intense development the CdTe thin film solar cell is now becoming a benchmark in photovoltaics. Modules based on Cadmium Telluride are manufactured on an industrial scale and new ventures are popping up. The father of this technology is Dieter Bonnet. Already in the 1970s, he made the first cell of the configuration still used today. He contributed to the fully automated module factory built in Arnstadt in 2000. The award of Becquerel Prize can be considered a compensation for the many set backs and the refusal to fund the development even at a time when the potential was clearly visible.

Today at the age of 72, he is still on the move: lectures, consulting, and acquisition. Dieter Bonnet could have retired 10 years ago like so many do in the affluent society and lead a comfortable life in his house near Bad Homburg in the Taunus and paint, take photographs and walk the forest with his wife and dogs. Yet the white-haired gentleman did not hang his job and connections at the nail in the wall. In February this year, he joined ROTH & Rau, a young dynamic company offering equipment for manufacturing silicon solar cells, after take-over of the CdTe know-how aiming at manufacturing turn-key CdTe thin film plants.

Having spent 40 years under the auspices of thin film solar cells, Dieter Bonnet is aware of the varied stories involved being a key player and eyewitness to the labyrinth ways the CdTe solar cell has taken to maturity.

Power for the World by W. Palz
Copyright © 2011 by Pan Stanford Publishing Pte Ltd
www.panstanford.com
978-981-4303-37-8

CdTe thin film solar cells did not find the kind of acceptance they deserved from the beginning. His personal involvement started in 1968, when the space age started and electric power was needed for satellites. A key event was the moment when the USA put the first synchronous satellites into orbit, demonstrating the business potential of space technology in telecommunication and television. The European nations did not have rockets strong enough to put a heavy communication satellite into the 36 000 km orbit. So they asked the Americans for "a lift" counting on the broad heart of the Americans shown so often after the war. The response was negative. So what? An interesting idea to solve the dilemma was to use the Euro-rocket to reach the minimum orbit (ca. 100 km) and use electric propulsion to spiral the satellite into synchronous orbit. Electric propulsion should be powered by solar cells. But silicon solar cells (the only cells available at this time) would not survive the job, being destroyed by the particle radiation of the van Allen belts in which the satellite would spend a long time. So the idea seemed to be doomed, if not for a group of physicists who had said: "Let us make thin film cells which are less susceptible to the particle radiation." Thought and done, a call for proposal was issued and two contracts were made in Germany for the development of thin film solar cells. (France also started a substantial effort under the guidance of Wolfgang Palz.) At Battelle Institute in Frankfurt, a young physicist with his friends considered II-VI compound semiconductors for use in solar cells and finally chose CdTe, using the concept of a graded CdS-CdTe hetero-junction solar cell. His name was Dieter Bonnet, his age, 30. The project was granted at the same time as a parallel project at Telefunken building on the work at Clevite Corp, in the USA. Together with a young engineer, Dieter Bonnet set to work at Battelle enthusiastically and 18 months later they had small cells of 8% efficiency in their hands. But at that moment the Americans changed their minds and offered a ride into synchronous orbit to the European space agency ESA. This vaporised the motivation for new thin film cells in Germany and also in France some time later. Logically enough the projects were discontinued. The group members nearly lost their jobs, but they produced, based on their fresh know-how, other applications for thin II-VI films like vidicon targets (using the photoconductivity in CdS) and ultrasound image converters (using the piezoelectric effect of CdS) found sponsors and waited for the energy crisis. Dieter Bonnet made a presentation at the American photovoltaic specialist's conference in 1972 in Washington. It drew the attention of Richard Bube from Stanford University who involved Dieter Bonnet into a discussion on thin film solar cells based on II-VI compound semiconductors. A few weeks later, the US National Science Foundation asked Dieter Bonnet to review a proposal from the Bube group on a new thin film solar cell, based on CdTe. What could he do but give a good rating. While the CdTe solar cell crossed the Atlantic and became a major field of work in the USA, he had to stay inactive, his first great disenchantment.

Between 1972 and 1975 work on CdTe/CdS thin film solar cells has been taken up by different groups after the high potential has been shown by the Battelle group, indicating that a temperature of around 500°C had to be employed during the film deposition to achieve good electronic properties. Light should enter through the CdS film, which acts as n-partner. Later on it was learned how to reduce the influence of the lattice mismatch at the junction interface by suitable "activation" procedures. Broad academic studies of high scientific quality at Stanford University (Bube 1977) started 1973, shortly after the initial publication of Battelle. Also basic work using single crystals was initiated at Matsushita Laboratories (Yamaguchi, 1977). This group deposited CdS by a PVD process onto p-doped CdTe single crystals and achieved efficiencies of around 10%. Martinuzzi in France also studied junctions prepared by high vacuum deposition of CdS onto p-CdTe single crystals around 1976 (Martinuzzi, 1976). Another French group (under Cohen-Solal) at CNRS made shallow junction CdTe solar cells by depositing p-type CdTe onto n-CdTe single crystals, achieving 10% efficiency under AM0 radiation (Lincot, 1979). A group at Kodak Co. in the USA prepared an all-thin film structure using the technology of close spaced sublimation and in 1982, reported the achievement of the "magical" value of 10% efficiency under terrestrial illumination conditions (Tyan, 1982).

A new promising technology for low cost manufacturing was used by a small industrial venture in El Paso Texas ("Photon Power Inc."). As early as 1964 (Chamberlin, 1964), this group had studied deposition of CdS by a chemical spray technique and used it for the preparation of CdS/Cu$_2$S cells. Around 1987, Photon Power studied the same technology for depositing CdTe and achieved promising results of 9% efficiency (Albright, 1992). In 1994, Photon Power was taken over by a subsidiary of the COORS company and re-opened in Golden, Colorado (USA) as "Golden Photon Inc." around 1990. A pilot plant was set-up, but, as some stability problems could not be solved fast, the venture was abandoned in 1996.

A group at the Southern Methodist University in Dallas (USA) led by Ting Chu in 1980 started work on depositing CdTe by reaction of the elements on a neutral substrate (Chu, 1981), then changed the technique to close spaced sublimation and made CdS/CdTe n-p solar cells of 10% efficiency in 1987. The group moved to University of Florida and, after Firesides joined the group, in 1992 achieved the long-standing world record for CdTe thin film solar cells of 15.8% (Britt, 1993).

Around 1989, the National Renewable Energy Lab. (NREL, formerly SERI, Solar Energy Research laboratory) also entered the field of CdTe cells by in-house R&D. Not so much aimed at record efficiencies, but at a thorough understanding of basic processes and materials questions, NREL has provided valuable inputs to the later industrial efforts, besides administrating project funding on behalf of the US Department of Energy. This work finally led to a new record efficiency of 16.4% in 2001 (Wu, 2001).

In the 1990s, a number of academic groups have contributed to the presently available know how on the CdTe thin film solar cell: ETH Zurich (Tiwari, 1992), University of Durham (Durose, 1992), Georgia Institute of Technology (Rohatgi,

1992) Colorado State University (Sites, 1992), University of Queensland (Morris, 1992). A group at the Brookhaven National Laboratory have contributed to the health risk assessment of CdTe solar cells in numerous publications. They provided massive data on the environmental benign properties of CdTe cells and modules.

A new industrial involvement took place in 1984 when British Petroleum plc bought the know-how and patents from Monosolar Inc. This company, based in California had brought the CdTe to 10% efficient cells by electro-deposition processes (Basol, 1981). In Sunbury-on-Thames, close to the huge BP research laboratories a small unit was established ("BP Solar"), which set forth to upscale and optimize the procedures. Around 1990, modules of 30×30 cm^2 have been made at efficiencies close to 10%.

During this time (1984–1999) the technology of electrodeposition at BP Solar matured by intensive work and in 1997 was transferred into the USA, where a partly installed production plant for a-Si modules was for sale. This plant subsequently was transformed into a CdTe plant by BP. This factory in Fairfield, CA, was laid-out to a capacity of nominally 10 MW. In 2002, the management of BP decided to close the factory.

In 1990, the US-based company Glasstech Inc., which at this time operated in Wheatridge Co. commissioned a study aimed at identifying the most promising thin film solar cell for the future. Under direction of Peter Meyers (formerly with AMETEK Corp.), the best cell was identified to be CdTe and the technology of choice was to be close spaced sublimation. Under the guidance of the former chairman of Glasstech, McMaster, including private funds, a new company Solar Cell Inc. was established, aiming at large scale production at a capacity similar to those of BP Solar and ANTEC Solar, assisted by funding from NREL. In 1996, a 3/4 m^2 module was measured by NREL at an aperture efficiency of 9.1%.

Under new investors the company (under the new name of First Solar) has been restructured and the pace of development accelerated. First modules at 7% efficiency have been made on a pre-production scale and sold. In the years after 2000, under very lucky hands, First Solar became the world leader in thin film technology and presently is expanding its capacity at a tremendous rate.

In 1993, the Battelle Foundation decided to close down the Frankfurt lab in which the CdTe thin film cell in the meantime had been brought up to an efficiency of 11%. Four engineers from Battelle at that moment founded the new start-up company ANTEC (Applied New Technologies), and took over four fields of work from Battelle, with the aim of transferring them into the industrial stage, among them the CdTe cell technology. Dieter Bonnet was the head of ANTEC's R&D group. With marginal but essential funding from the European Union, very soon 10×10 cm^2 modules of >10% efficiency were made. At the same time funding for an industrial venture was sought and finally in 1996, ANTEC Solar GmbH was established in the eastern part of Germany. A factory with a capacity of 100 000 m^2 of 60×120 cm^2 modules was built, and commissioned in 2001. The production equipment has been installed as a fully automatic deposition line of 160 m lengths

(Bonnet, Harr 1998). First modules have surpassed the 7% mark. In 2002, due to a market collapse for PV modules in Europe, ANTEC Solar went insolvent, was closed and after extended negotiations with interested parties, was taken-over by wind strom frusta GmbH. Production was been re-started in May 2003. This new company also went into insolvency; the plant was later bought by a new owner, who put it back into operation and is selling modules today.

Why did First Solar experience the breakthrough, that was denied to ANTEC? When Antec had the first unit running, the manager of the First Solar asked to see the facility. They saw a 160-m long integrated fully automated production line churning out a module every few minutes. For First Solar, this was the right moment to go for an IPO and collect money to power up the operation.

Split opinions about the use of cadmium telluride solar modules still can be heard in the solar industry. Dieter Bonnet sees clearly the dangers of cadmium for human being but "cadmium telluride is no cadmium", explains the physicist. It's all well wrapped. The cadmium is so strongly bounded to the tellurium that both elements are thermally split only at a temperature over 1200°C. There are numerous studies, experimental and theoretical, which show that the risk is virtually nil.

And again he is on his way to work. In the Frankfurt S-Bahn, he once almost sold a factory to a Chinese photovoltaic broker. Not the final ironic trait in his life. Maybe his Huguenot heritage? He recently founded an association having all groups as members who work in the field of CdTe, some 30 members in total: SOLARPACT eV = Solar Power by Advanced Cadmium Telluride.

Dieter Bonnet

Chapter 8

Photovoltaics in the World Bank Group Portfolio[1]

Anil Cabraal

Advisor, World Bank/IFC Lighting Africa Program and Consultant (Rural and Renewable Energy) Potomac, MD 20854, USA

The World Bank Group comprises of the World Bank, the International Finance Corporation (IFC), Multilateral Investment Guarantee Agency (MIGA) and the International Centre for the Settlement of Investment Disputes (ICSID). The World Bank comprises of two development institutions owned by 186 member countries: the International Bank for Reconstruction and Development (IBRD) and the International Development Association (IDA). The IBRD aims to reduce poverty in middle-income and creditworthy poorer countries, while IDA focuses on the world's poorest countries. IBRD and IDA funds are lent to governments. The IFC, created in 1956 and owned by 182 member countries, fosters private sector investment in developing nations. The World Bank and the IFC have been the primary

---

[1] This paper is based on several reports, among them are: Cabraal Anil, "Strengthening PV businesses in China: A World Bank renewable energy development project," Renewable Energy World, May-June 2004; International Finance Corporation, "Selling Solar: Lessons From More than a Decade of IFC's Experiences," IFC and GEF, July 2007; Martinot, E., A. Cabraal, and S. Mathur, "World Bank Solar Home Systems Projects: Experiences and Lessons Learned 1993–2000," The World Bank, 2003.The World Bank, "Catalyzing Private Investment for a Low-Carbon Economy: World Bank Group Progress on Renewable Energy and Energy Efficiency in Fiscal 2007, The World Bank Energy and Mining Sector Board, November 2007; The World Bank, "Designing Sustainable Off-Grid Rural Electrification Projects: Principles and Practices," Operational Guidance for World Bank Group Staff, The World Bank Energy and Mining Sector Board, November 2008; and World Bank Implementation Completion Reports for solar projects in India, Indonesia, and Sri Lanka.

Power for the World by W. Palz
Copyright © 2011 by Pan Stanford Publishing Pte Ltd
www.panstanford.com
978-981-4303-37-8

supporter of photovoltaic investments in developing countries within the World Bank Group.

Today, more than 300 million urban and rural households around the world remain without electricity access. Of the 260 million unserved households in rural areas, those most difficult to serve usually live in isolated communities far from the national electricity network in so-called "off-grid areas." These off-grid communities are typically small and dispersed and consist of low-income households — characteristics that often discourage potential private-sector energy providers or even government electrification programs that must prioritize the allocation of scarce resources. Since the 1990's the World Bank Group has supported projects that use photovoltaics to provide electricity services in developing countries.

1. World Bank Group Photovoltaics Projects

Today, the World Bank and the IFC constitute a major financier of PV in developing countries with total project costs of more than US$1.4 billion that benefit over 15 million people in about 30 countries in Africa, Asia, and Latin America, and support to PV manufacture (see Table 1). They range from support for the installation of over 1.25 million solar home systems in Bangladesh, 400 000 Photovoltaic (PV) systems in China, and financing solar lighting for teachers in remote areas of Papua New Guinea, to the provision of lighting and basic electricity services for households and essential public facilities in rural Tanzania to a grid-connected 1 MW PV system in the Philippines that demonstrates the value of conjunctive use of PV and hydropower. IFC had its first involvement with investing in PV manufacturing in 1989 when it made a US$3 million investment (debt and equity) in Shenzhen YK Solar PV Energy Co. Ltd., a solar PV manufacturer in China. Although the investment did not meet its original expectations, it established an important precedent for investing in solar PV businesses in frontier markets by not only demonstrating confidence in viability of the PV technology, but also by focusing on commercializing the deployment of this technology in the developing world. More recently, IFC has begun to re-engage in investing in PV manufacturing, including silicon production with projects in China, India and the Russian Federation.

2. Business Models for Off-Grid Service

World Bank PV electrification projects principally aim to improve electricity access for populations in remote areas who are unlikely to be reached by grid extension within a reasonable time frame. Intertwined with this goal are the objectives of having players other than the government implement the work, mobilizing additional human and financial resources, and reducing pressure on already

Table 1.   World Bank Group Photovoltaic Projects (as of November 2009).

Country	Project	Systems	PV Capacity (kWp)	Total Project Costs # ($ millions)
World Bank Projects				
Argentina	Renewable Energy in the Rural Market (PERMER)	8300	1521	18.3
Argentina	Permer Renewable Energy Additional Financing	16130	3042	36.5
Bolivia	Decentralized Energy, ICT for Rural Transformation	60000	2600	38.6
Ecuador	Power and Communications Sector Modernization and Rural Services	2200	110	1.5
Mexico	Renewable Energy for Agriculture*	1000	700	5.0
Mexico	Rural Electrification	36000	3900	7.9
Nicaragua	Off-grid Rura Electrification*	7500	2100	23.0
Peru	Rural Electrification	7500	8000	5.3
Burkina Faso	Energy Sector Reform*	8000	300	3.0
Cape Verde	Energy and Water Project	4500	129	2.5
Ethiopia	Second Electricity Access Rural Expansion	40200	407	20.5
Ghana	Solar PV Systems to Increase Access to Electricity Services	15000	750	15.0
Madagascar	Energy Services/Delivery/ Integration of Renewable Energy*	15000	625	7.5
Mali	Household Energy and Universal Access Project*	10000	420	5.0
Mozambique	Energy Reform and Access Project	9600	1012	12.5
Mozambique	PV for Schools and Health Clinics	200	84	1.0
Senegal	Electricity Services for Rural Areas*	10000	420	5.0
Uganda	Energy for Rural Transformation (APL)	90000	6300	67.7
Tanzania	Energy Development and Access	16200	810	12.2
Zambia	Increased Access to Electricity Services	10250	410	6.2
Bangladesh	Rural Electrification & Renewable Energy Development	198000	9900	91.4
Bangladesh	Grameen Shakti Solar Home Systems — Carbon Finance	1000000	50000	320.0
Bangladesh	IDCOL Solar Home Systems — Carbon Finance	250000	12500	80.0

(*Continued*)

Table 1. (*Continued*)

Country	Project	Systems	PV Capacity (kWp)	Total Project Costs # ($ millions)
Bangladesh	Additional Financing to RERED Project of $130 million			4.0
Cambodia	Rural Electrification and Transmission	10 000	400	144.9
China	Renewable Energy Development	400 000	10 000	24.0
India	Renewable Resources Development	45 000	2500	3.8
Indonesia	Solar Home Systems	8500	425	1.3
Laos	Southern Provinces Rural Electrification	4000	160	5.2
Mongolia	Renewable Energy and Rural Access	50 000	520	16.5
Pacific Islands	Regional Sustainable Energy Finance	21 000	630	2.2
Papua New Guinea	Solar Lighting Kits for Teachers	2500	100	107.0
Philippines	Rural Power	135 000	9000	7.8
Sri Lanka	Energy Services Delivery	19 400	776	28.3
Sri Lanka	Renewable Energy for Rural Economic Development (RERED)	85 000	3400	25.0
Sri Lanka	RERED — Additional Financing	50 000	2500	
International Finance Corporation (IFC) Global Programs				
Philippines	1 MW grid-tied PV	4000	1000	6.0
Kenya, Morocco, India	PV Market Transformation Initiative (PV MTI)@			19.0
Global	SME Program Loans to PV Companies through in Bangladesh, Vietnam, Honduras, Dominican Republic	15 800	732	6.3
Global	Solar Development Group			16.0
Global	Renewable Energy and Energy Efficiency Fund			
PV Manufacturing	Manufacturers in China, India and Russian Federation			202.0+
Joint World Bank/IFC	Lighting Africa Program	500		
	TOTAL	2 666 280	138 MW	1405

\# Cost: Includes only total investment cost of PV component/applications.

* Estimated kWp or costs in these projects. ⁺ Estimated number of consumers served by PV power at 30 kWh/month

@ PVMTI: Shows commitments to date. PVMTI investments seek to prove the viability of a number of different business models, ranging from partnerships with industrial players to leverage an existing distribution network to joint ventures between PV companies and finance institutions to provide credit for solar customers.

overextended utilities; alternative players could include private-sector companies or individuals, non-governmental organizations or community-based organizations. The key is to develop a system of incentives sufficiently attractive for these players to do business in off-grid areas.

The business models for commercial PV dissemination may be classified as (1) dealer model (also known as direct-sales or open-market model) and (2) fee-for-service model (also known as energy-service-company model). The main distinction involves system ownership. In the dealer model, the consumer purchases the system either with cash or financing. Beyond warranty service, the consumer assumes responsibility for all operational and replacement costs. In World Bank projects, the dealer model often features microfinance assistance, which addresses the issue of high upfront costs. An exception is the China project, which lacked rural-credit facilities; in this case, consumers were used to paying cash, and no micro-financing was introduced. The issue of high upfront costs was addressed by driving down costs in various ways, a very competitive marketplace that drove down costs, using "plug and play" systems that required no installation and focusing on smaller, more affordable units. Initially, consumers bought small systems (10–20 Wp); subsequently, as their incomes rose, they bought larger ones (40–100 Wp). In the fee-for-service model, the consumer is provided only service, the level of which depends on system capacity. The company, which retains ownership of the equipment, is responsible for maintenance and providing replacement parts over the life of the service contract.

An early fee-for-service example is the concession model applied in the Renewable Energy for Rural Markets Projects (PERMER), initiated in Argentina in 1999. Franchise rights to rural-service territories were granted to concessionaires that required the lowest subsidy to provide households and public centers service in the areas bid out. Although concessionaires could choose from a wide range of off-grid technologies, PV was determined the most cost effective for many remote areas with dispersed customers. This model was considered suitable, given Argentina's long experience with concessions for concentrated electricity markets. Thus, the requisite regulatory framework and procedures for dispersed markets could be easily added to the existing system.

The Senegal Rural Electrification Project, initiated in 2003, used a similar concession model with exclusivity rights. But in this case, the total subsidy was predetermined. The winning bidder for a concession area was the firm that offered to provide the most connections in the first three years; the firm was also required to make a minimum number of connections beyond 20 km from the grid.

The model used in the World Bank–supported Decentralized Infrastructure for Rural Transformation (ERTIC) Program in Bolivia can be viewed as a hybrid of the above-mentioned models. Known as the Medium Term Service Contract (MSC), this model adds mandatory local-market development and 2–5 years of operation-and-maintenance services to the dealer-model requirements for participating companies. The model can also be considered a revision of the traditional

ESCO concession scheme, whereby the exclusivity term is reduced to only 2–5 years and opened to a broader menu of ownership options.

The sustainable solar market package (SSMP) used in the Philippines combines a tendered contract for institutional installations with incentives and the non-exclusive opportunity to sell PVs to households in the area. The SSMP is a contracting mechanism that provides for the supply and installation of PV systems, along with a maintenance-and-repair contract (five years with an option to extend) in a defined rural area. Applications in schools, clinics, and other community facilities are bundled with requirements and incentives for commercial sale to households, businesses, and other non-governmental customers. By bundling applications in a defined area, the SSMP approach addresses key affordability and sustainability issues of past PV projects: standardization, reduced transaction costs, larger business volume, and reduced risk. In the Philippines, SSMP contracts benefiting 278 villages are currently being implemented. In Tanzania SSMP contracts to benefit 80 villages are under implementation in Rukwa Region, with SSMP contract packages are being prepared for villages in five regions.

In Peru, three distribution utilities are beginning to offer PV off-grid services at regulated tariffs to customers whom they cannot cost-effectively reach through grid extension. This is the first instance in a World Bank supported project that a regulator has established an off-grid PV tariff on the same basis as life line tariffs for low income consumers are set for grid-based electricity service provision.

Variations on the above-described models include the leasing model — pioneered by Soluz in non-Bank projects in Honduras and the Dominican Republic — which falls between the two categories. This system is provided to the consumer via a direct lease or lease-to-own agreement.

There are no clear-cut rules for determining which PV-dissemination model is appropriate for a given project in a particular country. The dealer, ESCO, and MSC models have their comparative advantages and disadvantages. The dealer model is easier to launch, requiring only the accreditation of several participating dealers and establishment of a microfinance support system, as needed. Competition in all phases of implementation could, in theory, lead more quickly to cost reductions and better service for consumers. Conversely, because the model is fully market driven, the pace of coverage is hard to predict or control. The ESCO model has the potential to achieve faster coverage and obtain lower equipment costs due to volume transactions. At the same time, it requires more complex regulatory procedures that are often hard to establish in many countries. Lack of competition once territory is acquired may suppress innovation and lead to lower-quality service (unless well regulated) and cost savings from volume procurements may not be passed on to consumers. The regulated distributor model or the MSC hybrid model combines most of the above-described advantages of the dealer and ESCO models, while avoiding some of their limitations (e.g., via emphasis on post-sale maintenance requirements and reduced exclusivity term). Like the concession model, the regulated distributor model and MSC hybrid model requires that a competent, transparent, and effective regulatory system be in place to assure service quality.

Country conditions are important determinants of model choice. In countries where the potential PV market is economically attractive in terms of scale and geography and where there are enough qualified prospective competing companies, the dealer model (with attention paid to after-sales service) may be appropriate. Where universal access is the national goal or the market is unattractive (e.g., small and highly dispersed, consisting of uniformly poor households located in difficult-to-access terrain), the ESCO or MSC hybrid model may offer a better approach. In many countries, PV market features are mixed, which may call for a combination of models.

3. Key Lessons of Experience

Lesson 1: Project designs must remain flexible and adaptable to address issues of affordability, risks and other market constraints. Affordability varies among market segments (such as relative income levels or market applications), and it remains a challenge for PV companies to sell to the niche market segments even where PV is the least-cost energy alternative for the consumer. High first costs, lack of financing, limited awareness are among the constraints faced. Project designs must therefore be able to quickly adapt to market conditions so that responsive solutions are created to address barriers. Four such examples are projects supported by the World Bank and GEF:

- Bangladesh Renewable Energy for Rural Electrification project where 30 000 solar home systems are being installed monthly.
- The Off-Grid Rural Electrification Project in Nicaragua brings electricity services to households, public centers, and productive uses by facilitating access to micro-financing and strengthening the institutional capacity to implement a national rural electrification strategy.
- Papua New Guinea Teachers Solar Lighting Project where financing for solar lighting systems for teachers posted in remote areas is provided through the Teachers Savings and Loan Society.
- The Sustainable Solar Market approach first introduced in the Philippines under the World Bank and GEF-assisted Rural Power Project and now being implemented in Tanzania, Zambia and planned for Liberia.

In these and other solar projects, it was imperative during implementation to adapt to changing market needs, such as by extending the eligibility of PV system sizes to smaller, more affordable systems, such as solar lanterns, or by introducing loan guarantee facilities or capacity building support to entrepreneurs and microfinance institutions.

Lesson 2: PV must be considered as one of several options for rural electricity provision. All consumers prefer access to unlimited supplies of electricity at low prices. As grid-based rural electrification is often a highly valued political tool, significant government resources are employed in grid extension, even when it is a drain on government budgets and where economic justification is weak. Subsidies for off-grid populations are justified on social-equity grounds; that is, the need for remote or poor dwellers to achieve a level of parity with households in concentrated areas that benefit from subsidized grid-extension infrastructure costs and lifeline tariffs. There is also the expectation that the welfare gains from off-grid interventions are higher than the long-term costs. As a result of these subsidies the cost of grid electricity to the end user is significantly less than alternative options (see Tables 2 and 3 that compare subsidies given for grid and PV in selected countries).

Table 2. Subsidies for Grid Connected Electricity Services.

Country	Grid-connection subsidy level (% of construction and connection costs)[1]
Costa Rica	20–30
Chile	70–80
Honduras	85
China	85–90
Mexico	~95
Tunisia	100
Philippines[2]	100 (plus a portion of fuel and operating costs)

Sources: Various World Bank reports.
[1] Excludes subsidy for lifeline tariffs below the marginal cost of electricity supply.
[2] Diesel gensets.

Table 3. Subsidies for PV Applications in World Bank Projects.

Country	Project	PV system size (Wp)	Approximate subsidy range (% cost)
China	REDP	15–500	15–22
Bangladesh	RERED	20–70	12
Argentina	PERMER	50–100	up to 50
Tanzania	TEDAP	20–50	13–21
Sri Lanka*	RERED	10–60	10–25
Philippines	RPP	20–100	20–60
Mexico	IESRM	50–100	up to 90

Sources: Various World Bank reports.
* In Sri Lanka, the capital subsidy for micro-hydro minigrids is US$400 per kW or about 15–20 percent of investment cost.

However, there is now an increasing appreciation by governments and rural electrification authorities that PV is an economic least cost option for rural electricity provision, especially where the alternatives are small generators, batteries or kerosene lighting. With oil prices exceeding US$75 per barrel, and kerosene subsidies becoming untenable, PV's attractiveness is increasing. Various countries — for example, Bolivia, Laos, Nepal, Papua New Guinea, Philippines, Tanzania, and Zambia — provide subsidy support through rural electrification or rural energy funds that transparently cover the subsidy portion of electrification costs. Both grid and off-grid investments are eligible to receive support. The Philippines, for example, has a formal rural electrification planning process that demarcates villages where grid extension is not viable. Where solar PV is least cost, market and willingness to pay studies must be conducted to confirm that consumers will indeed demand these systems before PV financing programs are established. Subsidies to buy down first costs, access to financing through rural banks or microfinance organizations may be necessary to enhance affordability.

Lesson 3: Private equity is not the most appropriate financial mechanism for investing in PV rural electrification businesses in developing countries. An important lesson for the IFC was that, investments in PV manufacturing to serve the developed country markets along with private equity and venture capital, is promising. As IFC investment experiences from China, India, the Philippines show, there are profitable opportunities for entrepreneurs in developing countries to supply PV modules to developed-country markets that are highly subsidized. However, investments by IFC in PV for rural electrification have typically resulted in returns that are less than they could obtain in other ventures. PV for rural electrification in many developing countries requires patient capital where returns may not be commensurate with perceived risks to private equity investors. Nevertheless, China is an exception, where PV retail dealers depend primarily on private equity to finance their businesses.

Lesson 4: Good government relations and support are strong success factors. In some countries, duty structures bias consumers against off-grid technologies, encouraging further consumption of kerosene and other less suitable alternatives that may be subsidized or exempt from the value added tax and other duties. Such countries as Sri Lanka, Kenya and Tanzania have recognized the value of off-grid technologies, such as solar PV, and have exempted them from import duties. Since certain components of off-grid power systems have multiple uses (e.g., batteries), fiscal authorities are sometimes reluctant to grant duty exemptions, which can be abused. One option for governments to consider is to grant exemptions only for off-grid equipment that has met prescribed quality standards.

Although there are examples of companies able to establish successful PV rural electricity ventures without government support (Kenya is a good example), those companies fortunate enough to operate with such support (or with some form of subsidy or market aggregation support) tended to be more successful than companies operating without explicit government support. Given the value of mobilizing private enterprises to extend PV electricity services as an economic alternative to grid extension, public-private partnerships for risk sharing or to buy down the first cost of PV systems can be effective. Such commercial retail market opportunities do exist in developing countries, as experiences in Bangladesh, China, India, Kenya, and Sri Lanka, among others, have demonstrated.

Lesson 5: Quality must not be compromised. The most expensive form of electricity is a power supply that is not working. Quality of systems and ability to obtain spare parts and repair services must be an integral part of any PV electricity program. The World Bank was a pioneer in introducing quality requirements enforcement beginning with a project in Indonesia in 1997. All World Bank–funded projects now rigorously enforce quality requirements. In both Indonesia and China, the project specifications eventually were adopted as national standards. In China, additional support was provided for technology improvement and they introduced a "Golden Sun" quality mark to help consumers identify quality-certified products.

The World Bank worked closely with the Photovoltaic Global Approval Program (PVGAP) to introduce quality standards for solar lanterns and to undertake testing of solar lanterns. The World Bank supported the PVGAP in developing training courses for assisting PV companies to obtain ISO 9000 quality certification, improving quality of PV testing laboratories, and establishing certification programs for PV technicians and installers.

One option to enhance affordability without compromising quality is to provide smaller, lower-power systems that offer a lower quantity of service (e.g., reduced hours of lighting). For example, a solar lantern costing US$50-75 can provide 3–4 hours of lighting daily. A 50-Wp PV costing US$600 can operate four lights for 3–4 hours and power a radio or television for a few hours daily. Under the Renewable Energy Development Project (REDP) in China, where consumers had limited financial capability and lacked access to financing, most purchased low-cost 10- and 20-Wp PVs (US$80-160) initially and larger 45-Wp systems (US$400) after their incomes increased. In Sub-Saharan Africa, the Lighting Africa initiative builds on the philosophy that small, modern lighting products can be marketed at prices similar to or lower than those rural households typically pay for kerosene by limiting services to lighting, taking advantage of LED technological advances and cost reductions, and tapping into Africa's existing distribution and retail infrastructure.

Lesson 6: Access to financing is essential. Solar PV is capital intensive — both for manufacturers of key components, such as PV modules, and for consumers who purchase PV products. Therefore financing and financing vehicles that reach the target consumer are essential.

Subsidies might be complemented or substituted by encouraging or supporting microfinance institutions, commercial or development banks, or even leasing companies to offer consumer and/or trade financing. Such arrangements can increase affordability by spreading first costs over several years. Since financing off-grid electricity products may be unfamiliar to the financing entity, credit enhancement, such as a partial risk guarantee may help reduce the perceived risk to the lender. Some dealers have attempted to offer dealer financing; however, working capital constraints and lack of experience in credit-facility management have limited the success of such efforts. Successful off-grid lending programs involve a strong partnership between the microfinance institution and an energy company. The effectiveness of that partnership depends on a clear understanding of the roles and responsibilities of each partner and their competency and capacity.

On the supply side, financing is needed to increase manufacturing capacity, supply new materials that bypass the global bottlenecks caused by the limited supply of silicon and newer and higher efficiency solar PV materials and end-user devices (such as lighting with LEDs). Demand for larger utility-scale PV may also emerge, especially for reducing peak electricity demands.

4. Guidelines for Designing Sustainable Off-Grid Projects

Building on these project experiences, the World Bank issued guidance on the design of sustainable off-grid projects. To maximize the chances of sustaining operation of an off-grid project over the long term, fundamental project-design principles must be observed. These include the following:

- *The off-grid project must not be conceived of or implemented in isolation from the overall rural electrification plan for the region.* The project should not be influenced by such ad-hoc factors as one-time availability of donated renewable-energy equipment or pressure exerted by local politicians.
- *Both the government and officials of the implementing agency must take full ownership of the project.* The government must put in place light-handed regulatory measures that simplify operations for private-sector participants and limit the cost of doing business.
- *Because community ownership is essential, efforts must be made early in the assessment phase to maximize the awareness and involvement of target communities.*
- *For off-grid projects that rely on private-sector participation, the implementation design must be calibrated to fit the specific context in order to elicit the needed investment response.*
- *Project design must not be technology driven.* A cost-benefit analysis of technology alternatives must be carried out to determine the least-cost solution. Choice of technologies must be based on practical considerations. Data on energy consumption and income and willingness to pay across various sectors in the community should be collected upfront and factored into the technology selection process.

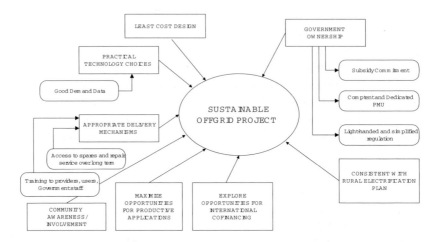

Figure 1. Elements of a sustainable off-grid electrification project.

- *Productive and institutional applications that complement the provision of household service must be proactively identified and supported as the income and benefits generated help those who cannot afford individual connections or systems.*
- *The simplest delivery mechanism or business model (or mix thereof) commensurate with local realities should be applied.*
- *Whatever business model is chosen, care must be taken to ensure that users have access to quality equipment and products and qualified repair service and spare parts over the long term.*
- *Opportunities for international co-financing should be explored.* Such funding sources might include the World Bank's Global Environment Facility (GEF) or Global Partnership for Output Based Aid (GPOBA), Climate Investment Fund, the Clean Development Mechanism (CDM) of the Kyoto Protocol or a future mechanism, bilateral donors, or a country's sectoral ministries (e.g., health or education).
- *It is important to obtain the government's upfront commitment to pick up the subsidy slack when grant co-financing ends to ensure that implementation momentum is not lost.* Grant co-financing by international donors for the cost of PV hardware is often provided on a declining basis and ceases at project closure. The assumption has been that, because of market growth, system prices would have decreased significantly by that time and would be affordable. But World Bank project experience over the past 4–5 years has shown this is not always the case. One possible solution is to create a rural energy fund to which off-grid projects have access.
- *Because competence of the local Project Management Unit (PMU) is critical to successful project implementation, commitment must be obtained from implementing agency officials that PMU staff will devote their time fully to the project.*

5. Future Support for Photovoltaics

The World Bank Group recognizes that PV is but one rural electricity option and therefore should be offered as one option on a menu of options to meet rural electricity service goals. Moreover, we recognize that electricity is merely an intermediate product and that it is the services rendered by electricity that matters for operating motors, lighting, refrigeration, communications, and so forth. To this end, the World Bank and IFC embarked on a joint project, Lighting Africa. It will catalyze local and international lighting-related companies into offering the unelectrified population greater access to modern and affordable off-grid lighting products while displacing fuel-based lighting products, such as kerosene lamps or candles, for which developing-country consumers spend about US$20-40 billion annually. PV technology is expected to be a principal source of power for such off-grid lighting applications.

Within the World Bank Group, in addition to normal financing available for its various institutions, special funding facilities for clean energy investments and for rural energy development can be tapped to finance photovoltaic projects. Among these are the following:

- Since its creation in 1993, the Global Environment Facility (GEF) has been the traditional co-financier of World Bank off-grid electrification projects through grants provided for renewable energy technologies that are ready for practical deployment but face market barriers.
- The Global Partnership for Output Based Aid (GPOBA) is another source of grant assistance. GPOBA's goal is to apply output-based approaches to support the delivery of basic services to the poor, not only for electricity but also for water, sanitation, telecommunications, transportation, health, and education. In addition to on-off subsidies such as to buy-down the initial cost, the GPOBA provides technical-assistance grants for project design and evaluation and disseminates lessons learned.
- A key challenge for off-grid project designers is potential service providers' lack of capacity, making training in such basic skills as business and financial management imperative. Such studies and activities may be eligible for grant financing provided by the Energy Sector Management Assistance Program (ESMAP). The World Bank–managed Asia Sustainable and Alternative Energy Unit (ASTAE) and Public Private Infrastructure Advisory Facility (PPIAF) could also provide grant financing for technical-assistance activities related to off-grid electrification.
- The Clean Development Mechanism (CDM) of the Kyoto Protocol and future mechanisms emerging from the forthcoming climate negotiations, may offer opportunities for enhancing the financing of off-grid projects through carbon credits for renewable-energy systems. Bangladesh signed contracts for the purchase of emission reductions to be achieved by its large solar PV dissemination program for remote off-grid areas. The program targets about one million

PVs installed by 2015, totaling more than 50 MW and avoidance of 84 000 tons of CO_2 per year at full implementation.

- Finally, Climate Investment Fund (CIF), approved by the World Bank Board of Executive Directors in July 2008, is a potential funding source for off-grid renewable-energy projects in developing countries. Under the CIF, two trust funds — the Clean Technology Fund and Strategic Climate Fund — totaling about US$6 billion are supporting clean energy projects. The Clean Technology Fund provides large-scale financial resources for projects and programs in developing countries that contribute to the demonstration, deployment, and transfer of low-carbon technologies. The Strategic Climate Fund, broader and more flexible in scope, will serve as an overarching fund for various programs to test innovative approaches to climate change. One program is the $250 million pilot, Scaling Up Renewable Energy in Low Income Countries (SREP). This facility supports low income countries transition to energy development pathways that will reduce their carbon footprints. The first SREP pilot projects are being prepared for Honduras, Mali, Ethiopia, Kenya, Nepal, and Maldives.

Having extensively evaluated not just its own experience, but also the experience of several important players in the solar PV business, the World Bank Group remains cautiously optimistic that it is not a question of "if," but "when," the goal of a self-sustaining solar PV market in developing countries will be reached.

Anil Cabraal

Until April 30, 2010, Dr. Cabraal was a Lead Energy Specialist at the Energy, Transport and Water Department, Sustainable Development Vice Presidency, The World Bank, Washington DC 20433, USA.

Chapter 9

Solar Bicycles, Mercedes, Handcuffs — PlusEnergy Buildings

Gallus Cadonau

Swiss Solar Agency, Waltensburg/Zurich, Switzerland

1. A Worldwide Unique Solar Decision: Tour de Sol

1.1 CO_2-free Hotel Ucliva in the Swiss Alps: 1st Solarcar Race of the World

On October 27th, 1984, a worldwide unique decision was taken in the CO_2-free, ecological Hotel Ucliva in Waltensburg, approximately 100 km north-west of St. Moritz in the south-eastern part of Grisons/Switzerland. In 1985, the world's first solarcar race, the Tour de Sol, shall take place. With a solarcar race from the Lake of Constance to the Lake of Geneva, the power of solar energy shall be publicly proven. The initiators were convinced that solar energy is not only useful in space, but can also drive vehicles.[1] The federal board of the Swiss Association for Solar

---

[1] The Tour de Sol decision of the federal council of the Swiss Association for Solar Energy (SSES) of October 28, 1984, in the Hotel Ucliva and its following realisation was taken by: Dr. Mario Camani (president), Dusan Novakv, Gallus Cadonau, Sven Frauenfelder, Herbert Marti, Thomas Nordmann, Dr. Anne-Marie Felkel, Ruedi Bühler, Urs Muntwyler, Ludwig Lübke, Ruedi Nüscheler, Dr. Jacques Dreyer, Franz Clauward, Felix Merki and Markus Heimlicher (minutes). The idea came from Hannes Ruesch and Josef Jenni, who employed Urs Muntwyler and trusted him with the realisation of the Tour de Sol 1985.

Power for the World by W. Palz
Copyright © 2011 by Pan Stanford Publishing Pte Ltd
www.panstanford.com
978-981-4303-37-8

Figure 1. The decision for the organisation of the first Tour de Sol was taken in the CO_2-free hotel Ucliva in October 1984. At the end of August 1991, the European Championship of Alpine Solarcars stops in front of the hotel Ucliva in Waltensburg, Grisons. The solarcars direct their solar panels towards the sun and refill their batteries with alpine sun.

Figure 2. The solar bicycles and solar cars of the world first Tour de Sol 1985. Left: Supplementary to PV solar energy, some of the solar cars were powered by pedals. Right: The winning solar car of Mercedes/Alpha Real won the first Tour de Sol from the lake of Constance to the lake of Geneva in 1985 with a velocity of 70 km/h.

Energy unanimously agreed to the regulations of the Tour de Sol and decided that the first Tour de Sol (TdS) shall take place from June 25th to June 29th, 1985.

1.2 *Tour de Sol 1985: Solar Bicycles and Mercedes Benz*

On June 25th, 1985, 29 pioneers with "solar vehicles without additional propulsion" (i.e. without pedals) and 26 "with additional propulsion" arrived in Kreuzlingen, sister city of the German city of Constance. The variety could not have been bigger: at the start were solar bicycles with two or three wheels, vehicles looking like soapbox cars and even a solar racecar of Mercedes Benz. Of course, most of the vehicles with solar panels also had to carry batteries. Many spectators were convinced that these "carriages" would never arrive at the finish line.

1.3 *First Solarcar Driving Past an Atomic Power Plant*

The technical regulations allowed a maximum of 480 Wp for the solar generator. Its surface was not to be bigger than 6 m². Shortly after the start in Kreuzlingen, the participants had the possibility to win a first award: Franz Hohler, a famous Swiss writer and cabaret artist, had sponsored a prize of CHF 100 for the first solar vehicle to drive past the atomic power plant of Leibstadt. In the evening, this and other awards were celebrated by all those who did not have to fix their solar vehicles for the next day.

1.4 *Geneva — Final Stage of the 1st Tour de Sol 1985: The Power of the Sun*

The 1st Tour de Sol went from Kreuzlingen via Brugg, Neuchâtel and Lausanne to the city of Geneva that had to be reached within four days. The popular success of the tour was overwhelming. Newspapers were reporting on the race on their front pages. And also radio and TV stations across Europe, Japan and the US reported on the Tour de Sol spectacle. The weirder the solar vehicles looked, the more enthusiastic were the people on the streets. And in many towns along the course of the Tour de Sol to Geneva the schools closed for the day, so that the children were able to attend the Tour de Sol. Mercedes Benz promoted its perfectly styled solar racecar as a European answer to the actual forest decline. And the company also used the opportunity to point to its environmental friendly car technology. From a total of 55 participating solar vehicles, only 27 were classified, the others had dropped out of the contest due to battery problems. However, all participating vehicles had one thing in common, they caused surprise and admiration! Mostly admired was the winning vehicle of Mercedes/Alpha Real with its solar peak velocity of 70 km/h. The stylish Mercedes solarcar managed the distance of 368 km to Geneva in 9 hours and 41 minutes.

1.5 *Tour de Sol 2 in 1986: Massachusetts Institute of Technology in the Roadside Ditch*

The 2nd Tour de Sol started in June 1986 in the solar friendly city of Freiburg in Breisgau, Germany. From there, it went to Basel, Biel, Berne, Interlaken, Lucerne and finally to Suhr in the Canton of Aargau. The entourage of the Tour de Sol was on its way for six days and travelled 432 "solar" kilometres. Since it also had to cross the Brünig Pass (1008 m) with a total difference in altitude of 1700 m, this tour became the "first climbing tour". There were many German teams participating. The Swiss School of Engineering Biel stood out with its solarcar reaching a peak velocity of 125 km/h. At the start were many universities and technical schools from different countries, amongst them also James Warden from the famous Massachusetts Institute of Technology (MIT), one of the leading universities in the field of technological research and science. The MIT vehicle was very original. At first sight it was not obvious if it had three or four wheels and what was front or back. Unfortunately, the MIT vehicle was not favoured by fortune on its way to Basel and ended up in a roadside ditch. The solarcar was all smashed to pieces and it almost looked like debris of a plane crash. The MIT student, however, arduously collected piece by piece of the car. With the support of the Tour de Sol organisation, he finally reached the first stage finish in Basel.

1.6 *Welding and Sweating Instead of Champagne*

Instead of celebrating stage victories, the teams often had to weld and sweat until dawn. The finish and parking areas of the different Tour de Sol stages looked like peaceful battle-grounds: the teams were hammering, screwing, disassembling and reassembling; invertors were replaced, batteries and wheels had to be checked. Often the teams had to hurry up, in order to be ready for the next day's mandatory briefing and start. Innovation and improvisation, enthusiasm and commitment almost had no limits. And on the second day, even the bleary-eyed MIT student James Warden was ready for the start. His solarcar was barely recognizable. The solar panels having looked like an engine hood on the first day were now installed at the back of the MIT vehicle. They were fixed with wires and cords and were slightly fluttering in the wind.

1.7 *Strong Solar Teams from Germany and the Swiss School of Engineering Biel*

At the end of 1985, Roland Reichel and his friends founded the "Solarcar Association of Erlangen". They intended to successfully promote PV utilisation also in Germany. The most famous solar pioneers of that time were the architect Rolf Disch from Freiburg as well as the team Trykowski with very good race rankings for several years. The solarcars continuously lost their soapbox image and looked the longer the more like real race cars: the prominent category I of the solar racecars was born. They were constructed professionally and built more solidly. Most were using continuously better mono-crystalline solar cells with an efficiency factor between 11% and 13%. Increasingly efficient solar generators and stronger motors increased velocity and rendered the solar race cars even more

Figure 3. The participating teams of the Tour de Sol worked day and night on their solarcars, on the way from Lake Constance to Lake Geneva, (figure left). The same was still the case, at the World Solar Challenge as shown in this picture (right) from 1993.

Figure 4. Dr. Fredy Sidler, head of the School of Engineering Biel at the left; the technical director René Jeanneret on the right.

attractive. The competition for the Swiss School of Engineering Biel was getting bigger, but in 1986 it won the 382 km long Tour de Sol with a total difference in altitude of 1851 m in 7 hours and 42 minutes. This corresponded to an average speed of 50 km/h. "Without the break-downs lasting up to an hour, the average speed could have been 57 km/h", reported the victory-addicted technical director and "solar king of Biel", René Jeanneret, to his happy students.

1.8 *Solar Cells for "Independency" or Terrestrial PV Utilisation?*

In the meantime, some tour participants had founded the "Driver and Constructor Association (FKVS)" in Berne. In September 1986, the FKVS appealed to the tour organisers with a widely discussed controversy: According to the tour regulations, "the solar electric cars did not imperatively have to carry the solar cells on the vehicle," if it was guaranteed 100% and unmistakably proven that each meter was driven exclusively by solar energy. However, the TdS organization was afraid that the race would lose its attraction and transparency if the solar cells were not integrated on the vehicles. The organizers were convinced that "the solarcars show their independency by carrying their solar cells along". Only the (everyday) solar production cars of category III did not have to carry the solar cells along. They were allowed instead to use solar cells with a higher energy output that were integrated in buildings and optimally facing South. This was the beginning of the triumphant success of efficient, terrestrial, building-integrated solar utilisation, contrary to the fear of the Tour de Sol organizers.

1.9 *Tour 3 in 1987: Huge Interest and "Chermobiles"*

The 3rd Tour de Sol started on June 29, 1987, in the German and French speaking city of Biel/Bienne and lasted until July 4. On each tour stage, an exclusive exhibition was organised for the media and other interested parties. If there were no repair works to be done, the participants had the possibility to go for a short solar "drive". The exhibition showed the newest technologies, invertors, lead-acid and nickel batteries as well as the best solar modules. For transparency reasons, the production cars of category III had to carry huge pictures of their solar energy feed-in. Furthermore, there were lively discussions between federal politicians, corporate CEOs, famous artists and VIPs on each day with focus on responsibility, sense or nonsense of nuclear power and combustion of fossil energy with damaging greenhouse gases. During the discussions about the use of PV on solar racecars or about solar energy feed-in with fixed and integrated solar cells, the emotions often went high, especially after the atomic disaster of April 26, 1986 in Chernobyl. Due to this, the electric cars without solar cells were named "Chermobiles".

1.10 *Huge International Media Coverage*

At the 3rd Tour de Sol, participants from foreign countries had the advantage to start first. Most of the solarcars came from Germany, France, Liechtenstein and the US. The technical progress was not only visual, it also lead to always faster solarcars. For speeding, rigorous penalties had to be imposed. "50 km/h with solar energy", reported the German newspaper Frankfurter Allgemeine already after the 2nd tour on July 21, 1986. Interest in and media coverage of the tour increased in Switzerland, Europe and overseas: "Tour de Sol: A total success. Tour de Sol: Victory for the Sun Tour de Sol 86 was a total success in all aspects. Solarcars almost as fast as conventional cars Race with watts and volts. The US participant James Warden was even accompanied by a TV team of seven people. They made a 60-minute report about the event that was broadcasted on 207 TV stations in the

Figure. 5a. Ursula Westphalen, media officer of Solar Agency, has a look at the remarkable media response for solar energy and the Swiss Solar Prize (1992).

Figure. 5b. The Japanese car industry and small and medium enterprises participated at the race with a solarcar looking like Mickey Mouse (third car on the left).

US in October 1987. The Swiss radio and TV stations reported daily about the Tour de Sol. And also the American TV program "Good Morning America" reported almost daily on the event and especially on MIT participant James Warden and other US participants.

1.11 *Tour Organisation and Its Regulations*

The Tour de Sol 1987 went from Lake of Biel to Emmen, Zurich and St. Gall. From there it crossed the Austrian border to the city of Rankweil. The following day it went to Liechtenstein and back into Switzerland to Chur, the capital of Grisons. Quite thrilling were day four and five from St. Gall to Rankweil and Chur: The

Tour de Sol was organised in a way that the chief of start was responsible for preparations, briefings and punctual starts. The chief of finish organised time keeping, arrival and parking for solarcars and their escort vehicles. A member of the organisational committee was chief of race and another was responsible for the total organisation. The chief of race always started first and drove ahead of the solarcars. James Schwarzenbach had to know the course best and all solarcars had to follow him. This way, it was ensured that the solarcars did not drive to fast in town and that they respected the traffic rules. On July 2, 1987 in St. Gall, the chief of race took the wrong way shortly after the start. Some solarcars followed him according to the regulations. Others did not follow and took the correct route towards the Austrian border. At the finish, there were lively discussions: Who was the winner? Who was to get the prize money?

1.12 *Tour de Sol Protests and Appeals in Court?*

The Panasonic team and others entered their protests. The appealing committee sat until deep in the night, listened to the protests and decided: The decision will be adjourned and the start on the following day in Rankweil will be on time. The protesting teams will also start punctually. The tour continued and tension increased. In Chur, the radio and TV stations showed less interest in the winners than in the pending appeals. The first teams consulted their lawyers and informed themselves about the ordinary courts and court proceedings. Approximately 30 minutes before the daily TV news, the race jury decided in Chur as follows: 1st: The "case St. Gall" will be adjourned and the court will continue its consultation at night. 2nd: The start on the following day will be on time and the decision will be announced in the morning at 9 am. The most important aspect of the adjournment was that the media did not get any "hot story" about a totally messy situation. In the following final decision of the jury the first part of the stage St. Gall-Rankweil was neutralised for all teams. The remaining parts were considered correctly and the teams were ranked accordingly. Thereupon, all protests were abandoned.

1.13 *Solar Mountain Race: Through 360 Curves to Arosa/GR*

The final stage of the Tour de Sol 1987 was a big solar challenge: After five hard racing days and therefore highly stressed batteries in some solarcars, the cardinal question of the tour and from the media was: "Does solar energy generate enough power for a hard mountain race of 30 km and with a difference in altitude of 1100 m? Will the solarcars starting in Chur on 600 m above sea level ever arrive in the famous mountain resort Arosa on 1720 m^2. The steep and winding final course on July 4th, 1987, caused suspense. For security reasons, the street from Chur to Arosa was closed for regular traffic for some hours. The street belonged all to the solar racecars and solar vehicles. Although this first solar mountain race lead to smoking electric motors, burned cables and empty batteries, the best solar racecars drove the steep mountain course of 30 km and with 360 curves almost in the same time as conventional cars. With 38 minutes and 42 seconds, the solarcar pilot Stefan Bräcker broke all records. Only a few, powerful passenger cars are able to equal his average speed of 46.5 km/h through 360 curves. Solar energy had passed its hardness test and had surprised everybody.

2. World's First Energy Feed-in System in Burgdorf/BE

2.1 *Tour de Sol in France — Solar Energy Instead of Air Pollution*

The wide range of solarcars was attracting more and more persons and institutions. Cooperation between interested parties became increasingly professional, interdisciplinary and international. In 1986, the SSES and the Swiss Greina Foundation for the Protection of Alpine Rivers (SGS) launched the Tour de Sol Foundation that became responsible for the organization of the Tour de Sol. Silvio Bircher, member of the Swiss socialist party and National Councilor of the canton Aargau, was appointed 1986 as first president of the Foundation "SSES-Tour de Sol". In his parliamentary motion of March 20th, 1987, he asked the Swiss government to "reduce air pollution and effectively promote the use of solar-cars". The Federal Council agreed to accept his motion as a postulate and to verify the possibility of "using solarcars (with or without integrated solar cells) in its administrative departments". From March 19th to March 22nd, 1987, the first Grand-Prix du Soleil took place in Colmar/France. It was also a preparatory race for the Tour de Sol 1987 which was won by the solarcar of the Swiss School of Engineering Biel.

2.2. *World's First Energy Feed-in System in Burgdorf/BE: "Grid Interconnection"*

In 1987, the Tour de Sol for the first time, introduced the category "grid interconnection". At the expert meeting, Winfried Blum pointed out that this category with "half a dozen participants in 1987" is getting more and more interest".[2] Many tour participants had a positive attitude towards the new category; others were still skeptic. The ecological center Langenbruck — founded by emeritus Prof. Pierre Fornallaz[3] — had been fiddling with solarcars and grid in-terconnection

---

[2] Winfried Blum, engineer and editor of the Swiss Association of Power Plants (VSE); "Solarcars in everyday life", edition 2, January 1988, was regarded with skepticism within the solar scene, because the interest in an efficient electric drive understandably was still bigger than in the substitution of nuclear power plants by photovoltaic installations.

already in 1986 and experimented with it at the Tour de Sol. The explanation was simple and technically convincing: If solar panels are integrated in a building and are optimally facing south, they have a bigger output than those installed on solarcars. Theo Blättler, director of the power station of Burgdorf (EWB), canton Berne, was commissioned in 1987 to write a report about the use of solar energy.[4] In 1988, he handed in an extensive report to the City Council of Burgdorf proposing solar energy feed-in. Subsequently, the power station of Burgdorf paid 1 CHF (0.75 € cts) for each kWh that was fed-in. By doing so, the solarcars were able to get the solar energy from the grid at night and to drive with 100% solar energy during the day. Since the solar feed-in took place during the day — when the energy need is biggest — and the batteries were loaded at night, the solar feed-in had a compensative, counter-cyclical effect.

2.3 *The Principles of Solar Energy Use: Best Technology or Self-sufficiency?*

The idea of a solar energy feed-in did not correspond to the overall principle of energy self-sufficiency saying that each unit is supplying itself possibly on its own and in a decentralized manner. For the Tour de Sol this meant: best possible energy self-supply of solarcars, apartment or business buildings, service or industrial buildings etc. Each unit that exploits its full technical potential of energy efficiency and is still under-supplied must provide for a central supply. Living comfort and sustainable productivity of society and economy shall not be cut back: This was the conviction of most of the tour participants. And after the catastrophe of Chernobyl on April 26, 1986, society developed as similarly innovative as solar technology. Citizens submitted initiatives, and financing models for the construction of photovoltaic installations or solarcars were presented.[5] On the

---

[3] Prof. Pierre Fornallaz vs ETH: In 1974, many years before his early retirement in 1981, Prof. Pierre Fornallaz, co-founder of the Société Suisse pour l'Energie Solaire (SSES) proposed an installation of solar cells on the roof of the Federal Institute of Technology in Zurich (ETH). The institute declined the proposal and justified its decision with the remark that solar energy was too expensive. Fornallaz: "This justification was wrong then and is still wrong today". Also today, 30 years later, the students of the Federal Institute of Technology are still waiting for incentives of their institute for a sustainable use of energy. The Federal Institute of Technology in Zurich has a professorship for nuclear energy, but 30 years after the solar intervention of Prof. Fornallaz it still does not have a professorship for solar energy.

[4] Theo Blättler was awarded the Swiss Solar Prize 2001 for his pioneer work of energy feed-in for photovoltaic.

[5] Financing models for decentralized photovoltaic installations from Gallus Cadonau with written statutes for the "cooperative solarcar construction", see also "Solarcars in everyday life", edition 2 of Jan. 30, 1988, p. 128-138. Subsequently, numerous associations and cooperatives were founded to build solarcars and to participate at the Tour de Sol.

Figure 6. Solar race car on the Julier Pass, GR.

political level, people were asking for "consequences". The canton Berne showed an increasing interest in solar energy (thanks to its left-winged Cantonal Councilor René Bärtschi) and was open to "solar experiments", although proposals for a solar energy use such as the one in Burgdorf were often criticised.

2.4. Tour de Sol 4 in 1988: PV Innovation and Financing are Getting Broader

On June 27, 1988, the 4th Tour de Sol started in Zurich with more than 100 participants. The tour entourage drove to Uster, then to Emmen/Lucerne, Liestal/Basel-Country, Solothurn, Estavayer le Lac and finally arrived in Etoy near Lausanne on July 2nd, 1988. In the mean-time, the solarcar regulations were accepted in Europe, the US and in Australia. Practically all countries had adopted the tour and regulations of the International Solarcar Federation (ISF) with only minimal adaptations. Category 1 defined the general regulations for the solar racecars. This category was complemented by the solarcars of the World Solar Challenge. Instead of 6 m², the race category newly admitted up to 8 m² of solar panels: the WSC category was born. Category 2 included the production solarcars, and category 3 the solarcars with energy feed-in. Solar installations were not only integrated in cars, but increasingly also in buildings. New was the category of commercial vehicles. And in order to finance photovoltaic installations, solar cooperatives were founded and cooperative statutes were elaborated.

2.5 PV on Land and on Water

On the occasion of the 4th Tour de Sol and its solarcar championship, the first solarboat race took place on July 1st, 1988, in Estavayer le Lac on the Lac de

Neuchâtel: In two categories 17 solar captains were competing for victory in solar raceboats and utility boats. The event had officially been announced as the first world championship of solarcars and solar boats. Both boat categories were tested in a "sprint race" as well as in "endurance". The solar raceboats had to carry all solar panels on board, while the usually much bigger utility boats for passenger and goods transport were also allowed to use energy feed-in from the grid.

2.6 *Solar Energy is Getting Increasingly Popular also for Groups and Managers*

A novelty in 1988 was that Migros, the biggest food company in Switzerland with a turnover of approximately CHF 15 billion (10 billion), supported the Tour de Sol as a main sponsor. Already at that time, the Migros cooperative stood up for a sustainable development. Also in the fields of logistics, Migros identified itself with the emission-free high-tech-movement and transported an increasing amount of goods by train or with gas-powered trucks. At the start of the Tour de Sol were the food company's presidents, Dr. Jules Kyburz and Eugen Hunziker, who also publicly took a positive stand towards a low-emission traffic circulation. The Tour de Sol was widely supported by "enlightened people" and "innovative managers committing themselves to sustainability". They were ready to do their bit for a sustainable development.

2.7 *Safety, Recuperation and Road Capability in Winter*

Safety and road capability of the solarcars were tested during winter events on the frozen Lake of Lenzerheide/Grisons. The solarcars proved to be suitable and reliable on snow; some even had a small heating installed. From the technical point of view, the use of photovoltaic became more practical and consumer-friendly: Safety and the rigorous compliance with safety regulations were important. New was also the recuperation of the gained solar energy. In 1988, most of the solarcars had a more or less efficient recuperation in order to regain up to 30% of the energy when driving downwards or when braking. Very positive was also the participation of professional companies from the industrial sector such as Horlacher AG of Max Horlacher from Möhlin/Aargau, a pioneer in building light vehicles (see Fig. 50). His solarcar drivers reported about "avowed admiration, incredible astonishment and applause" from tourists and about doubtful, but friendly policemen.[6]

---

[6] Second expert meeting, Werner Meier/Horlacher Sera-Mobil, "Solarcars in everyday life", p. 111, 30 January, 1998.

Figure 7. The solar-powered light vehicles of the Josef Jennis solar company in Oberburg (BE) were named "Sulki".

2.8. *Solar Power — A Friendly Alternative*

Goodwill for the Tour de Sol got larger and larger, not only from students who were glad to get out of the classroom for some hours to form a guard of honor for the tour. Also very popular were the solarcar drawing contests in public schools. At school and at home, the subject often caused fierce discussions about pollution as well as the use of solar, fossil or nuclear energies. There was a high amount of goodwill and sympathy; and cooperation with the authorities was very good. On October 20, 1987, the Chief of the Cantonal Police of Grisons wrote: "I was look-ing forward to the tour and will be happy to experience it again. For the Cantonal Police of Grisons it was a great experience".[7] On June 23, 1988, the TV magazine TELE reported: "In 1985, some idealists and home constructors were campaigning for the promotion of solar energy at the first Tour de Sol with their crazy vehicles. Three years later, the tour has become a spectacular event; it demonstrates that solar power can be a functional alternative on the roads".

2.9 *Tour de Sol: "A Hotbed for Solar-electric Mobility"*

The German solarcar pioneer, Roland Reichel from Erlangen, called the Tour de Sol "a hotbed for solar-electric mobility" and said: "Tenaciousness and endurance

---

[7] Dr. Markus Reinhardt, Chur/Grisons, Chief of the Cantonal Police of Grisons was responsible for almost everything around solarcars, tour routing in Grisons, the European Championship of Alpine Solarcars (ASEM), road closings, pass drives, permissions etc.

were enormous, as well as team spirit amongst the participants." Reichel refers to an "atmosphere of departure" that mainly prevailed in Middle Europe, Japan and the US. At that time, "batteries" and an optimal tuning between motor output and "energy economy" were not as essential as in today's discussions. The powerful electro-motors guaranteed high speed also uphill. The danger or weak point of the solar racecars lurked in a fast discharging of the batteries. After a fulminate start or a crazy mountain drive outdistancing one car after the other, the solar racecars often suddenly stopped at the roadside due to empty batteries. They had to face their solar panels towards the sun, and while they were "refueling", other solar-cars drove by with still well-charged batteries and arrived at the finish first. Pressure for new solutions increased.

2.10 *Car Makers, What Have You Done in the Past?*

Parallel to the Tour de Sol, expert meetings were held each spring and car makers were asked some awkward questions. The engineers of the Tour de Sol referred to the fast developments in aircraft construction, computer technology, telecommunication etc. and critically asked 1989 the car makers: "What have you done in all these centuries?" At that time, it was clear already that despite its broad use in Brazil in one third of all private cars, alcohol was not a real solution for fuel. In the new century, the world's meadows and fields will have to nourish twice as much people as today. And also our water and food reserves will not suffice to supply an extremely inefficient traffic sector with energy.[8] And not to forget air pollution, increasing CO_2 emissions and the climate problem in general. Do we really need 1.5 to 2 tons of material in order to transport a person of 70 kg? Where are the light vehicles? And where are the emission-free cars? What will people drive with after the decline of the oil century?

2.11 *Tour de Sol and the International Solarcar Federation (ISF)*

Shortly after the Tour de Sol 1987, the tour directors Gallus Cadonau, Urs Muntwyler and Jürg Schwarzenbach founded the International Solarcar Federation (ISF) in Berne.[9] The ISF regulations and its 26 articles were meant to be

---

[8] Tour de Sol 2000 with five theses from Gallus Cadonau: 1st — Solarcars are self-sufficient; 2nd — Solarcars do not need additional roads and complement public transport; 3rd — Solarcars meet the requirements of individual road traffic and are no source of noise; 4th — Load capacity of solarcars corresponds to approximately 50% of empty weight; 5th — Solarcars are no harm for the biosphere. "Solarcars in everyday life", edition 3 of 18/02/1989, p. 13–26 with 26 paragraphs on development of population and nutrition, energy, water, urban and regional planning, air and climate, the ideal solarcar etc.

[9] The International Solarcar Federation (ISF) was founded in mid October 1986 according to the statutes that had been elaborated by Gallus Cadonau, Urs Muntwyler and Jürg Schwarzenbach on September 27–29, 1986.

Figure 8. The first elected ISF president Gunter Topmann, member of the European parliament (SPD) from Germany (1987).

Figure 9. From left to right: Hans Tholstrup, president of World Solar Challenge, Takahiro Iwato, chief engineer of Honda Dream Team, Howard Wilson, former vice-president GM and Fredy Sidler, director of School of Engineering Biel. Picture: Gallus Cadonau.

applied internationally in order to allow safe and roadworthy solarcars to start possibly everywhere. Günther Topmann from Germany and member of the European parliament was appointed as first ISF president. In 1990, he was followed by Richard King of the US Department of Energy (DOE) who stayed in office for six years. After his demission, former GM vice president Howard Wilson was elected as his successor on November 3, 1996, in Darwin. Wilson was head of the GM Sunracer team that had won the first WSC in 1987. On the managing board were Hans Tholstrup, president WSC for Australia; Takahiro Iwata, chief engineer of Honda for Japan; Dr. Freddy Sidler, director of the Swiss School of Engineering Biel; Nancy Hazard, Tour de Sol US; Hennig Bitch for Denmark and Bettina Kosup (media) for Germany. Since 1987, Gallus Cadonau had been ISF

Figure 10. A new entraining possibility on double-deck coaches was proposed at the Tour de Sol expert meeting of February 18, 1989: While trains are getting faster and faster, the velocity in car traffic decreases. In double-deck coaches solar-powered light vehicles can be parked on the lower floor and recharged during the journey. At the same time, the passengers can enjoy a meal at the on-board restaurant.

secretary general as well as vice-president. He handed on his job as secretary general to Fredy Sidler, who shared the office with Iwata. The task of the ISF was to promote the Tour de Sol and WSC worldwide, to standardize the events and to arrange for international solar juries.

2.12 Rail 2000 and Solarcars in Double-deck Coaches

In 1988/89 within the project of Rail 2000, construction schemes for double-deck coaches were outlined for distances longer than 150 km. The project envisioned wagons with serial-production solarcars with a length of 2.5 m parked diagonally to the wagon on the lower-deck. The idea was that solarcars are charged with solar energy during the journey, while passengers can relax, dine, sleep or work on the upper deck. Hydrogen-fuel produced with biogas or renewable energies shall provide for emission-free air traffic for distances over 1000 or 1500 km. The emission-free century with ecological vehicles that do not take the daily bread from the poor shall begin.

2.13 The Ideal SOLARCAR 2000: Emission-free Traffic Circulation in the 21st Century

In 1989, the Tour de Sol expert meeting defined the guidelines for the "Tour de Sol 2000", and the "self-sufficient and ecological solarcar 2000"[10] was launched: The

---

[10] See note 8, Tour de Sol 2000 with the ideal solarcar.

ideal SOLARCAR 2000 produces most of its needed energy with its car surface (solar cells) and obtains the remaining part from photovoltaic installations integrated in buildings. Since 70% of all car driven distances are below 30 km, the solar energy supply of solarcars covers about 70% of all drivers. It makes sense to use solar energy in individual traffic for distances until 150 km or for drives with light vehicles of 1 to 2 hours. Distances over 150 km until 1000 to 1500 km are ideally covered with emission-free railway traffic. It is very important, however, that the railways are run with renewable energies or by solar electricity on the long term. Also aviation has to become more ecological for all distances.

3. The Solar Alternative in Road Traffic: World Solar Challenge

3.1 *The Solar Alternative in Road Traffic*

3.1.1 Two Hundred Years after the French Revolution: The Solar Revolution (1989)

Two hundred years after the social revolution of 1789 in France, the time was ripe for an energetic revolution in order to abandon the fossil-nuclear energy path of the 19th and 20th century with billions of tons of CO_2 emissions. No more radioactive waste that puts in danger millions of people for thousands of years. In Moscow, the general secretary of the Soviet Union, Mikhail Gorbachev, referred to "Glasnost". The French president, François Mitterrand, invited the world's chiefs of state to Paris. They celebrated the declaration of human rights from 1789 that included the principle of sustainability in its art. 4.[11] In Paris, the politicians aimed not only for technological, social, economic and other progress, but also set ecological goals — especially with a view to the international conference of Rio de Janeiro in 1992. There was an atmosphere of departure.

3.1.2 California's Clean Air Act, Zero-emission Vehicles, PV Program for 1000 kW Roofs

At the Tour de Sol expert meeting in Berne, Californian representatives spoke about California's tight Clean Air Act and Zero-Emission Vehicles (ZEV). They were aiming for an annual increase in ZEV of 2% and wanted 10% of all Californian cars to be ZEV by 2003. Ruppert Schmitt from Saarland/Germany informed about the roofs in the city of Saarbrücken with a capacity 1000 kWp of solar electricity. The minister for the environment of the Saarland, Jo Leinen, supported the possibility of a private grid connection in the Saarland. The German director for research and technology spoke about a photovoltaic program

---

[11] Art. 4 of the "Declaration of human rights" from 1789 includes a first principle of sustainability for future generations: "La liberté consiste a pouvoir faire tout ce que ne nuit pas à autrui." This means that today's societies shall not spoil the resources of the future generations and that they cannot leave the fumes and highly toxic radioactive wastes for thousands of years.

Figure 11. At the follow-up conference of Rio de Janeiro in 1985 in Berlin, Gallus Cadonau presents the Minister of Environment of that time, Dr. Angela Merkel, with a poster about the performance of solarcars in order to interest her in solar mobility in the traffic sector.

for 1000 kWp roofs.[12] The world's first parliamentarian from a green party and city councillor of Lausanne, Daniel Brélaz, advocated for more solar-powered electric cars. And Richard J. King from the US Department of Energy announced the first "GM Sunrayce USA" for 1990.

3.1.3 First Alpine Crossing with Solarcars: The Sun Conquers the Gotthard Pass in 1989

In everyday life it was soon clear that batteries and solar electric power were not the main problem. The problem was the massive inefficiency in the traffic sector. If vehicles of 1.5 t are used to transport a person of 75 kg, the total mass to transport is 20 times higher. In aviation, astronautics and computer technology, size and weight of the appliance, motors or chips were continuously optimized. The traditional car industry, however, contented itself with ever stronger motors. With its first alpine crossing in 1989, the Tour de Sol wanted to show the energy efficiency of light vehicles. On day two of the tour, i.e. on June 27, 1989, the Gotthard pass with an altitude of 2000 m had to be crossed. Solar electric power and batteries proved to be strong enough, and most of the solarcars managed the distance of 374 km on the steep road with 3890 m of ascent and 3830 m of descent without problem. After the hard pass ride, James Warden, the former MIT student who had to collect the pieces of his solarcar from the roadside ditch in 1986, drove through the town of Rothrist with 85 km/h. That was 35 km too fast and he was fined with a 10 minutes penalty.

---

[12] S. Roy Wilson, Board Member Air Quality Management, Los Angeles: California's Mandate for Zero-Emission Vehicles (ZEV); Dipl.-ing. Ruppert Schmidt was project manager at the city plants of Saarbrück for the program of 1000 kWp roofs; Dipl.-ing. Karl Wolling, Federal Ministry for Research and Technology in Bonn; Solarcars in everyday life, volume 5, Berne March 23, 1991.

Figure 12. Jo Leinen, Minister for Building and Environment of the Saarland, was interested in solar utilisation and solarcars from the beginning. He was committed for the realisation of the "1000 kWp solar roof program" in the Saarland and also acted as a model for the utilisation of solar energy in private with his "Solarcar Saarland".

3.1.4 Solarcar: A Danger for 150 pigs?

On July 1, 1989, the Swiss tabloid Blick published the following headline: "Short circuit after electricity theft: danger for 150 pigs". Obviously, a tour driver had been given a hard time with the alpine crossing. After having crossed the pass, Mr. A. A. drove to a pig farm in Hämikon/Lucerne in order to refuel his solarcar with electricity. The fuses blew and produced a short circuit. 150 pigs were caged without air-conditioning. As a consequence, the circumstances and the legal situation were clarified and the blameable participant was disqualified immediately. He was ordered to contact the farmer, to offer his apologies and to completely fix the damage.

3.1.5 Alpine Tests at the 1st European Championship of Alpine Solarcars (ASEM) in 1989

On August 30, 1989, another solar experiment started at the Vorder Rhine (Surselva) in Disentis/Mustér, a main town in the Romansh part of Switzerland[13]: At the first European Championship of Alpine Solarcars (ASEM) the solar electric drive, brakes and safety were not only tested once, but each day. After the start in Disentis, the race course followed the river Rhine for 30 km. The participants then

---

[13]Switzerland has 26 cantons. The canton Grisons in the south eastern part has a surface of more than 7100 km². It is the biggest Swiss canton and has three official languages (German, Romansh and Italian). The main town is Chur — the most famous resort is St. Moritz in the Romansh Engadin valley. The river Rhine rises close to Disentis at the Oberalp. It is the longest river in Western Europe and in parts it borders Germany and France. From Germany it passes through the Netherlands and ends after 1320 km at the sea in Rotterdam.

Figure 13. The hotel Ucliva was awarded as Europe's best ecological hotel by the Council of Europe in Strassbourg 1988.

Figure. 14. In the name of the Council of Europe, French minister and mayor of Strassbourg, Catherine Trautmann, presented the hotel Ucliva in 1988 with the award and the prize money of € 5000.

had to leave the Rhine plain and to overcome a difference in altitude of 350 m. In Waltensburg, a sun terrace on 1100 m, the solar cars could "refuel" in the sun in front of the ecological hotel Ucliva, where five years earlier the first edition of the Tour de Sol had been decided. After the racy mountain drive, batteries had to be recharged in order to pass the second mountain test to Flims/Laax, stage finish of the first day (see Fig. 1).

3.1.6 Bea Vetterli's Solarcar in the Mountains: Downhill with More Than 100 km/h

After the lunch break at the ecological hotel Ucliva, all ASEM drivers got on their racecars with recharged batteries. Beatrice Vetterli, a young electrical engineer, was one of the participants to drive her solar racecar on the curvy road from Waltensburg on 1100 m down to the Vorder Rhine on 750 m. Suddenly, her

Figure 15. Jo Leinen and Gallus Cadonau in front of the CO_2-free hotel Ucliva in 1992 — at the back the largest solar installation of the canton Grisons in 1987.

racecar got faster and faster. Recuperation that always had a brake function did not work any more! Bea's solarcar went already faster than 100 km/h. It was a wonder that she managed to avoid an oncoming truck and that she finally reached the main road at the Vorder Rhine. Everybody asked how this could have happened. The analysis then gave the explanation: The batteries had been completely recharged and therefore were not able to absorb any recuperation energy. Furthermore, the braking effect was not sufficient for such mountain roads — although the brakes had been tested in advance. Fortunately, the Tour de Sol and ASEM organisation had been spared from the first heavy accident with a solarcar.

3.1.7 St. Moritz: 1st ASEM Finish and 1st Electric Light in Switzerland

On August 31, 1989, after the night stop in Laax/Falera, the ASEM entourage drove down to Chur (difference in altitude of 400 m) and then up to Lenzerheide/Valbella on 1500 m. The third stage went to Tiefencastel and then accross the Julier pass to St. Moritz on 1800 m. The sophisticated and well-known ski resort in the Engadin valley is one of the most innovative towns of Switzerland: The first electric light was not turned on in Zurich, nor at the Federal Institute of Technology (ETH), nor in Berne or Geneva, but in St. Moritz in 1878.[14] St. Moritz also had the first electric tramway and the first electric train — the Bernina train. Until today, it still runs from St. Moritz across the Bernina pass to Poschiavo.

---

[14] After his visit to the world exposition in Paris in 1878, Hotelier Johannes Badrutt installed the first Jabochoff lamps in his hotel Kulm in St. Moritz. Consequently, the first electric plant of Switzerland came into existence in this Romansh mountain town of 1800 m.

Figure 16. The European Championship of Alpine Solarcars (ASEM) leads from Disentis at the Vorder Rhine to St. Moritz; Felix Schlatter, hotel director in St. Moritz, Bea Gaudenzi, director of the tourist office of Grisons, and Martin Berthoud, vice-director of the tourist office St. Moritz, at the ASEM in front of participating solarcars.

3.1.8 The British and St. Moritz: Inventors of Winter Tourism

St. Moritz is known worldwide for its invention of ski and winter tourism. Hotelier Badrutt convinced his British summer guests to spend wintertime in the alpine St. Moritz. He promised them holidays for free, if they did not like the three winter months. The British winter guests stayed. Together with the local population they invented new sports and winter games, skiing and sledging, ice skating, bob sleighing. Also the legendary and quite risky Cresta run was developed. Until today, it is managed by a British staff only. The courageous steer their sleigh down the ice canal with more than 120 km/h. The British winter guests were delighted and are visiting St. Moritz each year until today.

3.1.9 Clean Energy St. Moritz: The Overall Energy Concept

When the Swiss population rejected the solar initiative in 2000, the town of St. Moritz and the Association Clean Energy St. Moritz[15] decided to realise an overall energy concept for a complete change to renewable energies. The innovative and alpine spirit

---

[15] A committee of 180 persons launched a solar and energy initiative in 1992. Amongst the initiates were innovative companies and tourism representatives of the canton Grisons, e.g. the tourism director from St. Moritz, Dr. Hanspeter Danuser. After the rejection of the solar initiative, G. Cadonau developed an overall energy concept for St. Moritz. In 2001, the Association of Clean Energy St. Moritz was founded. Town, canton and swiss Federation constructed in cooperation with Rätia Energie AG and EW St. Moritz the solar installation at the Corviglia gondola and at the façades of the gondola station as well as on the top station "Piz Nair". Today, every third journey is powered by the sun ($\approx 50\,000$ kWh/a). It is the goal to change completely to renewable energies in St. Moritz.

Figure 17a. Prof. Stephan Behling, senior architect at Foster & Partners, London, was steering the winning solarcar "Spirit of Biel" for Federal Councillor Ruth Dreifuss to the Swiss Solar Prize Award in Berne in 1994.

Figure 17b. "Solar architecture is not about fashion, it is about survival" — Lord Norman Foster's commitment for sustainable architecture, solar and renewable energy, at the first European Conference for Solar Architecture 1993 in Florence.

of St. Moritz, however, is not alive everywhere.[16] In 2002, the renowned British star architect Lord Norman Foster created the "Chesa Futura" in St. Moritz, an inventive 10-family-house with wood shingles and exemplary heat insulation. The author of this article and Lord Normen Foster met in the hotel Laudinella after a cross-country ski training and Foster developed an optimally integrated photovoltaic installation for his Chesa Futura. Several requests — even from Lord Foster personally — to integrate this solar installation on the roof of Chesa Futura were rejected by a very rich co-owner from Zurich. In 2010, the same people agreed to install a 23 kWp-PV-installation for the reduction of CO_2 emissions at sports events in St. Moritz.

---

[16] In 1934, the resort of Davos/Grisons was the first ski resort to inaugurate a T-bar. In the second half of the 20th century it was also well-known for its tuberculosis sanatoriums. Between 1989 and 1995, however, the resort had rejected participation at the emission-free solarcar events.

3.1.10 Last Tour de Sol in 1991

In 1990, the Tour de Sol took place from June 25 to June 30. It focused not as much on high-lights such as the crossing of the Gotthard Pass, but more on endurance und use of solarcars in everyday life. The last Tour de Sol took place from June 30 to July 5, 1991.

3.1.11 Solarcar World Record: 148 km/h at the ASEM 1995

After 1989, 1990 and 1991, the last ASEM took place in 1995.[17] It started with a world record in Ilanz, the first city at the river Rhine. On August 24, 1995, the "Spirit of Biel" established a speed record of 148.16 km/h on the cantonal road along the river Rhine. On the whole, the solarcars managed their tasks in the alpine region superiorly and after 1989 without severe problems or accidents.

3.2 *World Solar Challenge in Australia and the US*

3.2.1 First World Solar Challenge in 1987 — 3005 km across Australia

In fall 1987, Hans Tholstrup, originally from Denmark, but living in Sydney, launched the first transcontinental World Solar Challenge (WSC) with a race distance of 3005 km across Australia. Already in 1982, he had been the first person to cover the distance of 4130 km from Perth in Western Australia across Southern Australia to Sydney in a solarcar within 20 days. The WSC 1987, however, followed the route of the Scottish pioneer John Mc Stuart who had been the first man to traverse the Australian continent from North to South. Until today, the 3000 km long highway from Darwin to Alice Springs and Adelaide in Southern Australia is named after him. The 1st WSC started on November 6, 1987, in Darwin. On day 5 and after 44 hours and 54 minutes, the Sunracer of General Motors (GM) crossed the finish line in Adelaide with an average speed of 66.9 km/h. The GM team consisted of 91 persons from 13 different automobile and technology companies. Amongst them were top engineers and they were using gallium arsenide cells with an efficiency factor of 20%! On second place was "Ford Motor Corp" with a race time of 67 hours and 32 minutes and an average speed of 44.6 km/h. The "Spirit of Biel" of the Swiss School of Engineering Biel under the leadership of the ambitious and hot-tempered Prof. René Jeanneret arrived third after 69 hours and 58 minutes and with an average speed of 42.96 km/h. Considering that the budget of the Swiss School of Engineering was approximately 1.5% of the GM budget this was a great result.

---

[17] All four ASEM took place in the canton Grisons. At the ASEM 1990, the subject "energy independent Val Müstair" thanks to hydro-power and solar use was discussed for the first time. In schools, solar competitions were held, and the students helped to prepare the different subjects about solarcars and the race courses of the ASEM between 1989 and 1995. G. Cadonau and some of his Tour de Sol colleagues were responsible for the outcome.

Figure. 18. The Honda solarcar of 1990 at the compulsory media stop in the Australian outback north of Alice Springs. (Picture: Gallus Cadonau, Nov. 1990.)

3.2.2 Japan's Waterloo at the 1st WSC — Detlef Schmitz Missed the Start

There were three German teams participating at the WSC. However, the renowned and successful Tour de Sol drivers Rolf Disch and Michael Trykowski had to give up since their solar modules bent in the Australian heat of 45°C and the electrical connection did not work. The third driver from Germany, Detlef Schmitz, brought his solarcar, type "HelioDet", in a suitcase. Although he was not ready on time, he took up the solar race. There were several strong Australian teams, as well as a team from Denmark and a solarcar from Pakistan named "Solar Samba". For the Japanese, the 1st WSC became a Waterloo: Behind three teams from the US, two European and six Australian teams, the team of the Hoxan Corporation only made it on rank 12. The Japanese team crossed the finish line in Adelaide after 17 days or 153 hours and 31 minutes with an average speed of 19.5 km/h and classified as last vehicle out of 22 vehicles started. Contrary to the European car industry, the Japanese, however, drew the consequences and learned from the debacle all according to James Joyce: "Mistakes are the portal to discovery".

3.2.3. The Second World Solar Challenge and Its Dangers in the Australian Desert

The 2nd WSC took place from Nov. 11 to 22, 1990, in Australia with start in Darwin and finish in Adelaide. Before the start, all solarcars were tested for their "braking ability and stability". On their way from Darwin to Adelaide, the solarcars

Figure 19. Start of the 2nd World Solar Challenge, on November 11th 1990 in Darwin; Honda solar-car in front left, "Spirit of Biel" at its right.

weighing between 140 and 250 kg encountered 16-axle road trains with three trailers and a total weight of 80–120 t. The danger was that kangaroos or emus crossing the Stuart Highway were hit by such road trains, and that personal cars or solarcars crashed with the carcass and were catapulted into the outback or on the oncoming lane in front of a road train with very long braking distance. All WSC participants were informed about the dangers. They were given an emergency ration of rice and several litres of water since many people die in the Australian outback (desert) each year of thirst or get lost. The 2nd WSC promised to be spectacular.

3.2.4. The "GREATEST RACE on EARTH, Creating a SOLUTION not POLLUTION"

With this slogan on its WSC support and travel van, billionaire John Ward Phillips, main sponsor and US real estate investor, called attention already in advance. The Japanese car industry with Honda at its leader had learned from the WSC debacle in 1987 and reported back for the race in 1990 with eleven teams among them Honda, Kyocera and Hoxan Corp. At the start in Darwin were also eight US teams from the most important universities such as California University, California State University, Western Washington University, University of Oklahoma, Crowder Collage and Michigan University as well as Queen's University (CAN). Europe was represented by only five teams: Grundfos

Figure 20. Minister of Justice and Federal Councillor Dr. Arnold Koller talking to the General Manager of Shell Solar Dr. Gosse Boxhoorn at the Swiss Solar Prize Award in 1998. Shell-Solar was sponsor of the Swiss Solar Prize that year.

Solvogen (DK), Helioo Det, D. Schmitz and M. Trykowski (Germany), Farrand (GB) and the "Spirit of Biel" (CH). At the braking and stability test, the Tour de Sol winner of 1989, Michael Trykowski (BRD) came on rank 1, having won the prologue with 103 km/h the day before. It was followed by the "Spirit of Biel" (100 km/h) and Honda on rank 3 with a speed of 94 km/h in the prologue. With its armada of engineers, mechanics, electronic technicians, weather and other experts, Honda was the odds-on favourite. It was said that some of the Honda engineers had driven and studied the Stuart Highway several times in summer 1989. "Light and energy efficient solarcars have the best chances to win," said the race strategists before the start.

3.2.5 International Solar High-tech Competition Across Australia

At the WSC, all solarcars are facing towards the sun before the race starts at 8 am. During the drive, the solarcars "refuel" in the sun. If there is not enough sun, they need to tap the solar energy stored in their batteries. Usually, the race takes until 5 pm with only two or three media stops — and this at temperatures of 45°C in the shadow. Accordingly, the drivers in their small and low-comfort cabins are quite challenged. After the first racing day, the university teams of Michigan and Western Washington had completed a distance of 480 km. What a surprise that Honda reached only 449 km and came on rank 4. Leader was the "Spirit of Biel" with 527 km. The following day, Honda's chief engineer Takahiro Iwata said: "Not today, but tomorrow we will catch up with the 'Spirit of Biel'." Suspense and emotions were guaranteed. On the third day, the "Spirit of Biel" was still leading. It

Figure 21. The "Spirit of Biel" at the World Solar Challenge 1990 in Australia is surrounded by media representatives.

Figure 22. In 1990, the Honda solarcar fell back and is starting after a media stop at WSC near Alice Springs, Australia. Honda's chief engineer Takahiro Iwata: "Not today, but tomorrow we will catch up with the 'Spirit of Biel.'

won the race on November 16 in Adelaide with a race time of 46 hours and 13 minutes. Its average speed was 65.2 km/h. The Honda team arrived in Adelaide on the following day after 54 hours and 58 minutes. Three hours later, the teams of Michigan University (57.2 h), Hoxan Corporation (57.3 h), Western Washington University (58.5 h) and University of Maryland (60.7 h) followed. Obviously, the Honda team had expected another outcome of the WSC 1990.

3.2.6 What Technology and Strategy was Responsible for the Victory?

About 150 journalists and approximately 50 TV stations expected the winner at the finish in McLarens on the Lake near Adelaide. They all asked one question: How did it come that the odds-on favourite Honda with a budget of approximately seven million dollars was beaten? Did Honda's technology fail? Did they have the

Figure. 23. McLarens on the Lake: In 1990, the Honda solarcar finished the race one day after the "Spirit of Biel". After congratulations to the WSC winner "Spirit of Biel", the Honda team threw chief engineer Takahiro Iwata into Lake McLarens close to Adelaide.

wrong strategy? The monocrystalline solar cells of the "Spirit of Biel" had been processed by Hans Gochermann in Hamburg, Germany. They originated from Prof. Martin Green from the University New South Wales (NSW), who attended the race with eagle-eyes and was as enthusiastic as Gochermann and the Spirit team. Even the Australians, who may probably still remember the air raid of Darwin and Northern Australia in World War II, did not conceal their happiness with Professor Green. A comparison of strategy between the "Spirit of Biel" and Honda revealed that Honda's battery capacity was only 2.3 kWh whereas Biel had used the maximum allowable capacity of 5 kWh. For weight reasons, Honda had renounced to a "Maximum Power Tracker" (MPT). Biel, on the other hand, had a 2 kg power tracker with an efficiency factor of 98.6%. Without driver, the "Spirit of Biel" weighed 182 kg; Honda only 140 kg. And the "Spirit of Biel" driver had the clear command of chief engineer René Jeanneret that the battery capacity may never go below 50% until the last race day.

3.2.7. "Spirit of Biel": 1.8 dl (Solar) Fuel for 100 km — 55 Times More Efficient

Wily Solarfox Jeanneret had calculated that the transport of 1 kg from Darwin to Adelaide cost 70 Wh. Since 1 kg of battery mass, however, produced 127 Wh, it remained a net energy surplus of 57 Wh per kg/battery! The "Spirit of Biel" needed 55 kWh for 3007 km, i.e. 0.018 kWh/km or 18 W/km ≈ approx. 1.8 dl full for 100 km! Apart from comfort, this meant that already in 1990 solarcars

Figure 24. Detlef Schmitz transported his solarcar in a suitcase to Darwin in Australia. There he assembled it and drove across Australia.

Figure 25. Detlef driving full power in his Helio Det suitcase solarcar across the Australian desert.

were 55 times more efficient than average cars. Compared to 1987, in 1990 all solarcars had environmental-friendly silicium solarcells. The only exemption was "Solar Flair" of California Polytechnic University/USA. It used (toxic) gallium-arsenide solar cells that are usually used in space. Nevertheless, the California Polytechnic University only made it on rank 11 with an average speed of 44.2 km/h.

3.2.8. Great Suspense and an Odd Cup in McLarens on the Lake

After the arrival of the Honda team and the US University teams, the air was tight. Honda's chief engineer Takahiro Iwata took his time to congratulate the team of the "Spirit of Biel". Expecting a clear victory of its team, Honda had brought along a meter-high, golden four-pillar cup. What should be done with it? At first, the cup was hidden. Then, the Honda chief congratulated the winning team, and his team members opened the champagne bottles. They celebrated with the other teams and threw their chief Iwata into the waters of lake McLarens (See Fig. 23). After lots of beer and champagne, the Japanese fetched their four-pillar cup and initiated a great solar party.

3.2.9 Detlef Schmitz — the Friendly "Suitcase Man"

The WSC participation terms specify that the solar surface of the solarcars may not exceed 8 m^2 and that they need to be storable in a box of 6 m $\times$ 2 m $\times$ 1.6 m. Detlef Schmitz's "HelioDet" from Munich was stored in a suitcase. At the first WSC in 1987, Detlef had to give up the race with its carbon suitcase solarcar due to the Australian heat after four days and 320 km. At the second WSC in 1990, Detlef again came to the start in Darwin with his suitcase. After 320 km he took a break and said: "This year, I did the same distance on the first day than I drove during the whole race in 1987." On day 9 and after 1980 km, his Helio Det I was hit by a tornado; it broke and Detlef had to give up the race like 17 others out of 36 starting teams that five days after the arrival of the WSC winner still had not completed the 3007 km to Adelaide. In 1993, Detlef's Helio Det II had new mono-crystalline solarcells with an efficiency factor of 14.5% — i.e. 30% more power. When it was criticized that his solarcar of 65 cm was 35 cm to low, he just put an Australian flag of 1 m on top of it and complied with the requirements. After 750 km, his front wheel broke and he had to go to the next repair shop. When he came back, somebody had stolen his new batteries and Detlef had to give up the race again.

3.2.10 Detlef — Veteran and Misadventurer at each WSC 1987, 1990, 1993, 1996, 1999

In 1996, the "Suitcase Man" came with a new solarcar that had hidden batteries beneath the solar panels. Detlef, the perpetual misadventurer, was celebrated euphorically by all WSC participants in Darwin. They all wished that Detlef would make it to Adelaide. The evening before the break and stability test, Detlef and many other drivers, mechanics and engineers were working at their solarcars. Suddenly, there was a big bang. Everybody run out of the reparation hall immediately. Only one stayed pale-faced and looked at his Helio Det III. Detlef's battery voltage was too high and one of the two battery bolts was blown in the air at the 6 m high ceiling. The technical director explained that such accidents often cause severe head injuries and that many people lose their lives each year because of such accidents. Detlef was shocked but uninjured. On November 7, 1996, he started with his new solarcar constructed of steel tubes. But the steel frame was not flexible enough and Detlef had to repair his no. 21 at several stops. After eight days and 1650 km Detlef had to give up again near Alice Springs. In 1999, 2001, 2003 and 2009 he started again. In 2003 and after 16 years of trying, he arrived in Adelaide after 2 997.8 km. Despite his bad luck, Detlef is not a "loser", but a "successful solar pioneer" knowing what solar technology and solar energy is able to give to mankind. Detlef: "Each race has its winner. But in the end, all participants win." Mercedes, the winner of the Tour de Sol in 1985, however, cannot say so: After its first Tour de Sol victory in 1985, Mercedes never participated again, nor at the Tour de Sol nor at the WSC and therefore never took care of an emission-free, environmental-friendly car technology.

3.2.11 World Solar Challenge 1993 — Japan Invests Millions in Solarcars

As suspenseful as the WSC 1990 was the race in 1993. Fifty-two of the registered 55 solarcar teams came to the start in Darwin on November 7, 1993. After its defeat in 1987 and 1990, the Japanese car industry was eager for revenge. With 20 out of 52 solar teams, Japan clearly proved to be a leader in the solar traffic sector. The US and Australia had 10 teams at the start each. Europe was represented by seven solar teams, but there was no European car brand present. According to the stability test, the Japanese had 8 of the 11 fastest solarcars. The winner of the WSC 1990, the "Spirit of Biel", got starting position 1 with 129.9 km, followed by Honda on position 2 with 125 km/h. The University of Michigan got starting place 5, all other leading positions in 1993 were for Japan. Honda did not want to show any weakness this year. Their budget for the Honda car was estimated to be between 20 and 30 million dollars. All other teams had smaller budgets, the one of the School of Engineering Biel being about three million Swiss Francs. ($\approx$ $US)

3.2.12 Honda Changes Its Strategy for the WSC 1993

Honda had completely changed its technology strategy from 1990. Instead of light batteries, Honda had installed silver-zinc batteries of 63 kg; the team of Biel having remained with 31 kg of batteries. After the start, the "Spirit of Biel" raced away. Honda seemed to have difficulties to follow. But not for long; they caught up and made 730 km on the first day. Having troubles with a wheel suspension, the team of Biel had to camp after 715 km already. This was the main reason for Honda's triumphant drive to Adelaide in the morning of November 11, 1993. The Japanese team needed only 35 hours and 28 minutes for the whole distance and had an average speed of 85 km/h. It won the World Solar Challenge 1993 with an advance on the "Spirit of Biel" of three hours (38, 30). On rank 3 was Kyocera, on rank 4, Waseda University (JAP), rank 5 Aurora (AUS), and rank 6 was Toyota Motor Corp. Several Japanese and US Universities as well as Nissan on rank 12 followed. Five out of the first ten solarcars were Japanese, two American, two Australian and one European team with the "Spirit of Biel". For chief engineer Takahiro Iwata one of the main reasons for the eagerly awaited and well-deserved victory was the change in strategy with considerably higher battery reserves. The second and probably even more important winning factor were the new Sunpower solarcells of Richard Swanson. According to Iwata, the solarcells' efficiency of 21–22% was the main reason for victory. Presumably, the Sunpower solar panel of Swanson/Honda was between 5 and 10% more efficient than the one of the "Spirit of Biel". Therefore, the "Spirit of Biel"s problem with the wheel suspension was not the decisive factor, and Honda would probably have won the race in any case because of its better solar performance.

3.2.13 WSC and Sunrayce in the US and Other Solar Races in 1996

At the WSC 1996, the winners were the same as three years earlier: Japan on rank 1 followed by the team from Biel and respectively Schooler. In 1996, even more classic Japanese car companies were found on the first ranks. After 1987, the US car companies barely participated; even so they supported several solarcar teams especially from universities. The first Sunrayce, organised by Richard J. King of DOE, took place in 1990 and was followed by races in 1993, 1995, 1997, 1999 until 2008 going from Texas to Canada. Also in Europe several smaller solar races took place in Berlin, Kassel, Hamburg etc.

4 Solar Prize, Handcuffs and PlusEnergy Buildings

4.1 *Swiss Solar Prize and Handcuffs*

4.1.1 Solar Utilisation: From Traffic to Building Sector

The first UN Conference on Environment in Rio de Janeiro in 1992 cast its shadow already in advance. In 1987, the Norwegian Prime Minister and physician, Gro Harlem Brundtland, from the Social Democratic Party was elected as chairman of the UN Conference on Environment and Development.[18] With the prevailing subjects of "forest decline" and "acid rain" Brundtland succeeded in globalising the term of sustainability and to sensitise the countries of the Conference of Rio 1992. With their participation at the Tour de Sol, World Solar Challenge in Australia and Sunrayce in the US, the leading Japanese car companies such as Toyota, Honda, Nissan etc. demonstrated that they had understood the environmental problems. After 1990, their emission-free solar cars were using 50 times less energy than conventional cars and it had therefore become high time to look at another sector with the biggest waste of energy: the building sector. The following facts shortly summarise the solar history. Whereas many citizens were fully committed to solar energy, there were others who perfidiously tried to obstruct further development of solar energy and its use.

4.1.2 "Solar 91 — for an Energy-Independent Switzerland" [19]

An atmosphere of departure was prevailing also in Switzerland. On May 22, 1990, the Swiss Greina Foundation, the Foundation Tour de Sol and the SSES launched

---

[18] Earth Summit — United Nations Conference on Environment and Development (UNCED) in Rio de Janeiro in 1992.

[19] Solar Agency Switzerland: The "working group Solar 91 — for an energy-independent Switzerland" founded in 1990 changed its name to Solar Agency Switzerland (SAS) in 1999.

the "Association Solar 91 — for an energy-independent Switzerland" (today called Solar Agency Switzerland, SAS). For the first time, a Solar Prize in the building sector was launched with a wide information campaign.[20] Juridical demands for changes in legislation, initiatives and motions for the promotion of solar energy in building legislation as well as articles of association were elaborated and implemented in several towns and cities. The most important goals of Solar 91 (i.e. SAS) were summarised in four languages in a solar manual of 150 pages with an edition of 18 000 books. Solar 91 came under the authority of the Federal Department of Energy with Federal Councillor and Minister of energy Adolf Ogi in Berne, and it was largely supported throughout Switzerland. The cantonal directors of energy and energy offices, the association of Swiss municipalities, the Swiss federation of trade unions and the Swiss trade and crafts association were also in the party.

4.1.3 First Solar Prize 1991 for World's Biggest Solar Surface per Inhabitant

It was the goal of Solar 91 to have a solar installation for the production of electricity or heat of 1 kW to 1 MW in each town. At least 700 solar installations until 1991 — the 700th anniversary of Switzerland — were aimed to reduce Switzerland's 80% energy dependency on foreign countries, especially on the Middle East. The world's first Solar Prize was awarded on October 4, 1991, by Federal Councillor and Minister of Energy, Adolf Ogi, in Brienz (Canton of Grisons). Brienz was awarded the Solar Prize because the alpine town recorded

Figure 26: The first Swiss Solar Prize winners were awarded by Federal Councillor and Minister of Energy Adolf Ogi on October 4, 1991, in the mountain town of Brienz in Grisons.

---

[20] Solar 91: The goals of Solar 91 in a solar manual, 4th edition with 18 000 copies in four languages.

a solar collector surface of 1108 m² per inhabitant already in 1991. It therefore substituted 45 kg heat oil per inhabitant. With 789 installations, the project goal of building at least 700 solar installations was outnumbered by 10%.

4.1.4 Federal Councillor Adolf Ogi: Initiative, Courage and Solar Installations

Federal Councillor Adolf Ogi inaugurated the first Swiss Solar Prize Award on October 4, 1991, in the public school of Brienz with the following words: *"Dear citizens of Brienz. You all have seized the opportunity and I congratulate you in the name of the Federal Council. I also congratulate all other winners in the categories of industry, producers and employees. For you all, a reasonable use of energy starts at your own and not at others. For you all, the use of renewable energies is an interesting challenge and not a pill that others should swallow. You all prefer to act instead of just react."*[21] The Federal Councillor and other representatives of associations and institutions awarded 16 Swiss Solar Prizes in six categories: Not only communities, but also companies, owners of solar installations, personalities/institutions and best integrated solar installations were awarded. It was the aim to produce between 1000 and 1200 MW by solar installations until the year 2000. It was planned to realise such an output, comparable to a big atomic power plant, by the utilisation of roofs, protective walls and façades.

4.1.5 Handcuffs, Excavators and Solar Electricity

The Swiss Minister of Energy also awarded Gottfried Girsberger with a solar prize.[22] In 1985, he had organised the start of the first Tour de Sol and arranged for a cheerful mood in the starting area with the Tour de Sol song. The "mood" also rose at the energy authority in Girsberger's hometown Altikon/Zurich. The smart businessman had extended his PV installation on the roof of his two-family house without permission to 64 m² and the solar-thermal installation to 30 m². He furthermore insulated his residential and business house with 40 cm of mineral wool and made it practically autarchic by 100%. However, he got into a problematic situation with the energy authorities that cut off his electricity on January 25, 1984. Essentially, he became the first energy-independent home owner in Switzerland. The cantonal authorities did not believe that his house could function without connection to the grid with solar energy alone. They suspected Girsberger had tapped the wire illegally, something that according to Art. 146 of the Swiss penal code had to be punished with jail or imprisment of up to five years. Police and prosecution therefore appeared at the Girsberger family home and surrounded it.

---

[21] Federal Councillor and Minister of Energy Adolf Ogi held the opening speech at the first Swiss Solar Prize Award on October 4, 1991, in Brienz/Grisons. Swiss Solar Prize 1991, p. 26.

[22] Swiss Solar Prize 1991, p. 33/34.

Girsberger himself was arrested, put him into handcuffs and chained to a police-man. He had to follow the police looking for electric mains throughout his house while an excavator was digging a ditch of 8 m length in his garden. However, the authorities did not find any grid connection, and in the end they had to release the energy autarchic innovator Girsberger.

4.1.6 René Bärtschi: "Most Successful Swiss Governing Councillor"

The canton of Berne, wrongly labelled as a "slow" canton, already disposed of a "governmental report on solar energy" in 1990 saying that "the sun could cover between 5 and 8% of the current energy potential". With a PV performance of 1.012 MW from 170 PV installations and 12 593 m^2 of solar collectors, the canton of Berne was definitely in the lead. And the Bernese governing councillor, René Bärtschi, was also ahead of time from the legal point of view. He relieved solar installations on the roofs from requirement of authorisation and committed him-self to the promotion of solar installations. The laudatory speech for René Bärtschi was given by Dr. Wolfgang Palz.[23] Bärtschi was awarded the Swiss Solar Prize for being the "most successful cantonal councillor". Unfortunately, he died far too early in summer 1992 after a short, but severe disease.

4.1.7 Four Times Too Much Solar Energy and a Winter Bathe

One of the first winners of the Solar Prize in 1991 was Josef Jenni, an electrical engineer, Tour de Sol initiator and builder of his 100% solar house in Oberburg/Berne. On a sunny day in the cold month of February 1989, he invited the media to take a bathe in his garden pool: His 5.6 kW PV installation was pro-ducing enough solar energy for his grid-independent batteries and this contrary to all expert opinions, computer calculations and warnings. With 84 m^2 of solar collectors, the Jenni family even had four times more solar energy in the 100 000 litre hot water tank than needed. Professors and experts from the traditional energy sector pulled a long face; Jenni, the journalists and the public smiled.

4.1.8 European Commission, US Department of Energy and Japanese Industry

According to the weekly newspaper "Weltwoche", the campaign of the Solar Agency has "a unique and large political support" and "even the Japanese gas industry and the US Department of Energy" are interested in.[24] In his introduction

---

[23] Prof. Dr. Wolfgang Palz, director of the Directorate-General 12, renewable energies, European Commission (EC) in Brussels.
[24] Christian Speich about Solar 91 and the declaration of São Paulo, "Weltwoche", October 10, 1991.

Figure 27. Solar installations that are carefully integrated into roofs and façades are efficient suppliers of energy for the buildings and without destroying any cultural land or building any additional infrastructure such as streets or power lines. On the picture is Federal Councillor and Swiss Foreign Minister Micheline Calmy-Rey in front of the solar installation of the ecological hotel Ucliva in Grisons with a solar surface of 100 m^2 (21 June, 2003).

to the 1st Swiss Solar Prize on October 4, 1991, the representative of the Directorate-General 12 of the European Community declared: *"Solar energy and all other renewable energies are of highest strategic interest for the further development of mankind and its well-being as well as for the preservation of our environment. The Swiss Solar Prize 1991 was unique, not only in its goal but also in its realisation. The idea of the Swiss Solar Prize is unprecedented and points the way for a social implementation on a large scale. To interest all communities of a country for an award that concretely presents projects on the spot is the right way to put political awareness into practice."* It was impressive to see *"how Switzerland succeeded to acquire the interest of all important leading forces of the country: federal and cantonal government, the association of Swiss municipalities, the trade and crafts association, the Swiss federation of trade unions, the professional solar associations as well as the professional energy industry all demonstrated their unity at the award in Brienz. Within Europe, Switzerland is one of the leading nations in the promotion of solar energy. More than somewhere else, the interest of the single citizen was the driving force. I hope that the Swiss model will be an example for other European countries and for the rest of the world."*[25]

---

[25] Introduction by Prof. Dr. Wolfgang Palz, Swiss Solar Prize, 4th of October 1991, p. 2.

4.1.9 Best Integrated Solar Installations — Without Overbuilding Cultural Land

Of great concern was the careful integration of solar installations into the building. Therefore, a prize for *"the best integrated solar installation"* was incorporated in the award regulations. It was agreed with the branch of solar and building technology as well as with representatives of homeland security and monument preservation to use only urbanized surfaces and no cultural land nor any green space for solar purposes. In contrast to hydro-electric, biomass, geothermal and wind production, the basic question of solar energy was: Why not produce energy where it is used — directly on the buildings? And why "collect" a strong, decentral energy of approximately 1000 kW per square meter in a central and distribute it later over hundreds of kilometres decentrally? The effect of this solar prize strategy is that contrary to other countries almost no cultural land is "wasted" for solar utilisation in Switzerland until today.

4.1.10 The Solar Mission of the Federal Minister of Energy

Minister of Energy Adolf Ogi, President of the Federal Council and participant at the Engadin ski marathon for many years, appealed to the public in Brienz to follow the trail it blazed: *"I can feel the spirit of the ENERGY 2000 program; where targets are set — without ifs and buts. With the ENERGY 2000 program we take our mandate seriously. And you take it seriously with your Solar 91 program. However, Solar 91 must be followed by Solar 92 and Solar 93 and so on; because only continuity leads to the expected results. A good politician is a long-distance runner and not a sprinter."*[26] The Solar Agency has taken seriously its assignment of Federal Councillor and Minister of energy Adolf Ogi. The 20th Swiss Solar Prize will take place on September 3, 2010 in Zurich.

4.2 *Solar Energy on the Rise*

4.2.1 European PV Conference in 1992 and Popular Initiative for Solar Energy

After the Tour de Sol and Solar 91, little Switzerland with its four different language regions had approximately 8500 active SSES members and therefore the biggest solar association worldwide — also in absolute numbers. The 11th European PV Solar Energy Conference was held in Montreux in 1992 and included also the Swiss Solar Prize Award. In the same year, a popular initiative for the promotion of solar energy (solar initiative) was launched by the Swiss Solar Agency, SSES, members of the Federal Parliment and other friends of the solar energy. It aimed at stipulating the terms for the promotion of solar energy in

---

[26] Adolf Ogi: Opening speech of Federal Councillor and the Minister of Energy on October 4, 1991, in Brienz/Grisons. Swiss Solar Prize 1991, p. 27.

the federal constitution with an energy levy of 0.5 cts. per kWh on non-renewable energies. Fifty percent of this levy was intended for the promotion of solar energy and the other 50% for energy efficiency and other renewable energies. It was the first popular initiative for solar energy in Europe, and probably worldwide. On March 21, 1995, the initiative was submitted with 114 824 authenticated signatures.[27] Federal parliament and government of Switzerland needed to comment on the issue of solar energy.

4.2.2 European Parliament: Swiss Solar Prize — Model for European Solar Prize

At the European PV Conference in Montreux in 1992 Dr. Wolfgang Palz declared: *"In the European Community we want to take the Swiss Solar Prize as an example. Within the scope of the development program for solar energy, for which the commission in Brussels is responsible, we have decided to spiritually and financially support regional initiatives for the organisation of Solar Prizes according to the Swiss model — be it in Saarland, Swabia, the Provence or in Andalusia. By doing so, we hope that the idea of the Solar Prize will soon find its expression in a European citizens' movement for solar energy that leads into the new century and multiplies itself there."*[28]

4.2.3 Bonn-Cologne-Brussels-Amsterdam: More Solar Electricity Than in Australia

For the promotion of the European PV Solar Energy Conference in Amsterdam in April 1994, the high-end solarcars "Honda Dream Machine" and "Spirit of Biel" as well as two electric cars started in Cologne and drove through snow flurries to Brussels and from there to the opening of the PV Conference in Amsterdam.[29] On the morning of April 4, 1994, the director of the "Spirit of Biel", Dr. Fredy Sidler, noticed a solarcell efficiency of 20.4% and this at minus 2°C close to Amsterdam (compared to 18.5% in Australia at 40°C in the shadow). On April 11, 1994, the Greek president of the European Minister of energys, Elisabeth Papazoi, drove a "solar lap of honour" at the wheel of the "Spirit of Biel" in front of the conference building of the PV conference in Amsterdam. The conclusions from the first European Solar Challenge were three: *First*, the energy consumption for 500 km was 4 kWh for solarcars (equivalent to approximately 0.4 litre of petrol), 45 kWh for electric cars and 500 kWh for private cars, i.e. 100 times more than for solarcars. *Second*, the fears that in cold April 1994 the solarcars were not able to get enough solar energy on the way did not come true. On the contrary, because of the traffic jams in Cologne, Liège, Brussels and other cities the solarcars' batteries

---

[27] Message of Federal Council of March 17, 1997, BBI *199* II 805, p. 809; BBI *1995* III 1218, 1220.

[28] Prof. Dr. Wolfgang Palz, director at Directorate-General 12, renewable energies, Brussels, Swiss Solar Prize 1992, p. 3. At the end of 1993, the European Community changed its name to European Union (EU).

[29] Ligier Solar electric vehicles: The Ligier electric vehicles of Andreas Manthey and Thomic Ruschmeyer used nickel-cadmium batteries.

Figure 28. President of the European ministers for environment, Elisabeth Papazoi, from Greece climbs into the solarcar "Spirit of Biel" — accompanied by pilot Paul Balmer — and drives a lap at the 13th European PV Solar Energy Conference in Amsterdam in April 1994.

Figure 29. The winners of the 1st European Solar Challenge (ESC) drove with solar-powered electric cars with nickel-cadmium batteriesfrom Bonn, Cologne and Brussels to Amsterdam. From left to right: Andreas Manthey and Thomic Ruschmeyer. The prize is awarded by Prof. De Luce, University of Madrid, and Gallus Cadonau, ESC project coordinator, in April 1994 at the European PV Solar Energy Conference in Amsterdam.

were completely full — even at their arrival in Amsterdam.[30] *Third*, the integrated mono-crystalline solar cells of Honda and Spirit of Biel with a surface of 8 m^2 were much more efficient than the experts had expected. The potential of solar-driven electric cars in the traffic sector is much bigger than generally assumed: What

---

[30] European Solar Challenge 1994 Cologne — Brussels — Amsterdam, see Swiss Solar Prize 1994, p. 75.

Figure 30. From left to right: Takahiro Iwata — happy chief engineer of Honda and winner of the World Solar Challenge in Australia in 1993 as well as co-winner at the 1st European Solar Challenge (ESC) with the "Spirit of Biel" on the way from Bonn to Brussels and Amsterdam — Dr. Wolfgang Palz and G. Cadonau in Amsterdam in 1994.

other car in this world can drive 500 km and arrives at the final destination with a full tank?

4.2.4 Federal Chancellor Vranitzky Awards 1st European Solar Prize in Vienna

In 1992–1993, the Solar Agency elaborated a project for the European Solar Prize on the basis of the Swiss Solar Prize with the same prize categories etc. In spring 1993, the European Parliament gave the green light for regional initiatives *(EC official journal C 119 of April 29, 1993)* and published the award information in cooperation with Systèmes Solaires in Paris, Eurosolar in Bonn, Denmark, Austria and more. For the first European Solar Prize, Denmark reported 50 solar installations, Germany 87, France 52, Italy 48, Austria 49 and Switzerland 190. The first European Solar Prize Jury chaired by Dr. Wolfgang Palz and representatives of Systèmes Solaires, Paris, Eurosolar, Bonn, Denmark, Greece and Austria decided about the first European Solar Prizes on the basis of the slightly adjusted Swiss prize criteria. The Austrian Federal Chancellor Dr. Franz Vranitzky inaugurated the first European Solar Prize Award on October 3, 1994, in Vienna.

4.2.5 Chancellor Vranitzky: "Central Europe Free of Nuclear Power Plants"

Chancellor Dr. Franz Vranitzky inaugurated the European Solar Prize with the following words: *"It is our goal to play a pacemaker role in establishing a Central Europe free of nuclear power plants (NPP)."* He pointed to Austria as a "European

Figure 31. Federal Chancellor auf Austria, Franz Vranitzky, awards the city plants of Saarbrücken with the solar trophy of the first European Solar Prize for its "1000 kWp roof program" in Germany; the grid connection had been financed by the city plants of Saarbrücken. From left to right: chairman Leonhard of the city plants Saarbrücken, Hajo Hoffmann, the Austrian Federal Chancellor Franz Vranitzky and Gallus Cadonau, coordinator of the first European Solar Prize on October 3, 1994, in Vienna.

Champion" in the utilisation of solar-thermal energy topping *"one million square meters of solar collectors"* in the current month. Subsequently, he then awarded the first 14 European Solar Prizes (ESP).[31]

4.2.6 First European Solar Prize goes to Successful Opponent of EDF

Besides the three Danish award recipients, two Austrians and two Swiss, there was the French International Centre of Renewable Energies (CIEN), Great Pyrenees, to attract attention: After four years of fighting against the electricity giant EDF, the CIEN got the permission for the first grid connection of France. Saarbrücken received the ESP for its "solar promotion for 1000 kW on city roofs", for "four public solar filling stations" as well as for its "wind park Saar". Jo Leinen, Saarbrücken's Minister for Building and Environment, who was very active in the field of solar energy, also decisively contributed to the acceptance of the Alpine initiative in 1994. A prize also went to the "association for solar promotion in Aachen" for the first "cost-covering compensation" in Germany of

---

[31] European Solar Prize: The first European Solar Prize was awarded in 1994 in the city hall of Vienna, senate room; see Swiss Solar Prize 1994, p. 3 ff.

the "actual costs of approximately 2.00 DM/kWh."[32] Four weeks later, on October 31, Federal Councillor and Minister of internal affairs Ruth Dreifuss, Prof. Stefan Behling, Norman Foster and partners, London, Yves Bruno Civel, Systèmes Solaires, Paris, and the association of Swiss municipalities awarded the Swiss Solar Prize 1994 in Berne.

4.2.7 City/Charter: Implementation of the Goals of Rio on Municipality Level

Parallel to the Swiss Solar Prize 1994, a German-French codified *"City/Municipality Charter"* of 50 pages was published. It is used on local and municipality level for the gradual implementation of the environmental goals decided in Rio in 1992 for a sustainable development. The charter is based on a point system with 40 energy articles for renewable energies and 186 juridical regulations. The more renewable energies are used in a municipality, the more points it gets. In juridical terms, the 186 regulations of the city/municipality charter were the legal basis for low-energy and zero-energy buildings, because the maximum points could be achieved almost only with solar and other renewable energies. This charter was supported by the federal office for energy and by the social-democratic Minister of internal affairs, Ruth Dreifuss. She was asking for more renewable energies incl. solar energy and less emissions with the CO_2 law. Unfortunately, she encountered a lot of resistance with her requests. The charter and its point system were later adopted by the Association of Energy Cities and partly implemented.

4.2.8 Breakthrough in Parliament in 1997: One CHF Billion for Solar Energy

In negotiations with the parliamentarians, the co-presidents of the Solar Agency, National Councillor Marc F. Suter and National Councillor Dr. Eugen David, were able to convince the great chamber of the federal parliament (National Council) on June 4, 1997, to promote solar energy with a levy of 0.6 cts/kWh on all non-renewable energies (oil, gas, coal and uranium). It was decided with 88 affirmative and 82 negative votes to use 50% of the levy for the promotion of solar energy and 50% for all other renewable energies. This would have given approximately one billion CHF for solar and other renewable energies.[33] However, the approval of the Council of States (senate) was still missing.

---

[32] European Solar Prize 1994: Austria: 2; Denmark: 3; Germany: 5; France: 2 and Switzerland: 2. The financing in the Saarland was provided by the city plants of Saarbrücken; see Swiss Solar Prize 1994, p. 66–76.

[33] Official bulletin of the National Council of June 4, 1997: Motion of art. 14bis, energy law, of National Councillor Marc F. Suter (FDP/BE) and National Councillor Dr. Eugen David (CVP/SG).

Figure 32. National Councillor and mayor of Zurich, Elmar Ledergerber; Gallus Cadonau, director manager of Solar Agency Switzerland; National Councillor Marc F. Suter co-president Solar Agency Switzerland; Dr. Fritz Schiesser, Council of States of the canton of Glarus; Dr. Eugen David, Council of States and co-president Solar Agency Switzerland; Andres Holte, CEO of Eternit in Niederurnen; Helen Issler chief editor Swiss TV; general manager of Shell Solar, Dr. Gosse Boxhorn; Dr. Willy Durisch and Kees Vischer, financial representatives of Shell Solar. They are all discussing the construction of the first solar factory in Bilten/GL on November 1, 1997, in the hotel Ucliva.

4.2.9 Ucliva Agreement: First European Shell Solar Factory in Switzerland

On November 1, 1997, the co-presidents of Solar Agency Switzerland, National Councillor Marc F. Suter from Berne and National Councillor Dr. Eugen David from St. Gall, met with Councillor of States Dr. Fritz Schiesser from Glarus, Councillor of States Bruno Frick, National Councillor Dr. Elmar Ledergerber from Zurich, solar scientist Dr. Willy Durisch and CEO of Eternit AG, Andres Holte. Also present were Dr. Gosse Boxhoorn, general manager of Shell Solar, Cees Vischer, CFO of Shell Solar, as well as Gallus Cadonau, who had arranged for this meeting. The different representatives analysed the grounds of a former meat factory and different important logistic and strategic aspects on the site. Afterwards, the participants of the meeting all travelled to the Hotel Ucliva in Waltensburg/Grisons for a discussion of the project and the different proposals. All participants came to an agreement with the general manager of Shell Solar: If the Swiss Council of States follows the recommendation of the National Council in fall 1997, Shell will build the first European solar factory in Bilten, Glarus, in 1998–1999.

4.3 *Mephisto & Co Against Solar Energy*

4.3.1 The Wisdom of Arthur Schopenhauer and Solar Energy

According to Arthur Schopenhauer, social progress passes through three stages: first the ideas were ridiculed, then they are violently opposed and finally they are accepted as self-evident.[34] At the end of the 21st century, after the successful utilisation of PV in space and its successes in the traffic sector, the time of ridicule was over. Now, it was time for an active opposition against solar energy. However, since solar energy was the most popular energy source, strong counter-arguments were needed and the opponents of solar energy had to perfidiously deceive the public without making the misleading obvious. An example for how this can be done on the highest public level of jurisdiction is the treatment of the solar bill in Swiss federal parliament.

4.3.2 J. W. Goethe and "a Very Good Dinner" — Instead of Solar Energy

In fall 1997, the Council of States had to decide about the solar promotion proposed by the National Council. The responsible commission and the president of the small chamber, Gian R. Plattner, did not attack solar energy frontally. They raised doubts and warned about a "structural carburisation in the sector of solar energy" (note that the solar part was less than 0.1% of the total energy need!). And the solar opponents — leading representatives of the Swiss trade and commerce associations — were quite frank: "Two days ago, I had a very good dinner at the "Bellevue" with Andreas Leuenberger[35] and other parliamentarians on behalf of the promotion of trade and industry"[36] said the president of the commission. He did not mention the outcome of the business dinner, but then quoted the supposed "father of the ecologic tax reform", Hans Christoph Binswanger, a professor of economics and apparently unsuspicious "ecologic witness". Apparently, the professor had told the president of the commission the following: *"You better make it right, because it is so easy to take the wrong way and regretting it afterwards!"* If one follows Binswanger's expert opinion, the federal legislator could only take the wrong way in the solar sector. This form of deception almost comes to the level of world literature: "As for this science, it is so difficult to avoid the wrong way" says disguised Mephisto to the student who was thinking about studying theology in J W Goethe's Faust. Professor and economist Binswanger, who likes to see himself as "father of the ecologic tax reform", is a strong opponent of wind turbines. At a meeting in Zurich in 1985, where the author was present as well, he declared: "I prefer a nuclear power station to a wind turbine."

---

[34] Arthur Schopenhauer, German philosopher, author and professor in Danzig and Frankfurt am Main (1788–1860).

[35] Representatives of economics: At that time Andreas Leuenberger was president of the Swiss association of economics, today called Economiesuisse. He and other "prominent business representatives," such as councillor of States Vreni Spörri (FDP) or former governing councillor Eric Honegger (FDP), were also in the board of directors of Swissair until its grounding in October 2001 leaving debts of 17 bio. CHF. Müller was responsible for the propaganda aganist the solar initiatives.

[36] Official bulletin Council of States, October 9, 1997, president of the commission Prof. Dr. G. R. Plattner about the decision of the National Council on Art. 14bis of the energy law.

4.3.3 Combat Against Renewable Energies

The operative economists of Economiesuisse, the president of the commission and most of the right-wing politicians saw to it that the Council of States rejected the solar-friendly proposal of the National Council of June 1997. A minority of the right-wing parties, the small associations of building technology as well as CVP (Christian Democratic party), SP (Socialist Party) and Grüne (green party) supported the energy bills for renewable energies. The right-wing parties FDP (liberal party) and SVP (popular party) as well as the operative economists combated the promotion of renewable energies and the solar initiative. With the statement that in the energy sector Switzerland was top of the world, they caused polemics and deception; and on the other hand they combated the energy bills by demanding "no additional taxes."

4.3.4 Millions for Deception of Citizens

In order to mislead the public in the matter of renewable energies, the economical functionaries and their assistants in parliament said that Switzerland "is top of the world in the sector of renewable energies. *As to the solar performance of each household, our country is in the leading group of Central Europe.*"[37] It is true that after the Tour de Sol and Solar 91, Switzerland came on top position in 1992 with a PV performance of 4.7 MW or 0.7 W/per capita. In 1992, this corresponded to a performance that was almost eight times higher than in Germany with 82 million inhabitants and 5.6 MW i.e. 0.09 W/per capita.[38] However, Switzerland's dramatic energy dependency of more than 80% from abroad, the billions of energy imports and the energy losses of 70–90% in the building sector etc. were suppressed. Additionally, the oil prices increased in summer and fall of 2000. With their "no additional tax" campaign costing approximately 10 million CHF and misleading propaganda, the opponents achieved their goal and only 47% of the Swiss electors voted for renewable energies on September 24, 2000, while 53% voted against it.

4.3.5 Economic War against Innovative Businesses

The right-wing parties and economical functionaries thought to combat the left-wing and green parties because they support solar and renewable energies almost worldwide. It is not proved, however, that the sun is shining "left-winged". It seems that it shines for everybody and this for free! An analysis of the

---

[37] Official bulletin of the National Council, November 28, 2000, p. 1274: Gerold Bührer was national councillor and is today president of Economiesuisse. He combated the solar measures.

[38] Pius Hüsser, IEA PVPS Trends Report 2004, see also Swiss Solar Prize 2005, p. 13.

facts shows: There is an economic war going on in the Swiss energy sector against sustainable and innovative small and medium-sized businesses — and against middle class. Ten years later, the rejection of the economical incentives in 2000 of 0.8 billion CHF for the promotion of renewable energies and a resulting added value of approximately six billion CHF had the following effect: In 1995, Switzerland paid *2.7 billion CHF* for energy imports. In 2008, the amount paid to Russia and the Arabian countries including the Bin Laden family was already *13.4 billion CHF*. At the same time, Switzerland has an unemployment rate of 170 000 — the highest in years. 13.4 billion CHF for fossil energy imports correspond to approximately 135 000 jobs. In Switzerland, thousands of jobs have successfully been obstructed by the right-wing parties FDP and SVP as well as the economical functionaries or "apparatchics" for years. These jobs are now created in the new high-tech countries such as Germany and Spain — and especially China (see paragraph 4.7).

4.3.6 Swiss Economical Functionaries: Best Work for the Chinese Communist Party

The irony in it is that with their attitude and conducts the right-wing parties FDP and SVP support China and its Communist Party with its violations of human rights in economic success! In 2008–2009, China invested billions of dollars in high-tech, wind and solar industry as well as in traffic infrastructure. In Switzerland, on the other hand, the right-wing parties and especially the liberal party FDP with Federal Councillor Merz successfully provided for billions of bonus payments and investments in racketeering of banks such as the UBS. If one reviews the voting behaviour of the National Parliament, it can be seen that fortunately there are always some exceptions in the right-wing parties as well, especially farmers and representatives of mountain regions. According to economists such as Dr. Werner Vontobel, trade and industry are losing extensive amounts of revenues because the money is invested in speculative businesses instead — one of the main reasons for the financial crises in 2008. As a consequence, the innovative small and medium-sized businesses of Switzerland have to pay higher interests and are left with nothing. Many other countries, however, promote innovation in the solar sector, and contrary to Switzerland, the governments and economists of these countries invest in a sustainable use of solar energy. The following text mentions just some of the most innovative and most sustainable Swiss companies that cannot or can only partially produce and realise their products in Switzerland due to politically obstructed business conditions. Not to mention the inventions and patents such as the crystalline solar cells of the University of Neuchatel leading to billions of revenues and many jobs in Asia. It is tragic that the innovative small and medium-sized businesses cannot carry their point in parliament against the well-trained economical functionaries. This becomes obvious by comparing the economic conditions in Switzerland with those in Germany. In Germany, about 300 000 sustainable jobs were created in the innovative sectors of small and medium-sized businesses. In Switzerland, innovative businesses need to export up to 90% of their products if they want to subsist. To end this fiasco, all small and medium-sized businesses need to cooperate and make sure that the influx of funds to economical functionaries and right-winged parties is stopped. By the way, such conflicts are not new: also the functionaries of the Soviet Union used ideology for justification. It was more important than law, constitution and jurisdiction of the Supreme Court. Rumours that the operatives of Economiesuisse have done a doctorate at the central committee of the Communist Party of the Soviet Union, however, have not been confirmed or denied so far.[39]

4.3.7 Do Authorities Harass Citizens that are Loyal to the Constitution?

The active combat of solar utilisation was supported by a negative attitude of the authorities in building procedure: On August 22, 1995, the council of the municipality of Hettlingen/Zurich gave the building permission for a four-family house replacing a farming house in the town centre. In a second building application, the owners (family Müller-Lüond) wanted to complement the wood heating with a solar installation. The municipality rejected the application and recommended a heat pump. Family Müller appealed and went to Supreme Court, paying approximately CHF 40 000 for the legal procedure. And lost! After that, the family called on the Solar Agency. Different experts were consulted. Solar Agency then proposed a complete integration of the solar installation into the roof according to the directives of the canton of Berne. This new building application included a solar surface that was twice as much as in the former application and therefore the circumstances were new. The municipality, however, regarded the new proposal as a provocation and rejected to treat it. In the following, family Müller appealed because of unconstitutional denial of justice — and won the case. In the meantime, the media were interested in the case as well, and the municipality now had to treat it.[40] It was also pointed out that the president of the municipality, who had rejected the solar installation and proposed a heat pump, was doing private business by selling heat pumps. The municipality surrendered and the solar installation is producing hot water for the four apartments until today.

4.3.8 Solar Energy Instead of Unconstitutional Bureaucracy

On the 21st European PV Solar Energy Conference in Dresden in September 2006, I heard that carpenter Paul Meury in Blauen, canton of Basel-Country (BL), had

---

[39] Dr. Ruedi Minsch: Today, Dr. Ruedi Minsch, chief economist of Economiesuisse is an exception and interested in energy efficiency and renewable energies. However, this does not have any effect on energy positions so far.

[40] Do authorities harass citizens that are loyal to the constitution?, Solar Agency Switzerland, Zurich 2000, 130 pages.

been called to tear down his solar installation that he had built in 1997 with the permission of the municipality council. The town inhabitants had signed an initiative against the demolition of the installation. The small businessman had defended himself several times against the decision of the authorities unsuccessfully because he could not produce any building permit. After about 18 proceedings and interventions, even the lawyers refused the case because Paul Meury had missed several time limits for appeal. If Paul Meury did not take down his solar installation of 4 m² on his own, police would do so in approximately three weeks. In fact, the situation seemed hopeless and it was energetically absurd. The author, however, looked into it and wrote to the government:

(a) He pointed out the principles of the constitutional and administrative law, according to which legal decisions are not revocable.
(b) The government was reminded that before the fusion of the Laufental with the canton of Basel in 1994, the municipality of Blauen had belonged to the canton of Berne. There, governing councillor Bärtschi in 1990 had abolished requirement of authorisation for solar installations. Although the municipality of Blauen came to the canton of Basel-Country in 1994 (and therefore under jurisdiction of Basel-Country), and although the solar installation had been permitted in 1997 only, the old material building law of the canton of Berne was still valid until it had been replaced by the new urban planning law of the new urban planning revision for Blauen in 2000.
(c) Paul Meury therefore could not dispose of any formal building permission. By building his solar installation according to the administrative law of Berne and with the permission of the competent municipality authority, he had acted in full conformity with the law. There can be no question of an insufficient or incorrect building permission.

In the following, the administration saw reason of not having sorted out the legal situation completely and fortunately revised its decision: The solar installation in Blauen is still substituting fuel oil and CO_2 emissions for carpenter Paul Meury until today.

4.3.9 Constitutional Right for Solar Building Permit — New Law within Three Months

The Solar Agency Switzerland has been fighting for unbureaucratic and fast building procedures for carefully integrated solar installations for almost 20 years. With the motion of co-president M. F. Suter and Dr. Eugen David, a new federal law for fast solar building procedures was accepted by the Federal parliament of Switzerland within three months and five days. The new legal clause has been in force since January 1, 2008 with article 18a (solar installations) of the Federal law for spatial planning as follows: *"Solar installations that are carefully integrated into roofs or façades in building or agricultural zones have to be authorised if they do not affect any cultural or natural monuments of cantonal or national importance."*[41] According to this article, the Swiss citizen therefore has a constitutional right for a solar building permit for solar installations that are carefully integrated into roofs or façades. Indeed, such a right seems necessary because the building department of Zurich under the directorate of city councillor Martelli (FDP) is Swiss champion in rejecting solar installations; and this despite the fact that Zurich's citizens accepted the guidelines for a "2000 W society" on November 30, 2008, with a majority of 75%. It is therefore against the law if officials of Mrs. Martelli's building department deliberately try to avoid the application of constitutional law.[42]

4.4 *Market-based Compensation for Renewable Energies*

4.4.1 Market-based Compensation for Billions of Fossil-Nuclear Subsidies

The objections of many economists and managers against renewable energies are either of ideological nature or based on lack of knowledge. The promotion of renewable energies just corresponds to a compensation for the high state subsidies and privileging of non-renewable energy sources (and their external energy costs). De facto and according to the decision of the Federal Court *"measures that intervene in open competition in order to privilege certain industries are not permitted"* (BGE 111 Ia 186).[43] Therefore, a compensation *"for the nuclear state liability of billions of CHF"* is only restoring the equilibrium in countries with nuclear power plants. It corresponds to a minimum compensation for centuries of cross-subsidisation of nuclear power plants and coal-burning plants or for fossil emissions in all social and economic sectors. If there are no political objections against billions of subsidies and state privileges in the fossil-nuclear energy sector, why should there be any objections against the promotion of renewable energies? There is a right for non-discrimination and furthermore, renewable energies correspond to the constitutional principles of sustainability.[44] Several interesting examples show that

---

[41] Art. 18a Federal law for spatial planning has been in force since Jan. 1, 2008; AS 2007, 6095 6107; BBl 2006 6337.

[42] The city official F. Klaus informed on March 5, 2008, at 03:13 pm, in written form how Zurich's building department planned to *avoid the application of federal law* for the *granting of building permits for solar installations* in the city of Zurich: *"To be treated as cultural monuments of cantonal importance are also those of communal importance* (i.e. objects that are protected by the city council or listed in an inventory)."

[43] CASH, March 3, 2000 www.cash.ch/archiv: In the energetic market and in economic debates Article 12 of the nuclear energy liability law (KHG) is suppressed: *"The Federation insures the liable party against nuclear damages up to one billion CHF plus 100 million CHF for interests."* Since the establishment of the federal state in 1848, state subsidies were never as radical and as partial in favour of a single branch as in the sector of nuclear energy.

[44] Swiss Federal Constitution: See Article 8, as well as Article 73 and 74 ff. of the Swiss Federal Constitution (BV).

small and medium-sized businesses are at the front of empirical research and that they achieve the best and most sustainable results for economy and society.

4.4.2 Prof. Dr. René Rhinow: Best useage of Revenues for Measures

According to the jurisdiction of the Supreme Court it is clear that state taxes are used for police, schools, hospitals etc. and are to be paid by all citizens. It was also clear that the solar bills had nothing to do with a state tax, but that the revenues of the levy were to be used for promoting energy efficiency and renewable energies exclusively. René Rhinow, professor for state and constitutional law at the University of Basel, summarised the classic and most efficient regulating measure of the solar bill in one sentence: *"The regulating goal is achieved best by using the revenues for measures that support the achievement of the regulating goal."*[45]

4.4.3 European Court of Justice 2001: Grid Feed-in is not Tax

Different studies of Prof. Dr. H. U. von Weizsäcker, the Swiss Federal Office of Energy and the trade and crafts association of Basel — as well as practical experiences — all showed the same result: From the energetic, economic and ecologic point of view there cannot be found any measures in the whole OECD region that are more efficient than the regulating measure described by Prof. Rhinow. This corresponds to the legal system of the solar initiative having a regulating effect that is eight times more efficient than the CO_2 levy system.[46] In 2001, the European Court of Justice said about grid feed-in: "A regulation that obliges electricity plants to pay a minimum price for electricity from renewable energy sources , *cannot be compared with state subsidies."*[47]

4.4.4 Democratic Decision of the Electricity Consumer on Energy Investments

Thanks to the European Supreme Court it is clear that grid feed-in is not a tax as claimed by the Swiss operative economists, but a market instrument for the

---

[45] Prof. Dr. René Rhinow, liberal press service, Oct. 16, 1997.

[46] Federal Office of Energy Economics (today Federal Office of Energy), statement on the solar-cent initiative, Berne 1996; Prof. Dr. H. U. von Weizsäcker, reduction of unemployment rate by 50% with investments in energy efficiency, 1999; appr. 270 pages. A survey of the craft and trade association of the city of Basel proved that an incentive of 1 CHF in window improvement and building services leads to investments of more than 24 CHF. On the average, 1 CHF initiated investments of 8 CHF in the building sector, because private persons used the energy incentive for energetic investments.

[47] EuGH, in the case C-379/98, PreussenElektra AG v. Schleswag AG, Schleswig-Holstein/Germany, March 13, 2001.

protection of equilibrium in a market-based world. All new investments in the traditional nuclear or fossil energy sector are made with the yearly profits from electricity prices paid by the consumers and not by the electric power companies. But why only invest in fossil-nuclear — i.e. dirty energy sources — and not in clean, renewable energy sources? Renewable energies only need temporary, market-based, compensatory measures[48] and not any nuclear state liability of several hundreds of billion CHF, nor do they produce any radioactive and highly-poisonous waste with half-life periods of 24 000 years. If the fossil-nuclear energy companies are not afraid of democracy, they let the electricity consumers of their area decide about the energy investments and in which energy sources they shall be made.

4.5 *Best Innovative Entrepreneurs for Sustainable Economy*

4.5.1 Small- and Medium-sized Entrepreneurs are the Most Innovative

Walter Schmid, building contractor from Glattbrugg/Zurich, was involved and invested in sustainable technologies and projects with good ecologic and economic sense for years. In 1975, he built several apartments with 60 m^2 of the first vacuum solar collectors. In 1988, he built the first façade-integrated solar

Figure 33. Walter Schmid demonstrates it: Organic waste is transformed into compogas and used as fuel for his vehicles. One kg of kitchen waste lasts for 1 km car drive.

---

[48] Sunset Legislation: G. R. Plattner, president of the commission, spoke of a *"structural cementation in the sector of solar energy."* Official bulletin of the Council of States, October 9, 1997. According to the "Sunset Legislation" the solar promotion would have stopped automatically after 25 years and foresaw only the "promotion of solar utilisation on urbanized surfaces" with exceptions "for especially energy-intensive businesses". Facts such as the nuclear state liability, costs for the disposal of radioactive waste for at least 24 000 years etc. were suppressed.

Figure 34. With 3 kg of banana peel, heavy trucks of 40 ton drive 1 km. In 2003, Walter Schmid was awarded the Swiss and European Solar Prize as Europe's most innovative and most sustainable entrepreneur in the traffic sector.

installation with an output of approximately 6000 kWh/a for solarcars. In 1989, he was the driving force behind the electric utility vehicle Solcar. For its operation, it only needed a roof or façade surface of 15 m², and the battery charge sufficed for 100 km. Solcar was awarded several prizes but owing to lack of demand, however, the project was stopped.

4.5.2 Biogas — Compo-gas: 1 kg of Banana Peel = 1 km of Car Drive

In 1991, Walter Schmid began to gain energy from organic waste with compogas. Instead of composting organic waste with a high need of energy, he developed a compo-gas power plant that worked as an organic disposer as well as an electricity, heat and compost soil producer. One kg of organic waste is sufficient for 1 km of car drive. In Rümlang/Zurich with 100 000 to 120 000 inhabitants approximately 15 000 tons of organic waste are collected each year. With this amount, between 4500 and 5000 m³ of biogas i.e. approximately 30 000 kWh are produced each day corresponding to an energy equivalent between 2000 and 2500 litres of fuel. Therefore, 10% of the fuel need in traffic is covered with compo-gas. Since the launch of compogas, Schmid has been known as the most innovative organic waste disposer worldwide. Schmid's plants are run in different European cities as well as in Japan.[49] The biggest compo-gas plant is actually built in Qatar, and Schmid is world leader in this sector. In 2003, Walter Schmid was awarded the Swiss and European Solar Prize.

---

[49] Kompogas plants are today run under the name of AXPO.

4.5.3 Solar House on the Federation Square: Built in 22 Hours

Max Renggli, head of the Renggli AG in Sursee, is one of the leading suppliers of highly-efficient low-energy houses. In 1998, his company built the first Minergy houses in Switzerland.[50] In 1999, Renggli also built the first Swiss passive house estate. After the European PV Conference of May 2000 in Glasgow and prior to the popular vote on the energy bills in September 2000, Max Renggli met with Gallus Cadonau in order to plan the construction of a solar house on the Federation Square in Berne. Renggli built the house in a record time of 22.5 hours in cooperation with 60 other small and medium-sized businesses. The planned solar house was practically able to cover its total energy need on its own. The pioneer company Renggli produced a video clip on it, but the liberal editor in chief of Swiss TV, F. Leutenegger, did not want to inform about the solar utilisation. He preferred to broadcast a report about health insurance for the nth time. Federal President Adolf Ogi, instead, visited the solar house on August 31, 2000, and was fascinated. The National councillors Regine Aeppli and Rosmarie Zapfel stayed in the solar house over night and confirmed the following morning that they had slept well.

Figure 35. "Solar house of federal parliament" on the Federation Square in Berne, in front of the parliament, built on August 31, 2000, in 22.5 hours by Max Renggli AG and 60 other innovative entrepreneurs. Together with other prominent speakers, Dr. Wolfgang Palz, GD 12 of the EU, was personally interested in this solar house that had been built in record time, and congratulated the business pioneers to their worldwide unique achievement of "a solar house built in one day".

---

[50] Minergy buildings have a maximum heat need of 42 kWh/m^2 (hot water and heating).

Figure 36. Wattwerk's PV installation has a capacity of 30 kWp. It produces 24 500 kWh/a, and only needs 13 800 kWh/a. The excess of solar electricity therefore is 10 500 kWh/a, enough to supply the company's solar-powered electric cars which are used by the employees to supervise the solar projects in the region. Wattwerk is owned by Heinrich Hollinger, a veteran and successful Tour de Sol participant. He was awarded the Swiss and European Solar Prize in 2004.

4.5.4 Swiss Solar Prize for First PlusEnergy Building

The first Swiss Solar Prize awarded those municipalities, companies or families that had the biggest solar collector installation per capita. After the first Solar Prize Award, Prof. Pierre Fornallaz and jury member Jung stated: The event in Brienz in 1991 was wonderful. But it would be even better if family X did not only have 50% of its roof surface covered with solar collectors, bud also had better windows. After 1992, the focus came more and more to energy efficiency in the building sector. The century of low-energy buildings began, and they got increasing attention. In 2000, a service building in Chur was awarded as the first PlusEnergy Building (PEB) in Switzerland with grid connection. In 2001, the first PEB single-family house was awarded the Swiss Solar Prize. It produces approximately 10 000 kWh/a, and only needs 7500 kWh/a. Its energy self-supply therefore amounts to 130%. In 2004, Heini Hollinger's enterprise "Wattwerk" in Basel-Country was awarded the Swiss and European Solar Prize. This PlusEnergy Company has a self-supply of 175% and employs eight people. With its solar electricity excess, the electric cars of the Wattwerk company are driven.

4.5.5 Shell's Solar Factory in Gelsenkirchen: "We Want to Earn Money"

If even the biggest oil companies shift to renewable energies, why still against renewable energies? Shell Solar built its first European solar factory with 25 MW in Gelsenkirchen, Germany instead of Bilten, Switzerland. The author was invited

Figure 37. Shell Solar opens Europe's first 25-MW solar factory in Gelsenkirchen/Germany (instead of Bilten/Switzerland) in November 1999.

to the inauguration of the first Shell Solar Factory by Shell CEO Jeroen van der Veer on November 16, 1999. Also several Ministers and about 150 media representatives were present. After the failed attempt of Shell to dump an old oil platform in the German Sea in 1995 and several consumer boycotts of shell service stations, the media representatives were not sure if they could trust the CEO of Shell. He was aware of that and suddenly asked: "But do you think we want to earn money?" — After deathly silence from the public, van der Veer added: "Shell has been active in energy business for 100 years now. If we want to earn money for another 100 years, which we want, we have to shift to solar energy; because as you all know, oil reserves will not last that long."

4.5.6 Lord Norman Foster on the 15th Swiss Solar Prize 2005

In 2005, Lord Norman Foster gave the following celebratory speech for the 15th anniversary of the Swiss Solar Prize in the Federal Institute of Technology in Lausanne: "The Swiss Solar Prize is truly unique. It is an indication of the unremitting dedication to solar energy and sustainable architectural technologies within Switzerland. Crucially, the prize not only considers the environmental performance of buildings, but also considers the essential problem of how sustainable technologies can be an integral part of good architectural design and practice. I have always argued for the importance of sustainable design. It is essential that we take a holistic design approach that considers all aspects of a project — from the totality to the smallest detail — and the effect that each has on the others.

Figure 38. Energetic refurbishment of old and cultural buildings: The Bundestag from 1896 was refurbished by Lord Norman Foster in 1998 and is now using exclusively renewable energies. CO_2 emissions were reduces by 95% (PV installation of 37.5 kWp and biomass plant of 2.35 MW). If such a historic cultural building can energetically be refurbished by Lord Foster, this is also possible for other buildings. Further examples are the refurbishments by Karl Viriden in Basel where the energy needs were reduced up to 93% (see part 5).

Sustainable design means doing the most with the least means. Following the logic of "less is more," it employs passive architectural means to reduce energy consumption, minimising the use of non-renewable fuel and reducing the amount of pollution. [51]

4.5.7. PlusEnergy Buildings for Alpine Resort: 175% Self-Supply

In spring 2008, the author invited the best architects and engineers, who had won several Swiss Solar Prizes in the past, to elaborate an energy project with a self-supply of 175% for a planned resort of 1 billion CHF in Andermatt. It was shown that with a standard of Minergy-P/Passive Houses the heat energy need can be reduced by 50% compared to the legal requirements. Correspondingly, there is a decrease in the investment curve for suboptimal building technology. These means are better invested in energy efficiency and photovoltaic. By doing so, the total energy need of an average building decreases to approximately 42 000 kWh/a. Self-supply amounts to 72 700 kWh/a, i.e. to 175%. CO_2 emissions drop to almost zero. If the whole resort was built this way, it would not be necessary to

---

[51] Swiss Solar Prize 2005, p. 3.

buy electricity of 5 GWh/a. Instead, approximately 1.5 GWh/a could be sold as solar electricity excess or used for the traffic sector. The slightly higher investment costs are compensated with energy efficiency and the selling of solar electricity.[52]

4.5.8 Energy-Intensive Industrial PlusEnergy Building: 125% Self-Supply

For a large construction of one of the biggest food companies in Switzerland, Solar Agency Research proposed to realise a PlusEnergy Building according to the standard of MinergyP. By realising this project, the former energy need of approximately 1.6 GWh decreases from 900 000 kWh/a to 70 5000 kWh/a (−22%). Yearly energy costs drop from 100 000 CHF to 76 000 CHF (−24%). Also CO_2 emissions sink, depending on the PV installation, to 0.0% g after two years.[53] The increase in energy efficiency amounts to 56% — and energy costs decrease by 56% compared to today.

If the *PlusEnergy alternative* with MinergyP standard, an optimum energy production and with Sunpower solar cells with an efficiency of 19.3% was built, the energy self-supply would *amount to approximately 0.9 GWh/a (125%)*. The part in input energy would decrease from 1.6 GWh/a today to *0.0 kWh/a*. Furthermore, the PlusEnergy Building would also be able to sell *192 000 kWh/a to the grid* or could use the same amount for the operation of approximately 130 solar cars.

4.5.9 Installed PV Performance: World Leader in 1992 — Last in 2008

World leaders in 1992 were not the operative economists, but the small and medium-sized entrepreneurs as well as dedicated citizens who built solar cars and integrated solar installations. Ten years after the realisation of the renewable energy law by its red-green government, Germany disposes of 12 300 MW with an

---

[52] Study on PlusEnergy Building: see AADC PlusEnergy study. February 2009

[53] CO_2-emission factor: Some electric power plants export between 80 and 99.3% of the hydropower. In total, Switzerland produces approximately 36 TWh/a of hydropower, but exports 51.4 TWh/a as "hydropower peak energy" and imports at the same time 50.2 TWh/a of electricity from the EU. Therefore and according to the UCTE (protocol of Kyoto), 535 g CO_2/kWh are listed as input energy. As to solar energy, 0.0 g CO_2/kWh are listed since all PV and thermal installations have generated their production energy after 3 to 30 months and thereafter produce CO_2-free energy for centuries (P.H.). CO_2 emissions for 1kg of fuel oil ≈ 10 kWh ≈ 3 kg/CO_2; 10 kWh of natural gas ≈ 2 kg; 10 kWh of nuclear electricity ≈ 1 kg, i.e. according to Federal Constitution 73/74 in addition to 100 g CO_2/kWh for the nuclear processing, the costs for the nuclear waste disposal incl. the costs for "final storage", future earthquakes, security, water inleakage etc. for at least 960 generations have to be integrated as well (see study of the University of Sydney, Australia 2006; German eco-institute and Jan Willem Storm van Leeuwen, 2005): URAN 235 half-live: 24 000 years ≈ 25 y ≈ 960 generations; radioactive deposit, Asse 2008–2009 etc.

installed PV performance of 119 W/per capita. This is almost 17 times more than Switzerland with 34 MW or only 4.4 W/per capita. In Bavaria, the installed PV performance with 307 W/per capita is even *34 times higher* than in Switzerland.[54] The innovative Swiss companies in the field of renewable energies have to export up to 90% of their products. The Swiss top position, often underlined prior to the popular vote, went up in smoke. Actually, the part in domestic and renewable energies on Swiss total energy supply sank from 36.7% in 1950 to 19.7% in 2007.[55] Almost in the same time, Germany multiplied its part in renewable energies by a factor of 5 from 20 TWh/a to almost 100 TWh/a and created 300 000 new jobs.

4.5.10 Sustainable Economy: Amateur Becomes World Champion

At the 20th European PV Solar Energy Conference 2004 in Barcelona the visitors amazingly looked at four or five small PV exhibitors with booths that would have found place in a large kitchen. "China is producing PV installations?" asked the conference participants. After 1998, Germany had made itself a name in the solar sector and had become market leader. The German states are getting increasingly independent from fossil-nuclear energy-imports. Germany is paying billions of € in its own states instead of paying the amount to foreign countries for fossil-nuclear energy imports. Billions of € therefore go into the domestic economy and generate local added value. This leads to millions of orders for thousands of small and medium-sized businesses and creates jobs and training positions. Today, exporting champion Germany is world's leader in the field of solar and wind energy with highly qualified jobs. In the meantime, the German energy law has been adopted by approximately 70 countries, including the People's Republic of China. During the financial crises of 2008–2009, it became obvious that China did not support the banking sector as massively as the West. Instead, China was investing in traffic and energy infrastructure. The consequences can be seen already: Since 2008, the high investments in solar and wind energy have led to price reductions in this field. It is fatal though, that the economic functionaries and representatives in parliament only refer to the negative circumstances such as coal-burning plants and pollution in China or the utilisation of nuclear energy in France. They cheat parliament and public with wrong figures.[56]

---

[54] Photon, Germany, 6/2010, p. 215; Swiss electricity statistic, 2008, p. 52.
[55] Swiss electricity statistic 1983: (1950: 70 200 TJ [hydropower and wood] with a total domestic energy need of 186,400 TJ; In 2007, the domestic energy production was 253 040 TJ with a gross energy need of 1 176 230 TJ Swiss energy statistic 2008, p. 11 and 16.
[56] Wrong figures: see part 4.3.4.

4.5.11 China could Outrun all — Economically and Ecologically

Reality is different: In the coming years, EDF — very nuclear-oriented so far — will build PV installations with a capacity of about 1000 MW with the American PV company First Solar Inc.[57] (since 1980, Switzerland has about 40 MW). In 1992 China has built about 0.5 million m^2 of thermal solar installations. With this, China's part in world's solar-thermal capacity is 50%, even though China's population of 1.4 billion people represents only 21% of the world's population. In 2008 China has built about 31 million m^2 of thermal solar installations; the EU $\approx$ 4.5 million m^2. Per capita it's about 3 times more than Switzerland with about 0.05 million m^2. By 2010, China would have installed 43 million m^2. The production capacity is about 135 million m^2 $\approx$ 81% of the world capacity. More than 3000 companies build and install thermal solar installations in China — and create jobs.[58] In 2008, China had a *PV production* capacity of 4000 MW and 450 PV companies. Three of them are amongst the ten biggest companies worldwide; the increase in PV production is 300% per year.[59] China is on its way to replace Germany as world's high-tech leader in the field of modern solar and wind energy. In only six years, the Chinese PV companies advanced on the first ranks worldwide and replaced Germany as exporting champion.

5. PEB Cover 75% of World's Energy Demand[60]

5.1 *From Solar Collectors to PlusEnergy Buildings*

5.1.1 Conclusion of Tour de Sol, WSC as well as Swiss and European Solar Prize

Races such as the Tour de Sol, World Solar Challenge *and* Sunrayce as well as other solarcar races have demonstrated that people can be moved in individual cars with only 2 to 10% of the usual energy need. Today, an average car with fossil drive has an efficiency factor of only 8 to 12%. European and Japanese electric cars prove that the efficiency factor can be increased by almost 80% without any loss

---

[57] Photon, 12/2009, p. 27.

[58] Systèmes Solaires, Paris, No. 195, Jan-Feb 2010, p. 22 incl. corrigendum.

[59] Systèmes Solaires, Paris, No. 195, Jan-Feb 2010, p. 25.

[60] Scientific American: A Plan for a Sustainable Future, How to get all energy from wind, water and solar power, Prof. Mark Z. Jacobson/Mark A. Deluchi, University of Stanford, November 2009, p. 58–61; Spectrum of science, 12/09 p. 80–87; According to BP Worldenergy Review 2008, the world's primary energy consumption 2008 amounts to a total of 11 295 million t of oil units — excluding wind, geothermal and solar energy production; see BP World Energy Review 2008, "Primary Energy Consumption 2008"; London 2009, www.BP-Review.uk. This corresponds to a world energy demand of approximately 133 000 TWh/a. The capacity is 12.5 TW, see Scientific American, Jacobson/Deluchi, University of Stanford and Spectrum of science, l.c. with regard to PlusEnergy Buildings (PEB), see part 5.1.3. item 3 ff.

Figure 39 Chesa Futura in St. Moritz was built by Lord Norman Foster in 2002. Thanks to its *envelope with 50 cm of insulation*, the big *energy losses are reduced* by 70 to 95%.

of comfort. For 100 km, the needed energy shrinks to 5 to 15 kWh. For an annual average of 15 000 km, the energy need therefore lies between 1000 and 1500 kWh/a. And even this energy need *can and must* be produced *ecologically*. No additional coal-fired, gas or atomic power plants are needed. An energy efficient building with a photovoltaic installation of 1 or 2 kWp is able to produce the annual energy need of the individual traffic sector without any problems.[61]

5.1.2 Energy Efficiency: "Sine qua non" of PlusEnergy Buildings

From the ecologic point of view, a building needs to cover its own energy demand first and can only then produce additional "traffic energy". The energy need for heating and warm water as well as electricity of all apartment and business buildings built in Switzerland and Middle Europe until 1990 amounts to approximately 20 to 25 litres of fuel oil per square meter and year.[62] That is the energy index of such buildings is between 200 and 250 kWh/m²a. It was the goal of the Solar Prize awards to increase energy efficiency in the building sector as much as in individual traffic.

---

[61] Energy need in individual traffic sector: The energy that is needed in individual traffic can be generated also with hybrid vehicles by using biogas or compo-gas motors or generators. In Middle Europe, approximately 950 kWh/a are produced with 1 kWp. In wintertime, the biogas motor and generator works as a combined heat and power station with a higher efficiency factor. For heavy trucks and for airplanes, hybrid or biomass solutions are more advantageous because the battery capacity is still limited.

[62] Conversion factor: The conversion factor for 1 kg of fuel oil is 11.63 kWh. In practice, for 1 kg of fuel oil, 10 kWh/a are assumed. 20 l of fuel oil therefore correspond to 200 kWh/m²a. This unit (energy need per square meter and year) is called energy index.

With Minergy-P/Passive buildings, some leading companies today demonstrate that energy efficiency can usually be increased by a factor of 10.[63] With today's level of technology, emissions and energy losses can be reduced by approximately 90% — or according to measurements in the building sector even more. The following examples are based on measurements on Solar Prize objects and other buildings during and after their first use.[64]

5.1.3 PlusEnergy Buildings (PEB) with a Self-Supply between 100% and 200%

PlusEnergy Buildings (PEB) are producing at least 1 kWh/m² annually, which is much more than their total energy need for heating, warm water and electricity.[65] In 2005, a SME with a self-supply of 170% was awarded the Swiss and European Solar Prize.[66] In 2006, the Solar Prize Jury was able to award a farming complex with an ideally integrated PV installation and a self-supply of more than 400%. On the average, the farming complex was producing 125 000 kWh/a and only needed 30 000 kWh/a.[67] In 2007, a service building with an exemplary integrated PV installation with amorphous cells had a self-supply of 100%. It was therefore awarded the Swiss Solar Prize. With today's best mono-crystalline solar cells, this

---

[63] Swiss and European Solar Prize from 1991 to 2009: See winners of the Swiss and European Solar Prize from 1991 to 2009. The efficiency strategy of Minergy-P/passive buildings also corresponds to the constitutional mandate of proportionality of art. 5, paragraph 2 of the Swiss constitution: Investing in heat insulation is much cheaper than investing in the production of energy for an equal energy and emission reduction. See also Minergy-P refurbishment in Basel: Reduction of energy losses of *170 000 kWh/a by heat insulation and self-supply of 37 200 kWh/a with solar heat (18 200 kWh/a), solar electricity (9000 kWh/a) and heat pump/environmental heat (10 000 kWh/a).* With the same investment, the heat insulation is four times more efficient and cheaper than the production of environmental energy (not forgetting the external energy costs especially of non-renewable energies). In the TEC21 magazine 5-6/2010, Prof. H. Leibundgut describes Minergy-P/passive buildings with advanced heat insulation as "nonsense" and "propaganda". However, his opinion does not comply with the above mentioned fact or with the principle of proportionality according to art. 5, paragraph 2 of the Swiss constitution. Leibundgut's energetically suboptimal buildings have to import solar energy from Spain and wind energy from Jura (constructing for the 2000 W society, city of Zurich/novatlantis/Hochparterre, Zurich, November 2009, p. 26/27)

[64] See FN 63 above, Swiss and European Solar Prize from 1991 to 2009.

[65] Requirements for PlusEnergy Buildings: On the annual average, the energy production index of a PlusEnergy Building has to be at least 1 kWh/m²a higher than the index of energy need. Legal and collective basis are the regulations of the Swiss Solar Prize with the mentioned and published definition.

[66] The Wattwerk in Bubendorf (Basel Country) employs between 8 and 10 persons and has a self-supply of 175%; see Swiss Solar Prize 2005, p. 32/33.

[67] Aeberhard farm in Barbarêche (Fribourg): 125 000 kWh/a correspond to the average measurements between 2006 and 2009; the forecast was 104 000 kWh/a. The actual production of solar electricity is therefore 20% higher than forecasted; see Swiss Solar Prize 2006, p. 32/33.

Figure 40.   Marché International, Kemptthal/Zurich, Swiss Solar Prize 2007, p. 32/33.

building would even have a self-supply of *at least* 200–270%.[68] In 2008, the Solar Prize Jury awarded a multi-family house with a self-supply of 160%.[69]

5.1.4 PV and Refurbishment of a 6-Family House: Energy Needs Reduced by 90%

With the refurbishment of this 6-family house and a complete PV-utilisation of the roof, the energy need of this building was reduced by 90%.[70] Exemplary is also the refurbishment of a 12-family house in the preservation area of the old town of Basel. The building dating from 1896 was refurbished according to the standard for Minergy-P/Passive buildings. Consequently, the energy need of the building was reduced by 93%. In Bennau, canton of Schwyz, a 7-family house with a self-supply of 110% rejoices its inhabitants. It was awarded the Swiss and European Solar Prize in 2009.[71] All these examples prove that according to the building

---

[68] Marché-International: Administrative/service building Marché-International, Kemptthal (Zurich) with a self-supply of 100% and a production of 40 000 kWh/a — thanks to amorphous thin-film cells (efficiency factor: 7%); with mono-crystalline Sunpower solar cells (offered in 2009) with an efficiency of 19.1%, a self-supply of 270% would be possible (see also Swiss Solar Prize 2007, p. 32/33).

[69] Two-family PlusEnergy Building in Riehen (Basel City) with a self-supply of 240% (production 25 000 kWh/a and Energy need 10,400 kWh/a) and a net electricity excess of 131% thanks to mono-crystalline solar cells, see Swiss Solar Prize 2008, p. 26/27 (measured in 2008/2009, W. Setz).

[70] Refurbishment of a six-family-house in Staufen/Aargau with a reduction of foreign energy by 88%: before the refurbishment, the house dating from 1967 was consuming totally 122 000 kWh/a. After the refurbishment and after integration of a PV installation of 15 kWp it only needs 15 000 kWh/a of foreign energy or 12.2%. Solar electricity is also used to operate the heat pump; see Swiss Solar Prize 2008, p. 24/25 (measured in 2008/2009, Erni).

[71] Refurbishment in the preservation area of the old town of Basel: Before the refurbishment, a 12-family house in Basel dating from 1896 was consuming 223 000 kWh/a. After the refurbishment and after integration of a thermal and PV installation on the roof, it only needs 15 800 kWh/a of foreign energy which is 7.1% of the former energy need; see PlusEnergy apartment building in Bennau (Schwyz) with a self-supply of 110%, Swiss Solar Prize 2009, p. 30/31 and 36/37.

Figure 41. With investments of about 100 000 CHF per apartment unit, the six-family house in Staufen, Aargau, from 1967 reduced its total energy need from 122 000 kWh/a to 15 000 kWh/a or by 88%. Increase in efficiency amounts to 62 000 kWh/a, whereas the 15 000 solar-powered heat pump and the solar installation produce approximately 45 000 kWh/a. The increase in efficiency could have been even higher if the refurbishment was done according to the standard of MinergyP/passive buildings instead of just Minergy; see Swiss Solar Prize 2008, p. 24/25.

technology of 2010, the apartment and business buildings in Middle Europe are able to cover their total energy need with integrated PV installations.

5.1.5 PV und Refurbishment of a 12-Family House: Energy Needs Reduced by 93%

Figure 42. With the MinergyP refurbishment of a 12-family house from 1896 in the preservation area of Basel, the former energy use was reduced from 223 000 kWh/a to 15 800 kWh/a or by 93%. Today's energy use of the refurbished 12-family house is therefore 60% lower than the requirements of the 2000-W society of 42 000 kWh/a. CO_2 emissions decreased from 72 450 kg to approximately 5100 kg or by 93%.

5.1.6 PV on PlusEnergy Buildings — the Level of Building Technology of 2010

Apartment and business buildings that were built intelligently and according to the level of building technology of 2009 in Middle Europe have a self-supply of at least 120% to 150%. After a refurbishment in 2009, the average building has a self-supply of 85% to 130%. These figures all correspond to measurements of Swiss and European average buildings (excl. energy-intensive businesses which only amount to approximately 1% of the 150 million buildings in Europe). Not considered for the solar energy production so far are the best mono-crystalline solar cells such as those of Sunpower with an energy factor of more than 19%.[72]

Figure 43. This PlusEnergy Building for 7 families in Bennau/Schwyz uses 62 000 kWh/a, and produces 70 000 kWh/a. This corresponds to a self-supply of 110%.

Figure 44. PEB refurbishment in Grüsch with a use of 15 000 kWh/a, and a production of 31 000 kWh/a. Self-supply: 207%. This PEB sells about 4500 kWh/a (130%) to the grid.

---

[72] Swiss and European Solar Prize winners from 1991 to 2009: The figures are based on measurements upon Swiss and European Solar Prize winners from 1991 to 2009. The best PEB were built after 2000 and especially after 2005. A total of 2800 buildings and installations were considered so far. It has to be taken into account that the insulation of buildings in Northern Europe is by far better than the insulation of buildings in Southern Europe. On the other hand, the solar radiation in the South is much stronger. With other renewable energies, such as wind energy, a renewable energy supply can also be guaranteed in the North. The wind potential is enormous and it will be used increasingly in the coming years everywhere.

5.1.7 Energy-intensive Business Buildings as PlusEnergy Buildings

Meanwhile, even energy-intensive businesses show sensational results: Shopping centres that seemed to be hopeless cases with energy indexes of 750 kWh/m²a can be refurbished into PlusEnergy Buildings with a self-supply of 120 to 150% by consequent heat insulation, with high-tech electric appliances, LED lights and mono-crystalline solar cells.[73] The developers of a building project in the Swiss Alps with investments of approximately 1 bio. CHF have the choice: With PEB and a self-supply of 175% they can sell approximately 1.5 GWh/a of solar electricity to the grid; or they realise the buildings conventionally as planned and will have to buy 5 GWh/a of foreign energy.[74] Experienced experts agree that from 2010 principally all European average buildings could be PEB with a self-supply of at least 120 to 150%.

5.2 PV-PEB Cover 75% of World's Energy Consumption

1st conclusion: PEB cover 75% of a family's total energy need Reduction of energy losses by 90%: If from 2010 on all new buildings were built according to today's level of Central Europe's building technology, they would have an energy self-supply of 100 to 200% — which is proved by different energy measurements. Refurbishments of apartment and business buildings lead to an efficiency increase of more than 90%. Additionally, the self-supply of refurbished buildings can be between 80% and 150% thanks to PV.

2nd conclusion: 100 to 200% PV energy gains: The massive, 90% reduction of energy losses in refurbished buildings and the energy gain in new and refurbished buildings of 100% to 200% allow the following conclusion: Thanks to PV, Central Europe's building park can averagely supply itself by at least 100%. *New PlusEnergy Buildings usually produce an electricity excess that can be used to operate solar-powered electric cars for the building inhabitants and generally also for the business employees. Excepted are energy-intensive businesses and heavy trucks if they are not operated with biogas or other renewable energies.*

3rd conclusion: Solar energy covers our energy need on the average only According to the OECD, apartment and commercial buildings make up for 46 to 50% of our total energy need.[75] Practically all national energy statistics furthermore

---

[73] Energy-intensive business buildings: For building plans with mono-crystalline Sunpower solar cells; efficiency of 19.1% for approximately 5700 CHF/kWp; Solar Agency Research, 2009.

[74] Solar Agency Research, AADC energy study, Zurich/Wil, March 2009; it depends on the ordering party whether PEB are partially realised or not.

[75] According to the OECD, the total energy demand of 46 to 50% includes heating, warm water as well as household and business electricity; see University of Cambridge/UK and Prof. L. Glicksman, MIT/USA, Swiss Solar Prize 2005, p. 10.

say that the building and traffic sector together make up for approximately 75% of the total energy need of an average family. The mentioned examples are not just theoretical cases, but existing Minergy-P/passive buildings and PEB that already cover the total energy need including traffic energy with today's level of technology. In Paris and Lyon as well as in other Central European countries the part in car owners has been decreasing since 10 years. More than 50% and in Basel even 65% of the urban households do not own a car. Empirical results show that future PEB could cover 75% of the total energy need of an average family. With PV-PEB, however, this can only be guaranteed for the annual average — without storage, but only if the sun is shining.

4th conclusion: PEB need regulating energy to guarantee 100% security of supply: The main argument against solar energy and especially PV is the unreliability of solar electricity. At night and when the sun is not shining, a solar energy supply is not possible. However, this is complemented by regulating energy that guarantees for electricity when the sun is not shining. The most reliable complementary energy and environmental-friendly regulating energy for PEB is water power. If PEB are built comprehensively, they depend on ecologic pump storage power plants.[76]

5th conclusion: Security of supply thanks to *ecological* pump storage power plants: In 2006, the Swiss Greina Foundation (SGF) asked the regional electricity company Rätia Energie AG to verify an environmental-friendly pump storage project at the Bernina. On one hand, because 16 000 km of the Alpine rivers are

---

[76] The Swiss Greina Foundation (SGS) for the protection of Alpine rivers was founded in 1986. Since 2004, it has been developing a project of ecologic pump storage power plants (EPSPP) in cooperation with the Swiss Federal Institute of Aquatic Science and Technology (EAWAG) of the Swiss Federal Institute of Technology in Zurich; In 2005, 93% of the members were supporting such a water protection and energy strategy if the following conditions are met:

(a) In the drainage zone measures for the reduction of hydro-peaking, detritus management and flood protection are needed.

(b) Besides the constitutionally "adequate" and ecologically sufficient minimum acceptable flows, dynamisation and seasonal graduation as well as flood waters will have to be considered as well.

(c) The produced energy shall principally be used as regulating energy for wind and solar energy or other renewable energies in order to make a significantly ecological contribution to the energy supply. Pump energy shall have a continuously increasing part of renewable energies, esp. wind and solar energy. The cheap wind energy excess must be used primarily to substitute coal-burning and nuclear power plants.

(d) Existing plants shall be used and optimised without harming additional rivers or protected landscapes. Therefore, the water shall be used in a possibly closed circuit in order to transform renewable energies into regulating and peak energy.

(e) A consistent implementation of these measures massively reduces the problem of hydro-peaking and guarantees adequate waters in the mountain storage plants in order to have "adequate residual waters" in all rivers as requested by the Swiss sovereign already on Dec. 7, 1975.

affected negatively,[77] and on the other hand because future buildings in Europe will only require 10% of today's total energy need; an amount that can be covered easily by an optimum solar PV-utilisation on roofs and façades. In order to guarantee a 24-hr security of supply, however, regulating energy is necessary. An ecologic security of supply (without gas-fired, nuclear or coal-fired power plants) with enough peak and regulating energy for Europe can only be guaranteed by storage power plants or pump storage power plants such as those planned and built at the Bernina Pass, Nant de Drance, Linth Limmern, Grimsel or in the Austrian Alps.

6th conclusion: Excessive wind energy is transformed into profitable regulating energy: The massive development of wind energy in Europe leads to big energy fluctuations in the European grid. In order to avoid a grid surcharge and excesses in wind energy, the consumer is paid 50 € for each MWh of energy consumption (negative price). In 2009, such electricity payments were made 18 times.[78] Due to the daily fluctuations of the stochastic wind energy, only between 15 and 30% of the wind energy production can be considered as "trustworthy." This is disadvantageous ecologically and also economically.

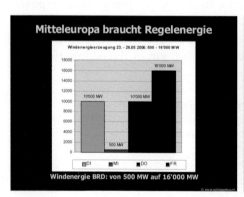 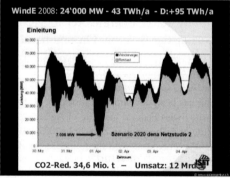

Figure 45. Central Europe needs regulating energy: The left chart shows the huge fluctuations in wind energy production during a week in May 2006 in Germany. The excessive wind energy is ideally used as pump energy in pump storage power plants. Figure 45 on the right shows that production fluctuations will massively increase in Europe and will probably amount to 60 000 MW by 2020. This ecologic, excessive and cheap wind energy has to be used.

---

[77] Dispatch of the Swiss Federal Council of June 27, 2007, on the popular initiative "living water", p. 5515. According to the Federal Council 15 800 km of our rivers are "highly affected" or drained — despite the request of 77.5% of the Swiss sovereign for "securing adequate residual waters" on Dec. 7, 1975.

[78] Frankfurter Allgemeine (newspaper): "When electricity prices get negative," December 10, 2009, p. 17.

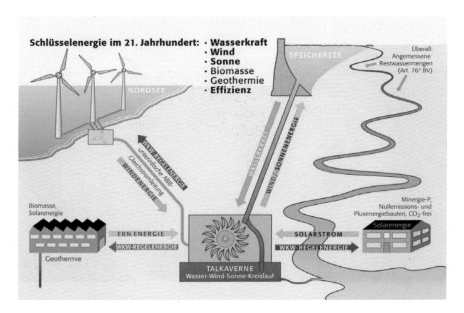

Figure 46. Upon initiative of the Swiss Greina Foundation, the first ecologic pump storage power plant has been planned and will be realised from 2013 on at the Bernina. Excessive and therefore cheap wind energy and in the future also solar energy will be used to pump the water from the bottom (valley storage) to the mountain. With such projects, existing storages can be operated in a better way and produce more regulating energy. The problems of hydro-peaking are reduced and „conforming residual water quantities" are guaranteed. On the average, PlusEnergyBuildings produce more energy than they need over the year — however, when the sun is not shining, they depend on regulating energy.

7th conclusion: More than 100% security of supply with wind, water and the sun. If the economic and ecologic disadvantages of wind energy are combined with PV and water power, the disadvantages are transformed into advantages: The excessive and therefore cheap wind and solar energy is ideal for pumping: with the cheaper wind energy — instead of nuclear or coal electricity — the water is pumped from the bottom (valley storage) to the mountain storage. On demand — when the sun is not shining — enough regulating energy for Minergy-P/passive buildings and PEB can be produced. Additionally, solar electricity exceeds of PEB are also fed to ecological pump storage power plants (see Figs. 45 and 46).

8th conclusion: Lake Binaco as ecological pump storage power plant. Figure 45 shows the feasibility chart of the newly planned Lake Binaco plant at the Bernina as an ecological pump storage power plant. Rätia Energie AG agreed to invest in wind energy for 150 to 300 MW: The pump storage power plant can only be labelled as ecological if the pumps are operated with wind or solar energy.

Figure 47. Proposal of the Swiss Greina Foundation of 2007. The figure shows the planned 1000 MW water power plant (actually 43 MW) in Poschiavo, Grisons, with a new power line of 17 km length between Lake Poschiavo (left/950 m) and Lake Bianco (right/2250 m). With this project, hydro-peaking (S-SV) is decreased from 1:24 (respectively 1:40 as planned in 1995) to 1:2. The concrete dam at the Bernina is only augmented by 4.3 m instead of 17 m, and conforming residual water quantities are guaranteed. The production of regulating energy increases from 120 GWh/a to 2500 GWh/a. Thanks to 20 times higher regulating energy reserves, a 100% security of supply for PEB can be guaranteed.

9th conclusion: PEB and ecologic PSPP secure Europe's energy future. New pump storage power plants (PSPP) have capacities between 600 MW and 1.2 GW (compared to older PSPP with capacities between 40 and 100 MW). If these new PSPP use the enormous wind potential in Europe for pumping, they become ecologic pump storage power plants (EPSPP). Despite 25% pump storage losses they are still able to guarantee the energy supply for the whole industrial and service sector with renewable energies.[79] Together, solar PEB, wind power plants and EPSPP are able to guarantee a 24-hr supply security for all European buildings and for individual traffic. Today, German wind energy production shows a daily

---

[79] In the North, there is an enormous potential of wind energy, which is increasingly used, especially after the decision at the beginning of 2010 of nine Northern states to establish an adequate North European electricity grid. A high-voltage power line with approximately 500 kV between the Netherlands and Norway is already in operation. Nevertheless, Germany's wind energy excess is increasing each year, which leads to negative prices (18 times in 2009 — with an electricity price of 1.50 € per kWh! (see newspaper "Frankfurter Allgemeine", December 10, 2009). The potential of renewable energies is therefore more than sufficient; see also Jacobson, University Stanford: Mark Z. Jacobson and co-authors assume an energy supply with renewable energies of 100% by 2030 (Spectrum of science 12/2009, p. 80 ff.). Considering that the sun is sending 10 000 times more energy to the world that it is totally needed, this is not astonishing. An energy transport with high-voltage power lines is only acceptable underground or along existing infrastructure such as highways, railways etc.

fluctuation between 10 and 15 GW. With the continuously increasing stochastic wind and solar energy production, Europe will have daily fluctuations between 30 and 50 GW until 2030. Until then, further EPSPP with a capacity of 20 to 50 GW will be needed in the French Alps, Austria and Switzerland as well as maybe in other regions such as the Pyrenees in order to level the production fluctuations from the energy production regions.

10th conclusion: PEB produce excess energy for cultural monuments, the traffic sector. It is not possible for all buildings or cultural monuments to use solar energy optimally. But there are, on the other hand, very powerful PEB producing 200%, 300% or even more than 400% of their annual energy need as shown by several examples on the country-side. Furthermore, there are also big halls and buildings in the cities that could be used accordingly. Since 2005, the farmer family Aeberhard is operating their solar farm with a self-supply of 400% or 125 000 kWh).[80] This is enough to supply four traditional apartments with a consumption of 27 000 kWh/a. After refurbishment, such apartments will only consume 1000 kWh/a, and the farmer family Aeberhard would be able to fully supply 125 refurbished housing units.[81] This is therefore another example to show the importance

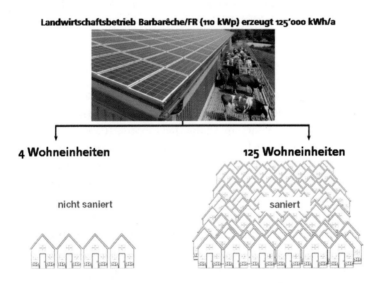

Figure 48. With a solar installation of 110 kWh, the agricultural business of farmer family Aeberhard in Barbarêche/FR produces on the average 125 000 kWh/a. This is enough to fully supply *four non-refurbished* housing units. According to the example of the the two-family house of family Spillmann in Zurich, the agricultural business would be able to *fully supply 125 refurbished housing* units (125 × 1000 = 125 000 kWh/a).

---

[80] See Swiss Solar Prize 2006, p. 32.

[81] Farmers and energy producers Elisabeth and Beat Aeberhard in Barbarêche/FR, Swiss Solar Prize 2006, p. 32/33 and family Spillmann in Zürich, Swiss Solar Prize 2009, p. 40/41.

Figure 49. SIG/SES-Société d'énergie solaire SA, 1228 Plan-Les-Ouates, Zurich/Basel, was awarded the Swiss and European Solar Prize in 2008. In 2008/09, the 571 kWp PV installation with Sunpower solarcells produced on the average 585 000 kWh/a. With a surface of 3395 m², this results in 172 kWh/m²a, which is enough to drive 15 000 km each with 400 solar-powered electric cars each year (see Swiss Solar Prize 2008, front page and p. 34/35).

of energy efficiency in the building sector (see next paragraph "the third dimension").

11th conclusion: SES-SIG supplies 400 solar-electric cars with solar electricity. Usine Solaire of SES and SIG in Plains-les-Quates (Geneva) was awarded the Swiss and European Solar Prize in 2008 for its optimally integrated solar installation with a surface of 3400 m². This PV installation produces 585 000 kWh/a, which is more than the solar installation on Mont Soleil using 20 000 m² of pasture and meadows. With a production of 585 000 kWh/a, this single solar installation is able to supply 390 electric cars with solar electricity. This means that 390 employees of Usine Solaire can drive a distance of 15 000 km each year with their solar-powered electric cars. In 2009, the PlusEnergyBuildings in Bennau/Schwyz and the PlusEnergyBuilding refurbishment of Züst in Grüsch were awarded the European Solar Prize. They have a self-supply of 110 to 160%.[82] Also these apartment and business buildings do not only supply themselves on their own, but produce solar electricity used for the mobility of the inhabitants. With such practical examples, the building and solar industry proves that the "Stanford plan for an emission-free world"[83] could be realised until 2030 according to today's

---

[82] See Swiss Solar Prize 2009, p. 32 and 33.

[83] Prof. Mark Z. Jacobson/Mark A. Deluchi, University of Stanford, Spectrum of Science, 12/09, p. 80–87.

technological standard. If renewable energies, especially PV, are combined with energy efficiency, between 80 to 90% of the world's energy need of all buildings and approximately 50% of traffic can be reduced and substituted by renewable energies.

5.3. *Stanford: "Clearly, Enough Renewable Energy Exists"*

1st PEB conclusion: PV-PEB supply 100% of the building needs

Due to the rejection of a sustainable energy policy, Switzerland has to import 80% of its actual energy needs. On the other hand, it is possible to enlarge several existing hydro-power plants to pump storage power plants for a sustainable us of wind energy excesses from offshore and coastal regions. Regulating energy could be sold "just in time" in everyone's interest to countries or European cities without peak or regulating energy (e.g. 1 kWh of regulating energy is sold for 2, 3 or 4 kWh of base load or excess energy). By implementing today's level of technology (with PEB standard for new buildings and Minergy-P/passive house standard for

Figure 50. In 1987/88 Max Horlacher, industrialist from Möhlin/Aargau had already built the most beautiful and best light vehicles — even for railway compositions and for industry. He also successfully participated at each Tour de Sol with his "solar fleet". It is very positive that Renault decided to build emission-free cars the first European car manufacturer to do so. However, Renault could have further developed and brought on the market such solar-powered electric cars already in 1987/88, i.e. 22 years ago.

refurbishments) on at least 1.5–2% of the buildings each year, today's energy need of 125 TWh/a in the building sector could be substituted completely between 2030 and 2050. Between 80 and 85% of the substitution can be done without any loss in comfort by just reducing energy losses; the remaining 10 to 15% can be covered with other renewable energies.

2nd PEB conclusion and critical remark to solar-powered "electric cars"

Critical remark: For the last 20 years it was not long-sighted pioneers such as Max Horlacher nor innovative students, engineers, physicians or chemists who were heading traffic development, but names such as Bob Lutz, Rick Wagoner of General Motors, Jürgen Schrempp of Daimler-Benz AG, Ferdinand Piëch or Martin Winterkorn of Porsche and Volkswagen AG. Bob Lutz, who went to work in Detroit in a helicopter each day and flew an Alpha jet and a Czech jet fighter weekly, was admired by the media. However, with their million-dollar-salaries and the production of hummers and other gas-guzzlers for the destruction of environment these guys steered the car industry into the ground. The American state had to save the car industry in 2009 with billions of dollars. It would be a nightmare if these "system cheaters" (Binswanger/TA-01/2010) and "bonus banksters" used the pension fund assets to invest against renewable energies and environment into coal-burning and nuclear power plants. Innovative small and medium-sized businesses have no chance for favourable investment and innovation loans, if casino managers gamble away the pension fund savings of others in their own interest. And if the tax payers have to finance another financial crises, innovative and sustainable products such as those of Max Horlacher are left behind.

3rd PEB conclusion: solar-powered "electric cars for all"

By changing from motors to solar electric drive in individual traffic, efficiency in the traffic sector would increase by 50 to 70%. Today, about 75 TWh/a are needed in the whole Swiss traffic sector. If individual traffic changed to electric drive, 25 TWh/a would be enough. In spring 2010, Renault asked for "electric cars for all" — and called upon car industry to focus on the human being again. If Renault wants a trustworthy implementation of its slogan, it should read "solar-powered electric cars for all." The wind, water and solar electricity excesses of the PEB are definitely enough to cover this additional energy need in the traffic sector; a fact that has been confirmed by the formerly mentioned examples already 20 years ago (see Fig. 50, Horlacher).

4th PEB conclusion: PV, wind and water cover more than 100% of the total energy need

The total energy need of the apartment and commercial buildings, according to the OECD making up for 46 to 50% of our total energy need,[84] will in the future be covered completely by PEB. If future apartment and commercial buildings have an average self-supply of 120 to 200%, the future need of 25 TWh for an electric traffic sector in Switzerland will be completely covered by the wind, water and solar electricity excesses of the PEB. Other renewable energy sources such as biomass and geothermal energy will be available as residual and reserve energy. All mentioned examples from the building and traffic sector confirm the study of Prof. Jacobson/Deluchi of the University of Stanford: "CLEARLY, ENOUGH RENEWABLE ENERGY EXISTS."[85] Instead of today's energy need of 200 TWh/a (buildings and traffic sector) and a continuous energy increase, for Switzerland this could mean a maximum energy need in the building sector and in individual traffic of 25 TWh/a to maximally 50 TWh/a by 2030–2050. Therefore, the former energy need can be reduced by 75% with energy-efficiency and renewable energies. Translated for other comparable countries this means:

5th PEB conclusion: In Europe, PV-PEB can cover 75% of the total energy need

Similar to China, Europe is consuming approximately 20 000 TWh/a.[86] Today, the average consumption of Europe's buildings amounts to approximately 10 000 TWh/a. PEB could substitute at least 75% of the average building consumption. Therefore, the building sector will in future only need about 2500 TWh/a and the traffic sector another 2500 TWh/a.[87] According to today's level of technology and depending on the general conditions, Europe's energy need for buildings and individual traffic will decrease to 5000 TWh/a by 2030–2050. As an additional

---

[84] According to the OECD, the total energy demand of 46 to 50% includes heating, warm water as well as household and business electricity; see University of Cambridge/UK and Prof. L. Glicksman, MIT/USA, Swiss Solar Prize 2005, p. 10.

[85] Scientific American Stanford Study by Prof. Mark Z. Jacobson/Mark A. Deluchi, University of Stanford, Spectrum of science, 12/09 p. 80–87.

[86] BP Statistical Review of World Energy June 2009, p. 40 ff. China is listed with 2002.5 million tons of fuel oil (17.7%) ≈ 22–25 000 TWh/a depending on the conversion factor in the Chinese energy sector. In Europe, it depends on which countries are included: with Russia and Azerbaijan it would be even 2965 million tons of fuel oil, i.e. 26.2%; for the EU — including all western European states — about 20 000 TWh/a are assumed.

[87] With an average self-supply of 100–200%, today's energy need in the European building sector of 10 000 TW/a will be substituted. In future, it will be approximately 2500 TWh/a for the building and individual traffic sector each, depending on the investment scenario.

reserve, Europe's cheap and increasingly excessive wind energy could be better used as regulating energy for PEB and electric cars with help of ecologic pump storage power plants. This renewable energy model with general conditions according to the demand-side management (necessary also from the electricity consumer's point of view) can be adopted for all continents with slight adaptations. If Europe is transformed into a "PEB building park", the total energy need of all heated buildings and of individual traffic could be completely covered with today's level of technology. Therefore, PlusEnergy Buildings cover 75% of the European energy demand.

6th PEB Conclusion: PV-PEB can cover 75% of world's total energy need

Since physical axioms about solar radiation, gravitation etc. are universal, what is possible in Europe seems possible also worldwide. Therefore, PEB can cover 75% of the world's energy demand. The measured values from housing and business buildings in Europe confirm the "plans for an emission-free world until 2030" in the study of Stanford by Prof. Jacobson.[88] With just one note: Presumably it is not 12.5 TW or even 16.9 TW[89] that are needed; according to the planned and proven increase in energy efficiency and substitution in the building and traffic sector 6 to 8 TW should be enough.[90] Instead of 3.8 million of wind installations and 1.7 billion of PV installations, a reduced number will be enough to hold today's live standard or to even optimise it by massively decreasing emissions in the cities worldwide. What is already possible in Europe today, might also be possible on other continents.

The **Stanford Professors confirm** that there are enough renewable energies (PV, solar-thermal, wind, hydropower and wave energy as well as biomass and geo-thermal energy) to supply 100% of the world's energy needs. A new and

---

[88] Scientific American loc. cit., Spectrum of science, 12/09 p. 82 ff. With a self-supply of 120% to 200% for new buildings and of 85% to 150% for refurbishments, Europe's building park will transform from a big energy consumer of approximately 10 000 TWh/a to an energy producer with a self-supply of 120-200% in all buildings by 2030/50. Depending on promotion and legal requirements, the rate of renewal amounts to 1.5 to 2% or more of all buildings. The worldwide energy need of primary energy and final energy is based on BP Statistical Review of World Energy, June 2009, p. 40 ff; International Energy Agency, Key World Energy Statistics, Paris 2009, p. 6 ff.; see Swiss Solar Prices between 2000 and 2009.

[89] U.S. Energy Information Administration, Scientific American, Prof. Jacobson/Deluchi, November 2009, p. 60, Spectrum of science, 12/09 p. 82. With a self-supply of 120% to 200% for new buildings and of 85% to 150% for refurbishments, Europe's building park will transform from a big energy consumer of approximately 10 000 TWh/a to an energy producer with a self-supply of 120-200% in all buildings by 2030/50. Depending on promotion and legal requirements, the rate of renewal amounts to 1.5 to 2% or more of all buildings. The worldwide energy need of primary energy and final energy is based on BP Statistical Review of World Energy June 2009, p. 40 ff; International Energy Agency, Key World Energy Statistics, Paris 2009, p. 6 ff.

[90] Scientific American University of Stanford Prof. Jacobson/Deluchi, November 2009 and all mentioned empirical results in part 1 to 5.

pleasing, but not surprising phenomenon are the very low future energy costs, that are indicated with 4–5 €cts.[91] However, the energy price is only important to "energy illiterates", because they have not understood that the annual energy costs are determined by two factors: price multiplied by amount. As the following example (5.3) shows, the price for solar electricity can be even 200% higher while the energy bill is still 80% lower. The amount of energy is as important as the price! Unfortunately, many "energy economists" have not understood this simple fact so far.

7th PEB Conclusion: The Dimension of PV is Unbeatable Economically and Ecologically

The example of today's total energy need of twelve apartments from 1896 in the old town of Basel shows what can be achieved economically and ecologically even in protected areas in European cities by an optimum PV integration and with high energy efficiency — despite "expensive PV":

a) Former energy need:	**223 000 kWh/a to 0.15 €cts ≈ € 33 500 €**	
b) Reduced to the following by energy efficiency:	53 000 kWh/a	
c) Today's PV-energy need:	**15 800 kWh/a to 0.40 €cts ≈ 6300 €**	

Today's energy need of the apartments is 93% lower than before the refurbishment, and despite "expensive solar electricity", the annual energy costs amount to only 19% of the former energy costs.[92] The difference of 27 200 € can be used for paying interests and amortisation of the energetic refurbishment investments without state subsidies. Such a difference in costs, however, is often not enough to finance all energetic refurbishment investments — especially due to cross-subsidisation of non-renewable energies. Therefore, compensatory payments are justified.[92a] It is important to understand that energy efficiency accounts for 170 000 kWh/a and solar power for 37 200 kWh/a.[93] In the building sector making up for 50% of the world's energy demand of 130 000 TWh/a, this combination of energy efficiency and PV energy production is unbeatable economically and ecologically.

---

[91] Scientific American loc cit. Jacobson and Deluchi, p. 64.

[92] Swiss Solar Prize 2009, p. 34/35 and Fig. 40

[92a] Market-based compensation for billions of fossil-nuclear subsidies; see part 4.4.

[93] Solar architecture in Basel If the inhabitants formerly paid 20 cts./kWh, they previously had annual energy costs of CHF 44 600. Today, with 15 800 kWh and assumed costs of 60 cts./kWh, they have annual energy costs of CHF 9480. With the difference of approximately CHF 35 000, the energy investments of approximately CHF 880 000 are financed; see refurbishment of multi-family house in Basel, Swiss Solar Prize 2009, p. 36/37. This generates a local added value and jobs for building examinations. It also leads to a decrease in oil, gas and uranium imports i.e. to more energy-independent countries.

In order to substitute these 50% or 65 000 TWh/a, about 8600 nuclear power plants with a capacity of 1000 MW each would be necessary worldwide — at the moment there are approximately 430 NPP still in operation worldwide.

8th PEB conclusion: The third dimension of PV is unbeatable

In summary it can be said that PV can be used like no other energy source in three dimensions:

- The *first dimension* relates to energy production: After their use in space, PV solarcells were increasingly also used for terrestrial electricity production — e.g. for solarcars and in buildings.
- *Second dimen*sion: Solar electricity production with an optimum integration of the installation on roofs and façades. During the last decade, innovative architects like Lord Norman Foster/Prof. Stefan Behling (London), Prof. Thomas Herzog (Munich), Beat Kämpfen (Zurich) and several building companies have realised award-winning examples.
- *Third dimension:* Solar electricity and integration of the installation on roofs and façades combined with energy efficiency in the building sector. The results are sensational.

9th PEB conclusion: Lord Foster is leading the way for architecture in the 21st century

The principles mentioned by Lord N. Foster at the 15th Swiss Solar Prize are valid also for the 20th Solar Prize and for architecture of the 21st century. "Architects, designers and planners cannot continue to ignore the damage our building inflict on the natural environment." As the consequences of our past inaction become ever more apparent, designing for a sustainable future becomes a necessity, not a choice. The way we shape our buildings, our neighbourhoods and our global lifestyles has now become even more important than ever — we must ensure that sustainability becomes as inseparable from our design processes as time, cost and quality."[94] The multi-functionality of integrated PV installations, which are replacing conventional roofs and protecting the building against rain, snow, wind or excessive sun and are at the same time also supplying energy, is unique and with regard to the material requirements also very economic and ecologic. Criticism of the French legislation about the "high aesthetic claim" is not justified at all and shows a lack of

---

[94] Lord Norman Foster, Swiss Solar Prize 2005, p. 3.

know-how and economic expertise.[95] Innovative PV businesses will one day replace the aesthetically ideal copper roofs of our national and UNESCO-protected monuments — and will even produce electricity. Therefore, we do not need to obstruct PV installations, but need to provide our workmen with better plans instead. With the almost CO_2-free Bundestag in Berlin, Lord Norman Foster has demonstrated that this is possible and has opened the way for solar architecture in the 21st century.

10th PEB Conclusion: Physics or Nature's Gift per square meter

Solar radiation in Central Europe brings approximately 1000 kWh/m^2 a. The sun is essential for the loop of nature and outnumbers the total energy need of the whole world by a factor of 10 000. The solar reserves will last for the coming four billion years. If we analyse the gifts of nature and physics per square meter from the viewpoint of the final consumer or according to the demand-side management, we note different heights of precipitation depending on the location, an annual increase in biomass and solar radiation. These three energy sources can be transformed into kWh/m^2 as follows:

- **Hydropower:** With precipitation in Central Europe, nature makes a water gift per square meter of ≈1 m height (1 m^2 × 1 m) i.e. 1 m^3 of water ≈ 1 t of water per m^2. If this ton of water is poured from a height of 400 m on a turbine, we get 1 kWh per year.
- **Biomass:** The annual increase of biomass in plants in Central Europe corresponds to an energy amount of 2 kWh/m^2 a.
- **Photovoltaic:** The optimally integrated PV installation of SES-SIG in Geneva produces a total of 585 000 kWh/a on a surface of 3395 m^2. This corresponds to 172 kWh/a, per square meter.[96] For additional floors, *façade collectors*

---

[95] Photon 12/2009, p. 24. The difference between "highly aesthetical" and simplified integration is justified, because it incorporates justified criticism on disfigurement of townscape and offers aesthetically unobjectionable solutions. Economically, the differentiation is also justified due to the mentioned multi-functionality of the environmental-friendly material (silicium). Furthermore, the needed energy is generated on the building itself and there is no need for a colonialist energy policy with depression or plundering of third and future generation's energy sources: Scientific American University of Stanford Prof. Jacobson/Deluchi, Nov. 2009 and all mentioned empirical results in part 1–5.

[96] An energy-efficient average apartment or business building needs between 25 and 40 kWh/m^2a. On a sunny roof in Central Europe, there is therefore four times more energy than averagely needed per year. Or in other words: a roof produces enough energy to completely supply four to five floors energetically. For additional floors, façade collectors produce between 70 and 130 kWh/m^2a (depending on the PV cells).

produce between *70 and 130 kWh/m²a* (depending on the PV cells.)[97] In Central Europe, a complete renewable energy supply is therefore feasible and undoubtedly possible.

The sun sends us per square meter in one year:

a) **Hydropower** = 1 kWh/a
b) **Biomass** = 2 kWh/a
c) **Photovoltaic** = 172 kWh/a

In their study, Jacobson and Deluchi from the University of Stanford compare today's energy need of 11.5 TW with renewable energy reserves amounting to 580 TW for solar energy, 40-85 TW for wind energy and 2 TW for hydropower.[98] If we look at these physical facts, how can we still talk of electricity or energy gaps? Either the so-called "energy experts" just ignore physics and the outstanding performance of building technology in Central Europe, or they need to fill in their "expert gaps" with the help of an expect for psychiatry.

11th PEB conclusion: What was Possible 25 years ago and is now Realised in Shanghai

If it is possible to refurbish energy-intensive businesses with an energy index of 750 kWh/m²a to PEB, why should it not be possible to realise all future European average buildings with ten times lower energy indexes of 40 to 70 kWh/m²a as PEB as well? Furthermore, if in Pudong at the Huang-Pu river of Shanghai apartment and business buildings with altitudes of 350 to 420 m are built in a weekly rhythm, why should it not be possible in European cities and towns to enlarge buildings or building groups by some floors? Adaptations of roofs and façades with an integration of PV installations allow an optimum use of solar energy. If 25 years ago, solar-powered light vehicles were able to drive across whole continents, and projects of CO_2-free hotels and business buildings were realised (see Figs. 1 and 50), these environmental-friendly technologies should finally be implemented on a large scale, especially in the building sector. Not only new buildings should be PEB, but also refurbishments should result in PEB by preserving the appearance of the townscape.

---

[97] Alpiq solar wall: In Geneva, the technological department of Alpiq presented a solar wall with micro-amorphous solar cells from Oerlikon-Solar and designed in co-operation with SUPSI. Installed on the façade, this wall produces approximately 85 kWh/m²a. With mono-crystalline Sunpower solar cells (offered in 2009) with an efficiency factor of 19.1%, an energy-production of ≈ 130 kWh/m²a would be possible (see also Swiss Solar Prize 2009, Bennau, p. 32/33).

[98] Scientific American loc. cit., p. 60; Spectrum of science, 12/09 p. 84 ff.

12th PEB conclusion: "Domestic investments for Jobs"

"Instead of sending our money abroad for buying fuel oil, we better create domestic jobs with the know-how of German engineers," declared the German Minister of Environment (and member of the CDU) Norbert Röttgen.[99] There is no other energy and job machine that substitutes more energy and creates millions of sustainable jobs at the same time than the implementation of a PEB program. The initiates of the CO_2-free hotel Ucliva acted according to Röttgen's motto on June 23, 1978, when they founded a hotel co-operative in Waltensburg/Grisons. They wanted to create jobs in their alpine town at the Vorder Rhine in order to stop migration (see Fig. 51 with representatives of the actual managing committee and participating entrepreneurs Riccardo Bertogg and Dora Veraguth, co-founders of the hotel in 1978). The city parliament of Zurich has provisorily carried a respective single initiative (single initiative Cadonau) for the implementation of the 2000 Watt goals on February 10, 2010: The single initiative is based on the German, French and Japanese legislation and provides for a remuneration for PV feeding of only 20 €$_{cts}$/kWh. On the other hand, an optimum integration will be remunerated with additional 20%, a better heat insulation with another 20%, Minergy-P/ passive building: +20%, 2000 Watt goals: +20%, and for PlusEnergyBuildings

Figure 51   From left to right: Gallus Cadonau, co-founder and first president of the Hotel Ucliva; Hansjörg Wehrli and Corina Issler Baetschi, members of actual managing committee; Swiss Councillor and Minister of Foreign Affairs, Micheline Calmy-Rey as well as representatives of the involved SMEs (Dr. Martin Pfister, surgeon and co-founder of the hotel in 1978; Riccardo Bertogg, baker; Dora Veraguth, host of Restaurant Post).

---

[99] Dr. Norbert Röttgen, German magazine "Der Spiegel", 53/2009, p. 31.

+20% — i.e. in total = 40€$_{cts}$/kWh. The initiative also foresees further competences for the government and exemption clauses in order to adapt the legislation to the respective conditions. See www.solaragentur.ch

13th PEB Conclusion: Creating Jobs with Ecological Behaviour

All these examples show that the opponents' arguments against renewable energies are based on lack of knowledge and ignorance of facts — to the disfavour of SMEs that have no lobby for PR campaigns. The main argument Bill Clinton used in 1996 in his successful election campaign is still the same today: „It's the economy, stupid." The main argument is jobs and then ecology. In 1978, a majority of Waltenburg's inhabitants combated the drainage of the Rhine and fought for its preservation up to Federal Court. Their action resulted in approximately 15 jobs (some of them part time) and the biggest employer in town. In 1988, the ecologic hotel was awarded by the Council of Europe in Strassbourg (see Fig. 14).[100] Thereupon it was also cited by the University of Bristol/UK as a model for the "development of rural areas".[101]

14th PEB conclusion of Socrates: Independence is the precondition for freedom

As millions of other examples from SMEs, the history of the hotel Ucliva shows that renewable energies create jobs and lead to a sustainable, regional ecology of public interest. Renewable energies and in particular solar energy can be used on all continents and even in far-away communities by the local population. The sun is shining for us all and supplies enough energy for free in order to be independent and to free ourselves from non-renewable energies that will come to an end in the 21st century — all in the true sense of Socrates' words: Independence is the precondition of freedom. Why do politicians and media representatives listen to those "energy experts" that know the use of renewable energies from hearsay only? Why do they not rely on empirical and scientifically proved figures? Why are they not interested in the measuring results of the past years? And think about the future generations and about the consequential costs of their energy behaviour? By asking these questions, the energy illiterates can be identified.

---

[100] Council of Europe, European campaign for the countryside: "Enhanced public awareness and changed attitudes towards mountain holidays" (result: Corporaziun Quaterfegl — today: Ucliva), project 119, p. 303. The prize money amounted to 5000 ECU — corresponding to approximately € 5000 today.

[101] Prof. Dr. Bernard Lane published several professional articles at the University of Bristol/UK on the subject. They were in parts also published in practically all Swiss newspapers and magazines; in occasion of the hotel inauguration in December 16, 1983, there were also several articles in German newspapers such as Frankfurter Allgemeine, Süddeutsche Zeitung, Der Stern, Berliner Morgenpost and others.

15th PEB Conclusion for Barack Obama: Yes, You Can Build a White PEB House (Open Letter)

Dear Mr. President, as President of the United States you visited Saudi Arabia in spring 2009. According to the media, you were asking your hosts to send less money to Pakistan and Afghanistan because it might be used for financing terrorists' actions. We have to ask ourselves if those who are obstructing the use of renewable energies in our countries and sustaining the payments of billions of $ and € for oil imports from Arabian countries are not financing terrorist actions as well? Due to such obstructing behaviour, the equity prices of most innovative US and EU solar enterprises (Sunpower, First Solar and Q-Cells, Connergy and Solarworld etc.) have decreased by 60% and more, whereas Chinese enterprises capture the most innovative high-tech market with their products and show increases in equity prices of over 400%.[102] In 2009, the Chinese government decided to invest 309 billion € in renewable energies and 436 billion € in new energy grids (smart-grids) instead of investing in the financial sector by paying huge bonuses to bankers.[103] With Sunpower, the US is producing the probably most efficient solar cells of the world, while Central Europe disposes of technological businesses that are building the most efficient apartment and business buildings in the world. Therefore, we propose you to refurbish the White House into a modern PlusEnergy Building and to set a clear and strong standard for energy, environment and job discussions in the US, Europe and worldwide. PS: In Europe, the best examples of PEB are awarded a Solar Prize amounting to CHF 100 000. A White PEB House would also be considered for the award. We therefore thank you for a favourable consideration.

5.4   *First European Award for PlusEnergy Buildings of CHF (≈$) 100 000*

In order to promote PEB around Europe, there will be a Swiss and European Award for Plus Energy Buildings in 2010. It is the goal of this PEB Award to support architects and engineers according to the actual PEB level of technology in order to realise PEB possibly nationwide. The most creative architects, engineers und businessmen realise increasingly efficient buildings consuming less energy at full comfort: PEB produce more energy than they averagely need for their total heating, warm water and electricity supply. They sell their electricity excess to the public grid and therefore reduce CO_2 emissions. The total prize money of the award in 2010 is CHF 100 000 or € 70 000.[104] All objects are first verified by

---

[102] Photon 1/2010, p. 51: The equity price of the Chinese enterprise Trina Solar has increased by 410.7% since January 1, 2009. In Europe, equity prices of solar enterprises have drastically decreased.

[103] Photon 1/EU, Joint Research Centre, Renewable Energy Unit, PV Status Report 2009, p. 101.

[104] PEB Award 2010 of CHF 100 000: www.solaragentur.ch

a technical commission and then awarded by an international jury according to the following categories:

1. **Best PlusEnergyBuildings Swiss PEB**
2. **PEB with best integrated solar cells** (Lord Norman Foster Solar Award)

Applications are accepted until **May 15 of each year, for the first time in 2010.** For further information contact Solar Agency Switzerland, P.O. Box 2872, 8033 Zurich; info@solaragentur.ch

---

Gallus Cadonau

Gallus Cadonau was born in Waltensburg/ Grisons. After school, in his hometown, he did an apprenticeship as machinist and worked as aeromechanic in Dübendorf/ Zurich and Payerne/Vaud (1965–1996); Higher School Certificate at the cantonal grammar school for adults (KME) (1972–1975); language stay in Bournemouth and Oxford/GB; study of law at University of Zurich, lic. iur. (1975–1981); traineeship at Registry of Deeds, at the cantonal and administrative court of Grisons and at a lawyer's office in Chur (1982–1985); popular election as constitutional council of the canton of Zurich 2000–2005). Photo on top: Gallus Cadonau with the Swiss Councillor and Minister of Foreign Affairs, Micheline Calmy-Rey, 22 June 2003.

- **1978–1991:** Environment and jobs: Creation of sustainable jobs in alpine regions, co-founder of the hotel cooperative Ucliva for sustainable tourism (1978–1991: CO_2-free hotel building with 72 beds and 100% renewable energy sources; award for scenery preservation in 1981; 3 EU awards; award by Council of Europe in 1988; Swiss eco-hotel of the year in 1997/98).
- **1978–1986:** Protection of environment, landscape and waters: Commitment against the destruction of alpine rivers; salvation of landscapes at the Vorder Rhine and on the Greina plateau; from 1986 co-founder and director of Swiss Greina Foundation (SGS; with approximately 50 National Councillors and Councillors of States as well as 102 000 members/benefactors); several sustainable amendments of constitutional law such as the "introduction of a landscape cent" for the protection of landscapes of national importance which was taken as a model in the UN year of mountains in 2002; several amendments of the constitutional law in the fields of energy, water and spatial

planning as well as water, landscape and environmental protection legislation; several voting campaigns from 1975 on; www.greina-stiftung.ch

- **From 1978:** Renewable energies and energy efficiency: Co-founder of the Tour de Sol and vice-president SSES, Foundation Tour de Sol, director of the Solar Agency Switzerland with Swiss and European Solar Prize Award, European Championship of Alpine Solarcars, European Solar Challenge, International Solarcar Federation/Steward World Solar Challenge (1985–1996); vice-president Euro-/Swissolar; European Solar Council/Paris/Steering Committee European Solar PV Conference. Bipartisan alliance for solar and an energy levy in the National Council and the energy legislation (EnG); president of Clean Energy St. Moritz, enforcing of legal claim for carefully integrated solar installations in 2007; K. H. Gyr Award of the culture foundation of Zug for cultural, scientific and social pioneering achievements; info@solaragentur.ch
- **Sports:** Judoka and coach of J. C. Dübendorf; multiple Judo Champion in Grisons and British Champion in 1971/BJA; ski and street marathons; member of the recourse commission of the Swiss Judo Association; initiator and co-organiser of the constitutional run in 2005 with sports associations with approximately 150 media pieces in 6 weeks.

Chapter 10

Photovoltaic Power Systems for Lifting Women Out of Poverty in Sub-Saharan Africa

Dr Dominique Campana

Director for International Affairs, The French National Agency
for Environment and Energy Management, France

One of the main results that people should remember about the Copenhagen climate conference in 2009 is the statement of Africa's position in the fight against climate change.

Representatives of Africa's nation states asserted how urgent it was to bridge the gap between development and environmental policy, the world's poorest populations being among those who suffer the most as a result of climate change. Currently, 3/4 of sub-Saharan African population has no access to electricity: this amounts more than half a billion people.[1]

What makes this such a paradox is the fact that the African continent is one of the world's major exporters of fossil fuels and yet, per capita energy consumption, is the lowest in the world. On average, it is seven times less than in Europe, with among the lowest electrification rate of all developing countries

If we are to meet the millennium development goals and win in the fight against poverty, then provision of energy services need to be developed, particularly services that can be accessed by the poorer members of the population for income generation. Many — who live on the outskirts of towns or in rural areas with low population densities where the cost of connecting them up to energy distribution networks is prohibitively high — do not yet have access to such services.

---

[1] Source : IEA 2007.

Power for the World by W. Palz
Copyright © 2011 by Pan Stanford Publishing Pte Ltd
www.panstanford.com
978-981-4303-37-8

Figure 1. Group of women in Lower Casamance (South of Senegal) using solar pumps for the development of garden activities (Fondation Energies pour le Monde).

Over and beyond the preponderance of poverty in rural areas, it is becoming increasingly prevalent among women. The figures speak for themselves: around 20% of the world's population of people over the age of 15 is illiterate. Ninety-eight percent of these people are in the South, and nearly 2/3 of them are women. The kinds of task which are usually carried out by women in society still carry with them many constraints and are physically very arduous. They leave women living in rural areas very little time for other activities, ensuring they remain caught in the energy poverty trap.

1. Solar Energy Against the "Energy Poverty" Trap

The days of thousands of women in rural sub-Saharan Africa involve three main activities: gathering firewood, drawing and carrying water, and preparing meals — which involves husking and grinding cereals beforehand.

In a family, women and young girls will spend all of their time on these types of tasks. The average African woman carries 40 tonnes of wood and water on her head per year, and carries up to 60 litres of water 10 km every day. This means that children — girls in particular — cannot be normally schooled, while the women have no time to do anything else, such as earn money, follow training programmes or get involved in the village's community and political initiatives. So they end up caught in the energy poverty trap, with any kind of development being impossible.

In these numerous regions which will not have access to national electricity grids for many years to come, solar power as an extra source of energy is a good means of providing local populations — women in particular — with development solutions.

The use of solar-powered pumps with photovoltaic panels shows how essential the role that women play is in getting populations to appropriate and share techniques that use solar energy. The very first installations date back to the 1970s — a time when the villagers found "solar water" bland, still preferring water drawn from wells to it. The women quickly realised the benefits of this "solar water". First of all, their children had fewer intestinal complaints. And then the fact that pumps could be installed in the villages themselves made their work considerably easier: no longer did they have to spend long periods of time walking huge distances and carrying back heavy loads or using their own strength to pump water. So they very quickly became enthusiastic followers of solar energy, so much so that in the 2000s, marriages could be more easily contracted where water pumps could offer better live conditions.

But "solar water" still has one major disadvantage: it has to be paid for. Indeed, water is often considered a divine gift in these villages, and so one that should be free. But solar pumps — although considerably cheaper than conventional systems — still need a minimum amount of maintenance to limit wear and tear due to sand, engine wear and any other problems with electronics. Otherwise, they simply will not last. Money also needs to be set aside to repair this tried and tested equipment, and to replace any costly components. This means that the service can only really be maintained over a long period of time in return for payment from its users. And this is a concept which — in rural societies — the women were the first to understand and accept.

There are two types of solar pump:

- Village pumps

  The pumps are installed over drilled wells, and used for a range of different purposes: to provide water for domestic purposes and cattle, and for watering small-scale market gardens with what is left over. Nowadays, drilled wells that are far away from where people live are avoided, with more pumps being erected in village centres. And so there are other benefits in addition to those mentioned above: if cattle are watered in the villages, then water can be sold to animal farmers who are passing through; and being able to water market gardens with what is left over means that families can develop income-generating activities.

  A solar energy project that was launched in the mid-1980s in the Sahel region and which is being run by the CILSS (Permanent Inter-State Committee for Drought Control in the Sahel) with support from the European Union has provided more than a million people in Africa's Sahel countries with access to drinking water: approximately a 1000 pumps were set up between 1990 and 2009 as part of the project.

- Irrigation pumps

   In Senegal's Ziguinchor region, in the heart of the Casamance, women have joined forces in order to run small-scale market gardens. But without any kind of a pumping or irrigation system, using buckets to draw water from the ground made their work slow and arduous. A pilot photovoltaic irrigation project concerning 14 small-scale market gardens was launched in 1997 by the French "Fondation Energies pour le Monde". Financed by France's Ministry for Foreign Affairs in conjunction with the ADEME[2] and EDF,[3] it is managed by some 700 women and has served as a means of demonstrating the advantages of photovoltaic power systems compared with drawing water by hand: twice as much land has been irrigated, the work has become much less arduous, yield has increased by approximately 30% because the women can now turn their attention to real agricultural tasks, and they are able to spend more time at home with their families and children, and can become involved in various other social activities. The project has now evolved with the implementation of a "drop-by-drop" irrigation system. Being able to save water means that production surface areas can be increased and more time can be spent on selling and transforming farming products.

   So increasing production leads to new sources of revenue and results in new activities: preserving (by keeping cold or drying), transformation (jam, etc.) and marketing. It's worth pointing out that providing people with access to energy does not necessarily lead to the development of income-generating activities that will create the kind of wealth capable of lifting the entire community out of poverty. In fact, a survey of villages that have been electrified for more than 10 years shows 20% coverage, with very few productive activities being initiated in them. Hence the necessity to define support measures for developing the capacity of the world's poorer populations to create and then manage their own productive activities.

   This project has served as a template for defining a development programme based on improving access to services through the provision of energy, a programme that is to be put into widespread use throughout the Casamance region.

More recently, the solar mill makes husking and grinding cereals considerably less arduous, with significant time savings for women.

   Millet flour, millet granules and couscous — all created by transforming millet — much sought-after food products. The grain transformation process (husking and grinding) is carried out by women and requires energy. Nowadays, as part of the drive to improve women's living conditions, diesel

---

[2] French National Agency for Environment and Energy Management.
[3] Electricité de France (French Power Supplier).

mills and huskers have been introduced. But given the fact that a number of villages are completely cut off, their operation is often interrupted as a result of fuel supply problems and — above all — the cost of fossil fuels. Solar mills have recently been introduced in Africa's Sahel countries in particular, and are a means of overcoming this constraint. They are not yet in widespread use, but they do represent an avenue for future exploration.

The last few years have seen the emergence of a number of mini-network projects, providing entire villages with electricity using photovoltaic facilities. The impact on the fight against poverty that these mini solar networks have brought includes:

- the development of sewing, welding and local craft activities;
- sales of ice cream and cold beverages
- facilities for conserving vaccinations so as to improve healthcare services
- an increase in school occupancy rates thanks to access to electricity.

There are still a number of other possible new applications for photovoltaic systems, such as the development of modern electronic equipment for use in communications and IT.

Other renewable energy sources can also be used, such as small-scale hydraulic power, wind power and biomass (combustion, carbonisation or gasification).

A number of projects have been set up that use hybrid solutions, combining photovoltaic or wind power with combustion engines. Diesel machines still have a number of advantages in rural areas: equipment is simple and can be easily maintained in the villages. What's more, a number of pilot operations have shown that some of these engines — rustic in their design — can also run on locally-produced biofuels (jatropha curcas oil, for example).

Using photovoltaic power and other renewable energies — as well as combustible fuels when they are justified — means that a wide selection of applications are now available that provide access to modern sources of energy, thus helping to meet the millennium development goals in rural and outlying districts that are not yet connected to the electricity grid.

But it is important to ensure that the technical resources to be implemented are based on an exact analysis of demand and that decisions made factor in the amount of power required, the dispersal of users throughout the region and their ability to pay for services, the technological environment and the potential of technically and economically mobilisable renewable energies, as well as the contribution that they can make to local economic development, combining energy efficiency and low carbon technologies.

In economic terms, until recently, photovoltaic power was only an advantage when used with very specific applications: to meet the low-power energy requirements of populations that were a long way from the national grid. But this paradigm is now changing as the cost of renewable energies falls and the cost of fossil fuels increases. Now, electricity generated from photovoltaic systems can be used for:

- basic services, such as lighting and the provision of drinking water, and access to income-generating activities for isolated populations;
- reliable and affordable electrical services for communities which previously had no alternative other than diesel as an energy source;
- an energy supply that is increasingly reliable and competitive in urban environments, particularly when integrated into housing, and when compared with the cost of power during peak hours

In the long-term, decentralised technical solutions based on local renewable energy sources will be more economically viable than conventional solutions, with lower annual operating costs. Conversely, the operating costs of conventional systems tend to increase over time as fossil fuel prices increase and they become increasingly scarce.

But implementing energy services which use renewable energy sources still requires a large amount of capital outlay. This means that financing them is still something of a challenge, particularly in terms of providing access to energy, insofar as the main goal is to secure a return on money invested, and so to maximise the number of connections — to the detriment of the long-term. So beyond technical innovation, how successful the deployment of renewable energy sources is will depend on the efforts made with regard to innovative financing models and setting up institutional partnerships.

2. In Conclusion

The results gleaned from these various projects demonstrate that there are tangible links between the use of a modern energy source, such as photovoltaic power and a set of related effects associated with development and the fight against poverty, thus helping to meet the millennium development goals:

- promoting gender equality and autonomy among groups of women;
- primary education for everyone, helped by the development of new energy sources, young girls being released from arduous manual work;
- reduction of extreme poverty and hunger. Time and energy savings mean that women can spend more time on income-generating activities, thus increasing and diversifying their income

Solar power is one of the means to reconcile equality, economic development and global environmental protection, a major challenge for the future!

Acknowledgment

With my thanks for their contribution to ADEME's experts, Michel Courillon who spent more than 30 years of professional life for the progress of energy access in Africa and to Hélène Sabathié Akonor who is now on the same track.

Dominique Campana

Dr. Dominique Campana is Director for International Affairs at ADEME, the French national environment and energy management agency.
In her career, she was involved since over 30 years in the field of energy efficiency, renewable energy and environment, beginning as a researcher in Ecole des Mines de Paris, and joining ADEME in 1992.

She was involved in bilateral and multilateral co-operation with many countries from around the world, and has personal knowledge of most stakeholders.

She had contributed to the international negotiations on climate change and sustainable development and to French and European initiatives with emerging and developing countries.

Dominique Campana holds a PhD in Physical Sciences from Lyon 1 University from 1978.
Her thesis was concerned with PV water pumping; in this frame, the world's first PV water pump was developed and set up in Corsica, France.

She has a long record of publications on energy management, bioclimatic architecture, and solar energy, including the editing and translation in French of Unesco's book "Solar Electricity" (by W. Palz) from 1978.

Dominique Campana was honoured in 2006 with the French order " Chevalier de l'ordre du Mérite".

Chapter 11

Solar Cell Development Work At COMSAT Laboratories (1967–1975)

Denis J. Curtin

Former Chief Operating Oficer, XTAR LLC, Rockville, MD 20855, USA

At COMSAT Laboratories in 1967, the Spacecraft Laboratory studied all non communications aspects of the communications satellites; that is, everything related to the satellite itself including structure, thermal attitude, power, etc. In my case, I was responsible for primary power development in the Electric Power Department of the Spacecraft Laboratory In 1967, all satellites were powered by silicon solar cells. We examined every major type of solar cell that was available in manufacture or under development including, silicon, cadmium sulfide, and cadmium telluride cells. We rapidly concluded that only silicon solar cells were a viable option at that time since the other devices were neither stable enough nor had good radiation properties. We thereupon characterized every silicon solar cell under manufacture that we could purchase doing all a full battery of tests to determine how well they would perform in a space environment. We also tested every type of transparent or translucent solar cell shielding material available that could be used to protect the solar cells from radiation in space including fused silica, Microsheet, Kapton and other plastics and even spray on coatings. At that time, the best material was fused silica. Finally, we also tested the various adhesives that were used to bond the shielding materials to the cells, both for stability and performance under radiation. From this data we predicted what solar cells, adhesives and shielding materials performed the best under space conditions. Finally, taking all of the data and recommendations we made predictions of how solar cell arrays of silicon solar cells with various shielding materials would perform in space over time.

During the course of the silicon solar cell studies, we realized that the industry standard efficiency of silicon solar cells was about 10.5%, compared to the

Power for the World by W. Palz
Copyright © 2011 by Pan Stanford Publishing Pte Ltd
www.panstanford.com
978-981-4303-37-8

theoretical efficiency of about 24%. Based on this, I approached the Head of the COMSAT Solid State Physics Laboratory, Dr. Joseph Lindmayer, to determine if the device efficiency could be improved. He informed me that silicon solar cells were of no interest to solid state physics with everything else going on with solid state physics developments in communications. After six months and numerous efforts to secure their assistance, Dr. Lindmayer finally agreed to at least look at the efficiencies.

In just six months, his laboratory achieved the 10.5% state of the art for silicon solar cells. My group worked as a team with the Solid State Laboratory. They did the solid-state physical development of the cells. My group worked to ensure the devices met industry standards and that they had the capability to be used on satellites. When we were satisfied we had a 10.5% silicon solar cell, equivalent to anything commercially available, we celebrated with a party! My wife even baked a cake with 10.5% efficiency written on it.

To my surprise Dr. Lindmayer informed me he felt there was a real possibility to improve the efficiency beyond 10.5% with the new solid-state techniques that were available. We formed another team, this time with his lab doing the device development and my group ensuring that single devices could be manufactured to industry standards and then conducting the tests to guarantee they became space-qualified solar cells.

In less than three years, the Solid State team developed a 13% device, the "COMSAT" solar cell, using very shallow junctions that captured more sunlight. Shortly thereafter, the team achieved a 15% device using shallow junctions and non-reflective surfaces — the so-called "Black" cell. My team in the Spacecraft Laboratory fully tested and qualified these cells. The COMSAT solar cell and later the Black cell became the standard for the solar cell space industry for the next 20 plus years, until the advent of multiple junction gallium arsenide cells. As with the earlier solar cells my team did radiation tests on the new cells, the solar cell shielding materials, and adhesives used to attach the shielding material to the cells and based on the data made predictions on how the shielded solar cells would perform in space over time. These predictions were used as one of the industry baselines for the next 15–20 years.

Interestingly, Dr. Joseph Lindmayer, left COMSAT Labs and started Solarex, which went on to become a major supplier of terrestrial solar cells.

In the late 1960s, communications satellites were spin stabilized with body-mounted solar cells. By the early 70s it became apparent that there was a need for larger satellites and antennas to meet increasing communications requirements. Three-axis stabilized satellites, with deployed solar arrays and large deployable antennas came under consideration to satisfy these increased requirements.

On spin-stabilized satellites, with the solar cells mounted on the body of the satellite, the cells experience a benign thermal environment operating around +20°C in sunlight. This performance is only slightly lower during the two 45-day vernal and autumnal equinox eclipse periods each year, when the solar cells are concealed from the sun for up to 72 minutes a day. This benign thermal environment is due to the larger thermal mass of the satellite body.

The thermal situation is quite different in the case of three-axis stabilized satellites with solar cells mounted on deployed solar arrays. In sunlight, the solar cells will operate at about 60°C. During the eclipse periods, the solar array will rapidly drop in temperature to the order of −180°C before rapidly returning to 60°C in sunlight. This is due to the deployed array having a low thermal mass.

Over a 10–15-year orbit the solar array will experience 1200–1300 of these thermal cycles. These cycles were found to dramatically stress the solar cell interconnections that were normally soldered in series and in parallel to achieve the proper satellite current and voltage. Often the soldered connections were found to fail after these cycles.

In response, our team at COMSAT Labs made a major effort to replace the solder with another approach and to substitute the high thermal coefficient of expansion silver used to interconnect the cells with a more thermally benign material. A number of tests and studies were done at COMSAT and numerous other facilities in the US and Europe. These studies led to the replacement of solder with electrical welding, and pure silver strips with silver with loops for more stress relief, or silver plated materials such as silver plated molybdenum or silver plated Invar. Our team was deeply involved in these studies and we did extensive testing under the Intelsat R&D program to support these developments.

The results of this work were applied directly to the Intelsat V program in the mid-1970s, where the satellite chosen was three-axis stabilized with lightweight deployed solar arrays. Along with a team of engineers from the satellite prime contractor, Ford Aerospace, we spent several weeks at MBB, the satellite solar array subcontractor, in Germany. We assisted the MBB team in the development of the solar array and, in particular, the welding approach used to interconnect the solar cells. We applied what was learned in the development programs.

Dr. Denis J. Curtin

Chapter 12

SolarBank

Michael T. Eckhart

President, American Council On Renewable Energy ACORE, Washington DC, USA

This is the story of developing SolarBank, which has not yet been finished.

Perhaps someday there will exist as a global source of low-cost capital to finance the solar energy revolution. SolarBank will support lenders all around the world by purchasing solar loans they have made to local companies and end users, and by providing training, technical standards, and insurance products to support the widespread financing of solar energy projects.

The idea for a SolarBank first came up in 1978, when I was working for the management consulting firm of Booz, Allen & Hamilton (now called Booze & Co.) in Washington, DC from 1976 to 1980. My group did studies on the emerging energy technologies like solar photovoltaic (PV), solar water heating, wind power, geothermal energy, hydropower, and ocean energy, biomass energy and fuels (ethanol was then called "gasohol"), cogeneration, synthetic fuels like coal gasification and coal liquefaction, shale oil, advanced nuclear power, energy storage, demand-side management, and such things. We conducted studies of markets, technologies, industries, economics, and public policy.

1. Landmark Solar PV Study in 1978

One of our landmark studies was the *"Assessment of Solar Photovoltaic Technologies, Markets and Industry"* that was conducted for the US Department of Energy's Office of Policy, but was funded actually by the White House under a directive from President Jimmy Carter that a series of rifle-shot studies should be done on each of the emerging technologies, but not published, only be given to DOE and the White House. He wanted a consultant to "tell the truth" about these new technologies. This was the first major national study of PV in the US

Power for the World by W. Palz
Copyright © 2011 by Pan Stanford Publishing Pte Ltd
www.panstanford.com
978-981-4303-37-8

Figure 1.

The "PV Study" was completed in November 1978, concluding that PV would be a "post-2000" technology that would take off after the turn of the century. We did a very good job with this study. It was not what the government program officials wanted to hear, but it turned out to be accurate.

One of the conclusions of the study was that the provision of low-cost capital would be a key success factor for solar PV because PV is the most capital-intensive power generation technology, so it is most sensitive to interest rates. This led to the first discussions about a "solar bank."

Indeed, a version of this idea of a solar bank was enacted into law by the Congress in 1981. But it expired with no implementation in later years.

2. Away from PV for 15 Years

For the period 1980 to 1995, I went off to work at General Electric Company as the strategic planner of their power business, at Combustion Engineering, Inc. as VP of Corporate Marketing, the VC firm Arete Ventures as a partner, and United Power Systems, Inc. as CEO. This work generally involved conventional power generation from nuclear power, coal, and natural gas, a range of high-technology business in computers, and then the development of independent power plants (IPPs).

Then, in 1995, after 15 years in the conventional power business, I decided to stop doing that, feeling that my life was being used by old and polluting business when there at the doorstep was perhaps the greatest invention of all time — solar

Figure 2.

PV — and I wanted to play a role in helping it succeed. So I rededicated myself to solar energy, and to solar PV in particular.

3. Return to Solar PV in 1995

I made a personal study of the field and was surprised to see that the PV community was essentially the way it was in 1980, still driven by research, development & demonstration (RD&D) funding by US DOE, still talking about driving down the learning curve, still talking about "commercialization" and still not talking about financing at all. The industry was stuck measuring their cost as dollars per peak watt ($/Wp). I used to say, how can we cut PV cost by 50% in one minute? They would say it can't be done. I would say, well, if we measure cost in terms of the end user's cost in ¢/kWh, then we can cut the cost by 50% in a minute by cutting the interest rate on the financing by 50%. This simple logic made a big impact in those days. This led to quite a time of giving speeches about the power of financing to help drive PV into the market and down the real cost curve which is what the end user pays for solar electricity.

4. World Bank 1996–1998

In January 1996, the President of the Rockefeller Foundation, Peter Goldmark, made a challenge to the President of the World Bank, James Wolfensohn, that their organizations should create a joint venture to finance a solar energy revolution.

Figure 3.

A joint study was commissioned under Richard Stern, Richard Spencer, Dana Younger, and others at the World Bank Group, and Mike Allen and Christine Eib-Singer at a new NGO called E+Co in New Jersey. I was retained as the PV industry expert, based on the 1978 PV study, and was given the mission of visiting every major PV company around the world to solicit their support for the financing initiative. I visited the PV companies in the US, Europe and Japan and found that they wanted the World Bank to provide financing, but were not willing to put funds into a common pool for fear that their competitors would benefit more than them.

Then, as the joint study team began to write up its conclusions, it became apparent that the people at the World Bank were sandbagging the study and setting up the outcome for failure. We were astonished. The bank formed the Solar Development Corp. and the Solar Finance Corp. but we could see that it was being set up to fail. What an eye opener about the World Bank!

I was deeply troubled by this and went to see a Vice President of the World Bank who was a business school classmate. He said: "Mike, we do many important things here at the bank, but I have to say that lending money to poor people to buy solar systems is not one of them. We finance entire economies, plan and build capital markets, and collaborate at the highest levels, not the lowest levels." I left the meeting disillusioned. I called a meeting with Richard Stern and presented him with a letter demanding that the World Bank return the name SolarBank. He was in complete agreement, saying "the World Bank will never do this job — you should take it back and do it yourself!" He signed the letter, and

I went on to pursue SolarBank myself and obtained a trademark on the name SolarBank.

The Solar Development Corp. and the Solar Finance Corp. were formed by the World Bank Group, and they did fail.

5. India 1996–2001

I attended the solar PV conference in Sun Valley Idaho back in September 1995, and met Bob Freling of the Solar Electric Light Fund (SELF), which was then run by Neville Williams. We got to talking and it led to my commitment to get to India where there were the beginnings of a financed market for solar PV — bank loans, not government funding — taking place under a new company called Solar Electric Light Company (SELCO) that was run by Harishe Hande. I went to India in January 1996 and met Harishe and his lieutenants — Thomas and Pi, who were selling and installing solar home systems from their motorbikes. Harishe's aunt (named Auntie), who owned the cable TV system in Mangalore, made the lighting equipment at night. Harishe was the original solar entrepreneur in the world. He still is.

For the next several years I continued to go to India on small grants from Winrock International and the Joyce Mertz Gilmore Foundation, and work for Shell International, continuing to develop the SolarBank idea.

In 1998, my company, called Solar International Management, Inc., won a $1.01 million grant from USAID to develop solar financing and the SolarBank idea in India. We worked with a marvelous man named Dr. B. S. K. Naidu at Winrock India. We employed Rahul Arora, Aesha Grewal and others in India to develop a training program, and we trained 1000 bankers on how to do financing for solar home systems. The program was led by a most wonderful man named K. M. Udupa, retired head of the rural bank within Syndicate Bank.

We also worked with the microfinance community to get them into the financing of solar home systems for their networks of women in poverty. I learned from many travels around India what poverty is. As Ghandi said: "The most expensive electricity is no electricity." Solar PV, seeming to be so expensive at $1 per kWh, was in fact cheap compared to the $10 per kWh (equivalent) that the poor pay for light from candles and kerosene lamps.

In a second element of the project, we worked for three years with the microfinance firm called Basix in Bangalore. We trained their staff, even taking them to Sri Lanka to see how the financing was being done there. In the end, after three years work, we all concluded that microfinance is not suited to such asset financing. They are small business working capital financiers, not asset lenders. The women would join together to borrow money to support each other's micro-business, but not to finance a PV system on only one roof.

Figure 4. **Washington 2010:** Mike Eckhart, President of ACORE, on the left, with Holzschuh, Vice-President of Morgan Stanley; standing: Locke, the US Secretary of Commerce.

The third area of work in India under the USAID funding was to create the institutional infrastructure of the SolarBank. We retained Arthur Anderson to do this research, paying the firm to develop organizational options and recommend a path forward so long as it was not based on what is called a "Non-Bank Finance Company" because there are regulatory issues that prevent foreign entities from taking money out of a non-bank finance company. In the end, Arthur Anderson recommended that SolarBank be a non-bank finance company, and therefore it cannot have any international ownership. We were stymied by India's tremendous regulations.

This work in India was a great learning experience and had a huge impact. We succeeded in training 1000 bankers and we are told that this continues to be done under USAID funding even today. We had made a great advancedment in the microfinance world. I came away believing that entrepreneurs like Harish Hande are the key to the success of renewable energy in India. There are many institutional forces in India keeping things from getting ahead, and it is the entrepreneurs who find ways of getting ahead.

For all this, I was given the award "Best Renewable Energy Man of the Year of India" by the Indian National Foundation of Energy Engineers.

6. South Africa 1997–2002

The late John Bonda, Executive Director of the European PV Industry Association (EPIA), become my friend and mentor in those days in 1995–1996 when the idea about SolarBank was taking shape. He planned a trade mission of European renewable energy companies to South Africa in January 1997, and could find no one to speak about financing! John called me in Washington DC and asked if I would go to South Africa to give the financing speech. I was thrilled, and it led to a three-week trip that led to 14 more trips to South Africa.

In 1996–1997, I went to South Africa about five times for Shell International, creating a joint venture with the national utility ESKOM, called the Shell-Eskom Solar Energy Company. The idea for it was hatched back on John Bonda's European trade mission when I met a South African by the name of Herman Bos, who had the idea of a company that installed solar home systems on the company's capital and charged a "fee for service" using prepayment meters. I suggested to Herman that we get Shell and Eskom to co-fund it as a joint venture.

We created the $25 million joint venture with equal funding from Shell and Eskom, which took a year, and enticed Herman Bos to be the chief executive. President Nelson Mandela heard about the project, which covered an area in the Eastern Cape where he grew up, and told his staff that he wanted to inaugurate the project. This became a massive two-day event, like a mini-Woodstock, out in his homeland village. Shell cut a 12-km road to the village to get the press bus in. People came from all around. There were marching bands, choir singers, men coming on horses, others coming to show their cow herds, women dressed in their finest, and children everywhere. President Mandela and the top officials arrived in their gunship helicopters. It was quite a scene.

The Shell-Eskom joint venture company installed some 10 000 solar home systems in the 1998–2000 period before the corruption set in. The central government took the Shell-Eskom idea and granted concessions to six other companies, each with a mandate for 50 000 systems. Then they added a government subsidy which never reached the poor end users. The whole enterprise of concessions that our Shell-Eskom joint venture spawned ground to a messy halt. It is still struggling to function today, but is not succeeding.

Also in the 1998–1999 period I went back to South Africa several times under grant funding, exploring the feasibility of establishing a SolarBank operation there. We met with all the banks such as ABSA, Standard Bank, NedBank, African Bank and others. We worked with the Department of Minerals and Energy (DME), the Development Bank of Southern Africa, the Independent Development Trust (IDT) and other institutions. My close colleague at IDT was Reinhold Viljoen, who became the national advocate of a SolarBank, and a trusted collaborator. This work continued until 2002.

My last effort in South Africa was to rescue $900 000 of funding that the US government had given to the South African government to renewable energy development but went unused. I had an idea. The US government has a program called the Development Credit Authority (DCA), providing loan guarantees for home mortgage lending in the developing countries. I presented a plan to the State Department — we would use the funds as the required reserve for an

$18 million fund to finance solar PV installations on rural businesses ("productive uses" in government talk). In September 2002, Dr. Griffin Thompson of the State Department went to South Africa with me for a week to set up the program. We succeeded. It was all ready to go. The chief economist in the US embassy was ready to issue an RFP and run the program. All of the banks we met with were ready to bid to the RFP. The DBSA was poised to join in if it worked.

Then we had our final meeting of the week with the Deputy Minister of DME before catching our flight home. We were thrilled with the results and reported that it looked good. With her go-ahead, the program would proceed. But we were stunned by what she said.

With considerable flourish, the Deputy Minister, with 6–8 staff in the room, folded her arms and said: "No! You will not do this. You will not help those banks loan the money to my people. That money belongs to my people and was taken from them over the years by those banks. No, you will give the money to me and I will take care of this myself!"

I looked at Griff and said: "It's your turn to talk." Griff said: "You're correct, so please don't say anything." He turned to the Deputy Minister and explained in very few words that the meeting was over and we would be departing. We left the room in silence and proceeded to the airport to catch our flight home. Years of work had been wiped out by the few words said in this meeting. America cannot do business in the presence of corruption.

I returned to South Africa with my family for a holiday in the summer of 2008, saw many old friends, and was saddened to see that the forces of bureaucracy and corruption had mired the solar PV strategy into doom. I drove my family through KwaZulu Natal and the Eastern Cape, and saw many homes still without electricity. It reminded me of a phrase we coined in those days: "solar electricity is the first step onto the path from poverty." With solar electricity comes light, and with light comes transformation into modern life, and modern life is based on education, reading and writing, and cash incomes. We could see this "connect the dots" from poverty to economic upliftment through solar electrification, but so many other forces always seemed to get in the way. We saw families still living in the dark in 2008, many years after they could have had solar lighting if the forces of bureaucracy and corruption had not prevented it.

7. Europe 1997–2004

Throughout this period I was traveling through Europe almost monthly. Hermann Scheer, the leader of the solar revolution and a hero to me in those days (and today), invited me every year to speak at the solar energy conferences sponsored by his NGO Eurosolar in Bonn. It was there that he once asked: "Mike, how can I get Wall Street to finance my solar energy revolution?"

I said: "Take the 5-year term of your feed-in tariff, and make it a 20-year commitment, and then stand back because the flood of capital will amaze you."

Hermann asked why is this, and I explained that lenders have a thumb rule that they will make loans for a term equal to no more that 75% of the term of the "assured revenues" of a project. So, the 5-year FIT will attract 3-year loans. If we want 15-year loans, as we had just been saying in the conference that day, then there must be assured revenue for 20-years.

Hermann Scheer went to the Bundestag the next day, introduced an amendment to change the rule from 5-years to 20-years, and it passed. Immediately began the stunning growth of the PV market in Germany.

There were other conversations in those days in Europe. I compare that period of time in Europe to being in Philadelphia in 1776 — it was a time when the principles of a revolution were taking shape. The most important and influential person of the time — the "Thomas Jefferson" of renewable energy policy leadership — was Wolfgang Palz, then head of renewable energy in the European Commission. I was nervous one day to be introduced to him for lunch in Vienna, he was so fearsome and important at the time, and then over the years he became my best colleague and most trusted friend, as we met in so many places around the world like Paris, Brussels, London, Valencia, Rome, Venice, Bonn, Berlin, Cape Town, New Delhi, Beijing, New York, Washington, and other places, always plotting the next move on policy, and always reserving one evening to smoke a Cuban cigar. He inspired me and all of us with his knowledge, wisdom and courage, and became a model to many, including me. We felt it at the time, that it was a time of important change, and that what was happening could lead to a better world.

One conversation with Hermann Scheer and Hans Josef Fell, both members of the German Bundestag, was about how the German feed-in tariff idea was like the way we regulated the price of nuclear power in the US. In those days in the 1970s when new nuclear plants were coming on line, the Public Utility Commission would calculate the "revenue requirements" of the new units, ensuring that the utility could cover operating costs plus debt service plus a return on equity. The beauty was applying the old nuclear power revenue requirement methodology to renewable energy technologies, and this led to the German feed-in tariff approach of calculating a different price for each technology based on what price is needed to cover operating costs, debt service and a return on equity.

8. ACORE 2001–Present

In June 2001, I was in Bonn Germany at one of Hermann Scheer's conferences where he was forming the World Council for Renewable Energy (WCRE). There were two Americans there, among the 200–300 attendees: Neville Williams and myself. Hermann asked Neville if he would be the US chairman of WCRE, but Neville said no, he was too busy with other things. Hermann looked at me, and I said that I'd be happy to serve in that capacity.

I came home from that trip and convened a breakfast meeting with Scott Sklar, Judy Siegel and Griff Thompson to talk about this new WCRE and how we should implement it in the US. We all agreed, there should be an American Council On Renewable Energy (ACORE) — and we each pledged an hour a day of our time to this new project. An hour a day soon became 100 hours a week!

We held the founding "steering committee" meeting of ACORE at my office on September 4, 2001, just one week before September 11. Had we started ACORE after September 11, it would have been quite different. Things took off from there, the hour-a-day commitment became an overwhelming demand on my time and eventually I became fully engaged managing ACORE while seeing the SolarBank initiative slip to the back burner and now on the shelf waiting to be restarted.

ACORE only got going because of a very talented staff, Jodie Roussell. This was the summer of 2003. I hired Jodie to help me conduct a study for BP Solar. When it was completed in December 2003, Jodie picked up duties on the ACORE side project, and was responsible for establishing much of the operational infrastructure that made ACORE successful.

For the next four years until January 2008 when Jodie left to attend IMD business school, we worked as a wonderful team to build ACORE. Her skills, dedication and personal integrity are without equal. Jodie defined the term "invaluable" and the renewable energy community owes her a place of honor for this remarkable effort.

9. Bonn 2004, WIREC 2008, and IRENA

In June 2002, I was attending Hermann Scheer's conference in Bonn when he called for the creation of an International Renewable Energy Agency (IRENA), which he first proposed in 1990, and which was actually formed in January 2009.

It was two months before the WSSD meeting in Johannesburg. We were angry about WSSD, taking place two months later in August 2002 because the deal was rigged — all renewable energy and energy efficiency proposals were being blocked by the Middle East oil-producing countries. Hermann Scheer, David Wortmann and I spoke in the hallway — that some country, with no affiliation with the UN, should call for a global meeting on renewable energy. We said, if one country comes, then they will all come!

Hermann Scheer wrote a speech for WSSD to call for such a meeting. But two weeks beforehand, UK Prime Minister Tony Blair announced that he was going, and then German Chancellor Gerhard Schroeder said he was going! Hermann Scheer challenged Schroeder to give his speech, calling for a world meeting. Schroeder gave the speech, calling for what became the Bonn Renewables 2004 world meeting. Then there was the Beijing 2005 world meeting. Then, ACORE organized the WIREC 2008 event in Washington, DC.

Finally, the German government supported the formation of IRENA, and it was inaugurated in January 2009 in Bonn.

This series of global meetings and organization on renewable energy gave great support to the development of markets for solar PV through this period.

10. SolarBank Looking Forward

When I began with the SolarBank idea seriously in 1978, there was no way of knowing where all of this was heading, but we felt that this technology could change the world. Then, in 1995–1996, I decided to devote myself to SolarBank. At that time the worldwide solar PV market was about 20 MW/year of installations and $200 million in revenues. SolarBank was too early, ahead of its time. Today, the worldwide PV market has grown to 7000 MW/year and $35 billion/year in revenues. Financing of solar is the hot topic of the day. SolarBank is very much needed.

We'll see what happens, whether SolarBank will stay on the shelf as an old idea that came and went, or a good idea whose time has come.

The solar energy revolution is underway now, and based on the deep commitments of United States, Germany, Japan and China, this industry will be stop. It has turned the corner. It has reached its tipping point. The solar energy revolution will happen. The remaining questions are how big, how fast and to what extent of the globe?

With a successful SolarBank, the impact will be bigger and faster but the question to what extent is another matter. Just providing low-cost capital is not enough to get solar energy implemented in the developing countries. I will never forget meeting a roughrider kind of man in South Africa who ran a rural lending business. He wore boots, and had a gun in each boot. He charged 35% interest rates to his good borrowers and up to 100% to risky customers. He said: "I don't want you coming into the rural areas lending money at 10% interest to my customers — you will make me look bad and put me out of business."

There are many complexities to the roll-out of solar energy that are not obvious at first look, and we can't see all the complexities until we try to implement our ideas. Over time, I believe we will overcome those complexities and make this very successful.

After all, we are dealing with an inexhaustible energy resource, the energy from the Sun. What greater accomplishment can humanity achieve than capturing the energy of the Sun to ensure the survival and sustainable life of all of us here on Earth? In doing so we will have connected our existence and prosperity to the energy of the Universe itself, beyond the bounds of Earth, and into the very hands of God.

Chapter 13

Will This Work? Is It Realistic?

Thoughts and Acts of a Political Practitioner with a Solar Vision

Hans-Josef Fell

Member of the German Parliament

1. My Way of Solar Thinking

In 1971, I had just gotten my high-school diploma. Shortly after, we had the first oil-price crisis and the Club of Rome published the *Limits to Growth*. As a young student I felt closely linked to the rebellious, relieving, and questioning ideas of the German student revolt of 1968 and I wondered how the problems of tighter resources and destruction of nature could be overcome. I realised quickly that one of the main dangers to our world was the use of the fossil and atomic resources. Suppression, exploitation, spoiling the natural environment, and conflicts about oil were already on the agenda in those days. I realised more and more during my studies of physics the problems associated with the use of the atomic energy; I could not simply believe in the many "atomic" claims expressed by my professors and put them into doubt. This was complemented in the early 80s by the awareness of the drama of global change.

In my first years as a student of physics I also learnt quickly that the Sun irradiates 10 000 times the energy on Earth that mankind needs and uses. Hence, I saw solar energy right from the early 1970s onwards, as an overarching comprehensive solution.

My purpose was always to work out overall solutions and not only partial ones. Not only to introduce cars with catalytic emission control, but to stop CO_2 emission, too. The success of the catalytic emission control of the cars since the early 80s and the sulphur gas cleaning of coal power plants has slowed down

Figure 1. House of Hans-Josef Fell. Power and heating for his house are provided 100% by renewable energies.

the dying of forests, but at the same time it could only make worse the climate change problem. Had we given preference to electric cars driven by green power and to wind- and solar power plants early on, we would not even need catalysers and sulphur cleaning and maybe we would not have to deal with a dramatic global climate change as well as classical pollutants.

At that time I did not understand why one keeps betting on the destructive fossil and atomic energies despite the availability of that abundant and unlimited energy from the Sun that does not destroy nature.

I followed with much hope the first industrial revolution on the Renewable Energies in the United States that was promoted by President Carter. When President Reagan later stopped brutally all support for Renewable Energy research and market promotion, I started to realise that the driving force behind the fossil and atomic energy system is made up by the economic interests of the energy monopolies that are sponsored by a complacent political patronage.

All this kept worsening until today. Governments of all nations are seeking despearately a way out of the looming climate catastrophe since the Global Summit of Rio in 1992. Nevertheless the global temperature increase kept accelerating. No solutions are found while the economic interests of the energy monopolies carry on regardless. For the world leaders, energy security is impossible without oil, natural gas, and uranium. Despite the fact that these resources stand for 80% of global Green House Gas emissions. The ongoing dependence on fossil and nuclear energy makes an effective climate protection impossible to realise. Moreover, leaving behind the "peak oil", neither today's energy supply nor that of the future can be fully secured with the conventional energies.

The current financial crisis was inter alia stimulated by an oil price of $150/bbl in July 2008; it actually gives already a taste of what kind of problems to expect for the global economy and all of our well-being as a result not only of global climate change but also of the limitation of conventional resources.

But instead of heading now resolutely at a 100% Renewable Energy supply, most of the world's leaders keep supporting the conventional energy monopolies. The market dominance of those has developed into the largest business worldwide. nine of the 16 largest concerns on Earth are oil companies. Six other technology companies, mostly those of the automobile industry, place on conventional energies.

And the profits of the oil companies are growing along with the oil price. For instance Exxon, the world's biggest concern, had in 2002 a profit of $12 billion while the oil price stood at $22; in 2008 it went up to $42 billion for an average oil price of $100 that year. The oil price developed in the meantime into an important instrument for steering money from the poor to the rich — but despite of this (maybe because of this) we still do not have a fully consistent policy for a rapid switch towards Renewable Energies.

This is the more surprising as it has now become visible for everybody how fast an industrial development of the Renewable Energies can be realised. Germany's "Feed-in" law EEG giving priority to Renewable Energy in the Power sector has stimulated growth rates that had still in the year 2000 been shrugged off as completely unrealistic and naïve. It did start an industrial revolution that can best be seen from the new jobs that were newly created. The 30000 jobs we had in 1998 in Germany's Renewable Energy sector have become 300000 in 2009, ten times as many.

2. Being Called a Solar Do-Gooder and Unrealistic Politician

But those who call for the decisive solutions, that is the 100% switch towards the Renewable Energies, are slandered and insulted as unrealistic, crackpots, or out of touch — no matter if in the past, it was the solar pessimists who got it all wrong. Unanimously, they had declared the great successes of the Renewable Energies, we have recently experienced as impossible. But to be called "unrealistic" is actually hiding a methodology of prohibiting the fast growth of the Renewable Energies. From the supposedly slow growth only for the renewables, they derive the authority for the ongoing investment into fossil and atomic energies. Deliberately they spread the fear that otherwise the lights may go out.

Many solar pioneers had the same or similar experiences like me. I myself was, in my whole life, was smiled at or even insulted as a stubborn do-gooder and dreamer. My analyses were rejected most of the time as incorrect, without discussing them with me in detail. My only possibility then was to confront myself with my own thoughts and to discuss with people who shared my ideas. That is how I was able to progressively sharpen my vision and strategy.

The developments of the last few years concerning the shortage of resources, global change, and the steep growth curves of the Renewable Energies did confirm

Figure 2. Hans-Josef Fell and his solar mobile in front of one of the largest solar panels in his constituency.

my earlier analyses and even exceeded my expectations to some extent. This is the reason why I could always stick to the fundamentals of my ideas of the 70s; I remained firm in following and implementing them into my private life and into politics. This persistence and the experiences gained were the reason for my successful political actions in favour of the Renewable Energies. In every phase of my activity I accepted critics on my ideas and analysed it thoroughly. Most of the time the critique was unjustified, but sometimes it improved important details of my world of ideas.

I remember many situations in my life where I was warned not to implement my ideas, where I was accused of a lack of realism and of crazy ideas and that it was to be expected that my plans were impossible to realise.

Nonetheless the triumphant advance of the Renewable Energies eventually would be realised by striving vigorously with the many companions of the Solar Revolution.

3. Some Important Steps of My Life Illustrate the Persistence of My Solar Way

I was told:
A home exclusively supplied with Renewable Energies were impossible —

My house that was built in accordance with solar criteria in 1985 is exclusively supplied with Renewable Energies since 1995. A PV generator of 1991 and a cogeneration plant fed with vegetable oils allow me to produce more electricity than I need. My heat requirements are met by a green house and solar heat collectors together with a wood furnace and the heat coming from the cogeneration plant. My two cars are solar powered since 1996 with my own electricity or respectively with vegetable oil.

I was told by the professionals of my community:
Solar heating of the public swimming pool is uneconomic and technically impossible to realise —

As a city councillor I was able to arrange a solar pool heating as early as 1992; it saves till today masses of heating oil economically.

Colleagues of the city council and lawyers told me:
A city regulation for "cost covering" tariffs for solar electricity is legally impossible —

The first decision at city level, worldwide, on the cost covering tariffs of solar electricity was arranged by me as a city councillor in Hammelburg, a small town in Bavaria. It also worked as a model for other communes such as Bonn, Munich, Nuremberg, and Darmstadt; eventually, thanks to the great success it had locally, it became the basis for a federal law in Germany, the EEG feed-in tariff law mentioned here above.

I was told by financing experts:
A cost-effective operating community for solar electricity is economically impossible; no financial support will be available —

I have created the first community of operators for solar electricity — the Hammelburg society for solar electricity. In only two years, we were able to collect the needed capital of 200 000 DM. It is profitable until today. Collecting that capital was easy despite the announcement that the investment was not secured as long as Bavaria had not yet approved the "cost-covering" tariff system. After tough negotiations with the Bavarian authorities we succeeded in 1996 to get the approval that many had thought impossible to reach.

Leading economists had told me:
A yearly installation of 80 MW of PV power that I had projected in 1995 by extrapolating our Hammelburg experience is unrealistic and wishful thinking; a solar industry of this size is outside any realistic financial scope —

In 2009 Germany installed 3800 MW new ones.

I was told by most of the federal and state politicians:
My demand for a German federal law of "cost-covering" tariffs for all types of electricity generation from Renewable Energies is nonsense. Still in 1998 it was rejected with a shrug by most politicians as unrealistic. This is the reason why it did not become part of the new Government programme that same year —

In the year 2000, the red/green majority of the Bundestag has adopted the EEG.

Even in 1999 I was told by some politicians defending the solar energy case: My proposal for 0.99 DM per kWh of solar electricity in the EEG was reasonable but unfortunately impossible to get through —

In 2000 the German Parliament adopted exactly this tariff in the EEG. Together with the "100 000 — roof" programme it became the breakthrough for the market introduction of PV in Germany; in accordance with the experience gained for the cost-covering tariffs in towns and cities, the principle of profitable PV tariffs became binding law in Germany.

Utility experts told me:
The goal of the EEG of the year 2000 to double the share of the Renewable Energies by 2010 to 12.5% is totally unrealistic —

At the end of 2009 the share of electricity from Renewable Energies in Germany stood at 16.1%.

I was told by the experts in the Research Departments of the German Government in 1999:
It is not worthwhile to invest further budgets into solar thermal power, into geothermal energy, other Renewable Energies, and electric cars. It would make more sense to devote this money to nuclear fusion —

Thereupon I threatened to vote against the whole federal budget in the Parliament as I thought that exactly these additional funds for Renewable Energy research were necessary. By this brute force, I succeeded in increasing instead of reducing the research funds for the Renewable Energies. This was the basis for geothermal power plants and a revival of solar thermal power.

I was told in the year 2000:
It is not worthwhile to promote the EEG in other countries; it could at best be realised in Germany as other countries give preference to atomic power and coal —

I did travel to many countries and reported on the great successes of the Renewable Energies in Germany and the legislation that made it happen. Today some 50 countries worldwide have adopted with various consistencies similar legislation following the model of the EEG. In many countries I was and still am consulted by Governments and Parliaments, by Universities and companies concerning the legislation issues; lately even by the US Congress.

3. And How Is It Today?

Today one encounters again the incredibility with the claim that by 2030 the world could switch to the Renewable Energies provided the world leaders agreed. It is opposed the same way as the demand to clean the atmosphere from much of the man-made CO_2, towards 330 ppm or so.

One could well think of industrial processes allowing for a 100% supply of the planet with Renewable Energies; also the CO_2 reduction of the atmosphere is possible, for instance via vegetation that absorbs the CO_2 and keeps it safely in the ground.

As usual there are the sceptics, the doubtful, and in particular the protectors of the interests in the fossil resources and the interests in conventional agriculture. They all prohibit such an effective and rapid climate protection.

But the positive force of the Renewable Energies will impose itself faster than the leaders of the fossil and atomic energy system anticipate. More and more people, companies, and political leaders realise the impasse of the conventional energies. They want to get rid of the dependency of the big energy monopolies that had given them a hard time with ever increasing energy prices. They want to liberate themselves from the need for wars about limited resources and fear increasing atomic dangers. And they want an effective climate protection.

I will never stop stressing the chances of the Renewable Energies' rapid growth while denouncing the deplorable state of the conventional energy supply and agriculture. I am going to promote the goal of 330 ppm of CO_2 — even though I will hear again that this is unrealistic and a crazy idea of an incorrigible visionary.

I learnt in my life that a lot more is possible than what most people believe in. This provides me and others with the strength to follow those goals.

The World has, Anyway, no Better Chances.

Hans-Josef Fell

Chapter 14

The IEEE Photovoltaic Specialists Conference*

Americo F. (Moe) Forestieri

NASA, Lewis Research Center, Cleveland, Ohio, USA

This collection of names, facts, data, anecdotes, and memories is my feeble attempt at documenting the history and growth of the Photovoltaic Specialists Conference. I began this effort about two years ago when I decided that much of the information on how this conference has grown over the past years was beginning to disappear.

1. Brief History of the US IEEE PVSC and the William R. Cherry Committee

At the beginning of the 1960s, a large number of organizations under the Department of Defense and NASA were carrying out and supporting R&D on energy conversion devices and systems. Representatives of these organizations held regular meetings under the auspices of the Interservice Group for Flight Vehicle Power (IGFVP). At a meeting in Philadelphia, Pennsylvania on March 7, 1961, the Solar Working Group (SWG) of the IGFVP decided that a broader meeting was needed which should include personnel from industry and the universities active in the photovoltaic device area. Minutes of that SWG meeting comprise seven pages and its report number is PIC-SOL, 3/1, dated March 16, 1961.

The Institute for Defense Analysis (IDA) was charged with organizing this broader meeting (The First PVSC) which was held on April 14. 1961. It took place

---

* June 1996.

Power for the World by W. Palz
Copyright © 2011 by Pan Stanford Publishing Pte Ltd
www.panstanford.com
978-981-4303-37-8

in a basement conference room in Washington, DC. The Power Information Center (PIC) of IGFVP was given responsibilities for information dissemination and resulted in PIC-SOL 209/1, a two-page Report on Findings of Photovoltaic Devices.

In February 1962, the Interagency Advanced Power Group (IAPG) held a Solar Working Group Conference (the second PVSC) in Washington, DC. The topics of discussion were "Radiation damage to semiconductor solar devices", and "Solar power systems calibration and testing". Proceedings of that conference were prepared by the PIC of the IAPG and distributed to all attendees. In April 1963, the 3rd Photovoltaic Specialists Conference, sponsored jointly by IEEE, AIAA, and NASA, was held at the Statler Hilton, Washington DC. Again, the proceedings were prepared and distributed by the IAPG PIC.

In June 1964, the conferences began using a numbering system along with the title Photovoltaic Specialists Conference. The 1964 and 1965 (4th and 5th) conferences were also jointly sponsored by IEEE, AIAA, and NASA; the proceedings were prepared and distributed by the IAPG PIC. Since 1967, (6th PVSC, Cocoa Beach, Florida) the conference has been sponsored solely by IEEE.

At one of the planning meetings for the 1980 conference (14th PVSC), shortly after the death of William R. Cherry, the conference committee agreed to establish the William R. Cherry Award.

Joseph Loferski wrote the following description:

Mature communities — be they nations, professions or groups of people working toward a common objective — identify and honor their heroes, those persons who through their vision, perseverance and commitment helped to launch the community during its delicate early years: Having been born twenty-five years ago, the photovoltaic community is such a mature community. It is time that it, too, should begin to identify and honor its heroes. It is for this reason that the IEEE Photovoltaic Specialists Conference Committee has decided to establish an award to recognize outstanding contributions to the advancement of photovoltaic science and technology, and to name it in the memory of one of the heroes of photovoltaic history, William. R Cherry. It was the intent of the organizing committee that an award be given at each subsequent specialists conference.

2. 8th PVSC: The 1970 PVSC in Seattle, Washington, by Joseph Loferski

The 1970 PVSC was held in August at the Olympic Hotel, Seattle, Washington, with Professor J. J. Loferski of Brown University as General Chairman. Henry Oman of Boeing Seattle was local arrangements' Chairman.

About 90 people attended, reversing a trend of ever increasing attendance at successive PVSC Conferences. The reason for the decline was a substantial reduction in NASA and DOD support and interest in photovoltaic cells. The "politically correct" position at that time was PV power in space would be limited to perhaps 1 kW; space power above this level would be provided by nuclear reactors in space; the SNAP program was the wave of the future! Dr. Werner von Braun, the rocket scientist, derided PV power supplies for large (1 kW and above) space systems as "elephant ears", an inelegant and unacceptable solution to space power. The PV community was pervaded by gloom; the 1960 decade of increasing investment in PV research and development appeared to have ended and future prospects for PV were very poor. In this disheartening situation, Bill Cherry and Gene Ralph proposed a symposium on terrestrial application of PV cells and solar power to be held on the afternoon of the last day of the conference. Cherry talked about PV "rugs" to be made by depositing thin film PV cells onto flexible reel-to-reel machines. The large area panels would be suspended above the clouds by "blimps" tethered to the ground. Gene Ralph introduced the PV community to the "learning curve" concept. He told us how industrial experience in manufacturing many products resulted in reductions of

Figure 1. Committee of the 11th IEEE PV Conference in Scottsdale, AZ, 1975.

price between 10 and 30% for each doubling of accumulated production. Application of this concept to PV cells led him to the conclusion that when PV accumulated production reached the level required for truly large scale solar PV power, the price could be expected to drop below the $1/US per peak watt. Cherry's and Ralph's ideas revived the flagging spirits of the PV community. Perhaps PV had a future as a terrestrial power source, as an alternative to nuclear power! A daring dream indeed!

A direct outgrowth from this symposium was the establishment by NSF, NASA and the White House Office of Science and Technology (OST) — a task force to explore solar energy as a national resource. Bill Cherry played a central role in that initiative which eventually resulted in a 1972 blueprint for development of solar power, including PV power. This blueprint was the "bible" for development of terrestrial solar power during the 1970s.

On the lighter side, the conference excursion was a visit to the new Boeing 747 plant which had just become operational in Everett, Washington. Boeing had just "downsized from about 100 000 to about 50 000 employees (so what's new) because orders were not coming in for the new plane. The Boeing plant was the largest building in the world. The preceding PVSC was held in Cocoa Beach, Florida where our excursion trip had been to the (then) largest building in the world, the launch building for the rocket which would place Americans on the moon. I predicted that the future PVSC would be held in Chicago where the Sears Tower was under construction; it would supplant the 747 plant as the largest building in the world. That never happened although other predictions of that 1970 Conference did eventually reach fruition.

3. 12th PVSC: 1976 Baton Rouge, Louisiana, by Americo Forestieri

I remember that the meeting notice for the 12th PVSC invited everyone to the new "Hilton at Corporate Square" in Baton Rouge, Louisiana, a convenient drive from historic and lively New Orleans. What I am trying to forget is the entire conference. But my "friends" keep bringing it up. I believe it has become the benchmark on how not to run a conference.

Some of the things that went wrong:

- Prior to the conference we were told by the hotel management that even though the hotel was under construction it would be completed in time.
- Construction was still going on in parts of the hotel at conference time.
- We were told there were adequate facilities for holding exhibits (a first for PVSC).
- We found that we had to remove the top of the elevator to transport the larger items to the exhibit floor which was on an upper level. Some of the exhibits had to be located in a tent in the parking lot.

- We were told that the hotel would prepare a special event for the conference attendees.

It turned out to be the infamous "Louisiana Luau" which started about an hour late and ran out of food before half of the people were fed. I was obliged to refund everyone's money. Several members of the conference organizing committee met with the hotel management to discuss the fiasco. After much discussion the hotel decided not to charge the conference for this event.

I found that about a year after the conference several of the hotel owners and managers were sent to jail for defrauding the public. By the way, all of the hotel arrangements made in those days were without a contract. Needless to say, after that, every detail was spelled out for future conferences. Perhaps I did a service to the PVSC community after all.

And, oh yes, the weather was terrible.

Prior to the conference I arranged for a committee dinner at a local Italian restaurant run by a friend of mine. I chose the menu for everyone and I don't believe that anyone will forget that delightful banquet. It's too bad we didn't have this after the conference. Maybe everyone would have forgotten the conference gaffes.

Americo F. (Moe) Forestieri

Chapter 15

Review of China's Solar PV Industry in 2009

Gao Hu

*Center for Renewable Development (CRED) of Energy
Research Institute (ERI) under National Development
and Reform Commission, China*

In 2005, China attracted the world's attention by issuing the *Renewable Energy Law*, which put wind, solar energy, biomass energy etc. in an unprecedented role in the national energy strategy. This is the first time that solar photovoltaic (PV) industry began to gather real attention from the market and to start an entirely different way in China. Prior to this milestone, solar PV production was considered too expensive to be afforded as the commercial power supply sources.

For instance, solar PV was only considered as one of the technically-efficient solutions to provide power to remote and poor areas where no electricity was accessed. China's government initiated the ambitious Song Dian Dao Xiang (Township Electrification Program) in 2003 to deploy the solar PV panels to households or build isolated solar PV stations (totally 20 MW) in seven western provinces by subsidizing the capital investment, in order to electrify all the townships. It is estimated that 1065 townships and about one million residents benefited from this program.

The electrification program built a sound basis for China's solar PV industry to grow up. Furthermore, the *Renewable Energy Law*, changes the potential roles that solar PV could play. Against the booming wind industry, solar PV strikes Chinese attentions by the wealth rankings. Dr. Shi ZhengRong, back from Australia and CEO of the solar PV enterprises, SunTech, was the Top No. 1 in the China's wealthy list, by successfully IPO in New York. After that, 13 China solar PV enterprises were successfully listed in oversea stock market and 11 in domestic stock market. Solar PV industry dominates and represents the emerging energy in China's capital market.

Power for the World by W. Palz
Copyright © 2011 by Pan Stanford Publishing Pte Ltd
www.panstanford.com
978-981-4303-37-8

The participation of the capital totally changed the growth road of solar PV industry in China. In 2000, China's production capacity of PV modules is less than 10 MW. According to the statistics of China Renewable Energy Society, in year 2009, China's gross output of solar PV cells has reached 4011 MW which accounted for almost 40% of the world's gross output, ranking the top position around the world for a third consecutive year. Notably, SunTech's annual production capacity was 704 MW, only after the USA-based First Solar. Moreover, Yingli, JA Solar, Trina, Canadian Solar and Ningbo Solar etc. from China mainland, as well as the Taiwan enterprises, i.e. Motech, GinTech, were all top-15 PV companies in year 2009.

A milestone and positive news for China's solar PV industry in year 2009 was that, after almost a 5-year preparation, the first large-scale grid-connected solar PV power project was commissioned. This project, size of which was 10 MW, located in GanSu province, western China, was the first to be selected during the long-list of candidates, all of which would be larger than 5 MW and initiated too in the coming years. The price of this pilot project was surprisingly low — at 1.09 RMB/kWh (roughly 16 US cents/kWh) through the intensive bidding process. Though it was beyond expectation and critics complained about the distortion of the bidding mechanism under the competition, this project represented the first step for China's efforts in harnessing the solar power at a large-scale manner. It would also serve as a benchmark in formulating future PV pricing policy.

The renewable surcharge mechanism, which was increased from 0.001 RMB/kWh to 0.002 RMB/kWh and further to the present 0.004 RMB/kWh in the 2009, and levied to all the power consumers since year 2006, was created by the Law, and was supposed to cover the incremental cost of all grid-connected RE power projects, inclusive of solar PV power projects.

Against the ground-mounted projects, the government of China released the subsidy policy in 2009 to support the penetration of building-integrated PV (BIPV) demonstration projects. As per this, over 2 US$/W of the BIPV projects' capital cost would be subsidized by the government. In addition to this, as a near-term measure (2009–2011) to speed-up the demonstration and the industrialization of solar PV power, the government of China initiated the *Golden Sun Demonstration Program* in 2009. This program would cover diversified pilot projects, key R&D activities during the industrialization process and necessary capacity building programs.

The year 2009 would prove to be a historical time for solar PV industry in China. During the United Nations' climate summit, Mr. Hu Jingtao, Chairman of the country, made the ambitious commitment to increase the contribution of non-fossil energy to 15% by year 2020 in the primary energy consumption. In line with this guideline and inspired by the booming industry, China's government made the draft to update the national RE targets. Regarding solar PV, the 1.8 GW target by year 2020 would be increased to an ambitious 20 GW.

It is noted that under the encouragement of the national policy of *Energy Conservation and Emission Mitigation*, solar PV, together with the wind manufacturing industry, was gathering extensive attention across the society. The risk of oversupply for poly-silicon was growing. The State Council indicated in its No. 38 document this year to seriously warn the six potential oversupply industries, inclusive of the poly-silicon, and made strong restrictions to highlight technology improvement and prevent duplicated expansion.

Despite the oversupply concern, the growing power that China's solar PV industry in 2009 is showing and has created sound basis for critics to have more expectations than ever. It was believed that solar PV power would never play a dominant role by the year 2030 for China, however, solar PV has began to become one of the most promising solutions for China to consider in fulfilling its long-term energy strategy. In year 2009, it makes this change and stride the first step.

Gao Hu

Gao Hu got his PhD as well as the bachelor degree of Hydropower Engineering from TsingHua University, China, and is with Energy Research Institute (ERI) since year 2002.

As the Associate Professor of ERI of the National Development and Reform Commission (NDRC), Gao currently works as the Deputy Director of Centre for Renewable Energy Development (CRED) under ERI, and currently focuses on the policy research of renewable energy inclusive of small hydro, biomass, solar and wind energy etc.

As ERI is the only comprehensive macro-energy related research institute in China, Gao Hu had got involved in the research work of the Renewable Energy Law, national renewable energy medium- and long term planning, and the western rural energy planning. GAO provided consultancy to the National Energy Administration, Office of the National Energy Leadership Team and the China Academy of Engineering etc. GAO also led some international cooperation projects financed by WB/UNDP/UNEP in renewable energy fields.

Chapter 16

Lighting the World: Yesterday, Today and Tomorrow

Prof. Biswajit Ghosh

School of Energy Studies, Jadavpur University, Kolkata, India

1. Light and Energy

Light removes darkness; improves visibility and acts as the source for inspiration for doing work. That's why, holy Bible stated "Let there be Light". The Sun supplies light to this world for a definite number of hours and darkness appears in its absence. As a result, mankind of the ancient world worshiped Sun God to have his blessed vision for coming out from dark to improve prosperity. From ancient age, the Sun was worshiped in the forms of either by chanting or by singing songs, describing its contributions to mankind.

Scientists and philosophers from the ancient age have tried to earn knowledge on the origin of light from many perpectives in order to understand its nature, use and applications for human benefits. These understandings led to the advancement of philosophy starting from classical to quantum conception explaining the fact of origin and conversion methodologies. In its true sense, its the origin of light quanta packets comes from the conversion of one form of energy to other through the devices. Conversion of solar radiation into electricity has been discussed in many books and reports. However, many aspects were neither covered, not mentioned or discussed in those literatures. The main aim of the present article is to place the facts and figures not covered to the readers for a better understanding of photovoltaics (PV) which was initiated in India since the early 60s.

PV is increasingly an important energy technology. Deriving energy from the Sun offers numerous environmental benefits. It is an extremely clean energy

Power for the World by W. Palz
Copyright © 2011 by Pan Stanford Publishing Pte Ltd
www.panstanford.com
978-981-4303-37-8

source, and the few power-generating technologies that have a little environmental impact. As it quietly generates electricity from light, PV produces neither air pollution nor any hazardous waste. Moreover, it does not require land and water when integrated into the buildings. Also, because its energy source, sunlight, is free and abundant, PV systems can offer virtually guaranteed access to electric power.

2. Path toward Initiatives on PV Research

There is no doubt that human conceived the direct conversion technology from the nature and that was from the electric eel. However humankind is still unable to duplicate the eel's capabilities in converting the chemical energy directly into high electrical voltage. In fact the eel is characterized by its lethal high voltage discharge, on the other hand man-made chemical devices still remain limited to the lower values than that to the eel's characteristics. It is interesting to note nature's contribution in direct conversion but one must pay attention to know how humankind is able to emulate nature and its success. Therefore, at the beginning of this article one has to scan the past and learn the lesson about the limitations on success.

In fact the conception on direct energy conversion appeared in 1802 by Sir Humphrey Davy who suggested that chemical energy released from the oxidizing of coal might be converted directly into electricity. Although he did not succeed at that time due to many constraints, his ideas led paths to the development of fuel cells. Much works initiated after that as soon as Se was discovered by Berzelius in 1817. Berzelius was also the pioneer in preparing the elemental Si, which took part in the semiconducting electronic revolutions.

The effect of light on the bulk properties of the materials came in 1839 when a French teenager Edmond Bequerel observed PV effect in liquid. In the same year, English chemist William Grove constructed a device which was able to convert the reaction of H_2 and O_2 into electricity. Discovery of photoconductivity in Se was observed in 1873 by Willoughby Smith led to the proposal for developing solid state light meter by Adams and Day in 1876 (Fig. 1). Seven years later the first Se PV cell was reported by Fritts who first made the simulation of human eye response by the combination of Se cell and color filters (Fig. 2). In 1904, photoresponsivity in Copper-Cuprous Oxide structure was studies by Hallwachs. In the same year Sir Jagadish Chandra Bose from Calcutta, India, reported on the use of metal semiconductor barrier properties in building up detector for the radio receivers.

The real era of PV came in 1954 when D. L. Chapin, C. S. Fuller and G. L. Pearson from Bell Telephone Laboratories reported about the successful fabrication of 6% efficient Si solar cell (Fig. 3). In the same year PV effect in CdS was reported by D. C. Reynolds, G. Leies, L. L. Antes and R.E. Marburger from

Aeronautical Research laboratory of Wright Air Development Center. After these research works were initiated from the many countries on these two aspects to have better knowledge and understanding on the behavior of the materials and systems.

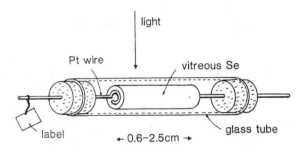

Figure 1. Solid state light meter as used by Adams and Day (1876). (After M. A. Green, 21st IEEE PV Conference, 1990.)

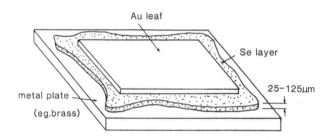

Figure 2. Simulation of human eye response with Se cell by Fritts (1883). (After M. A. Green, 21st IEEE PV Conference, 1990.)

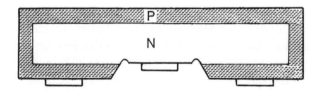

Figure 3. First 6% efficient Si solar cell from Bell Telephone Laboratory (1954). (After M. A. Green, 21st IEEE PV Conference, 1990.)

3. PV in India and International Scientific Cooperation

India is a Sun-rich country and is used to be habituated with blessed vision of Sun God. In fact, successful utilization of solar power in India has started during the 13th century when Sun Temple at Konark was built. It was in the east coast of India and is the world's first solar architecture when no one has even thought about the design of a solar home. Research work on the understanding the impact of solar radiation on human lives was started in India at early of 18th century. The first such work was initiated at Calcutta (now Kolkata) by Sir Jagadish Chandra Bose in 1904 while he reported about the carbon fixing process by leaves under the influence of solar radiation. In fact Bose's initial works on carbon fixing process indicate about the quantum conversion efficiency levels of the leaves. Depending upon the chlorophyll characteristics, leaves fix either six or eleven CO_2 and that of required amount of H_2O molecules either into glucose or sucrose. That's why the quantum efficiency of sugarcane leaves is higher than the ordinary tree leaves. He explained clearly the philosophy behind the utilization of solar quants in preparing hydrocarbon by the leaves. In fact his explanation on the activities of leaves is very much helpful in explaining the quantum efficiency phenomenon in solar cells.

In the late 60s research on solar power started at the several institution like the National Physical Laboratory (NPL), Solid State Physics Laboratory (SSPL), and Indian Institute of Technology (IIT) at New Delhi, Jadavpur University (JU), Raman Research Institute at Bangalore and Sri Aurobindo Ashram at Pondicherry. Researches on Si cells were conducted mainly at NPL and SSPL. Researches on thin film $CdS-Cu_2S$ PV cells were conducted at JU. Method for measurement of solar radiation using solar cell was published by the group of JU at IEEE Journal of Quantum Electronics in 1965. Later on this group projected about the successful applications of solar power for agro-irrigation by publishing another research paper in PV power generation international conference held at Hamburg, Germany, 1974.

In the late 70s and early 80s research works on amorphous Si cell was initiated at the Indian Association for the Cultivation of Science (IACS), Calcutta, and Photo-Electrochemical (PEC) cells at IIT-Kharagpur. Few more groups were generated at different IIT's and the universities for carrying out researches on the solar PV cells. For information transfer, solar Energy Society of India (SESI), and the Indian side of the International Solar Energy Society (ISES), first organized its national level convention at Jadavpur University, Calcutta in 1976. After that, SESI organized its yearly national convention at the different parts of the country. Having identified the need solar PV devices, a public sector enterprise, Central Electronics Limited (CEL) setup its plant at Shibabad near Delhi for manufacturing of commercial solar cells. Later, many public limited and private companies took part in the manufacturing of PV cells and systems. In addition to these, many small entrepreneur started manufacturing the PV components on smaller scale.

In the field of research, initial funding came from government organizations like University Grants Commission (UGC), Council of Scientific and Industrial Research (CSIR) and Department of Science & Technology (DST). The private sector enterprise like Tata Chemicals came forward for funding in solar energy research in India in the late 70s. With the initiation of the late Mr. D. S. Seth, the first private institution Tata Energy Research Institute. (TERI) — presently known as The Energy and Resource Institute — was set up but they were only able provide very limited funds to Jadavpur University and the National Physical Laboratory to carry out the research on the fabrication of thin film CdS-Cu_2S solar cells. Jadavpur University was able to fabricate 8–10% CdS-Cu_2S thin films cells with the innovative techniques like novel inexpensive front grid fabrication, magnetic field assisted encapsulation technique and novel H_2 plasma treatment on the Cu_2S surface. Major problem on stability of CdS-Cu_2S cells hampered further research on this type of cell structure. Research on amorphous Si cell was initiated observing the limitation on stability of CdS cells. Government initiated for more funding for carry out research on development of PV cell technologies. In the DST a separate division Commission for Additional Sources of Energy (CASE) was set up for nurturing solar research in India. Realising the need of more renewable energy technologies, a separate ministry named Ministry of Non-conventional Energy Source (MNES) was setup by the government in 1992. The ministry was later renamed Ministry of New and Renewable Energy Resources (MNRE). For propagating renewable energy to every corner of the country, the ministry setup state level organization Renewable Energy Development Agency (REDA) in each state of the country to implement renewable energy systems. This initiation resulted in the setting up of many small scale industries and entrepreneurs on solar power throughout the country.

Initial international scientific collaboration with the Indian researchers from funding by the various international agencies like the British Council, A. V. Humboldt Stiftung, German Academic Exchange (DAAD), Fulbright Fellowship and some individual fellowships from the various universities and research laboratories throughout Europe and USA. With the help of these fellowships, many Indian researchers sail their boats towards Europe and USA for carrying out researches on solar cells. In Europe, major research work was initiated at Germany and many researchers from India went to Stuttgart University, under the late Professor W. H. Bloss to carry out research on thin film solar cells.

4. Lighting the Remote

Among the solar PV products, solar street lighting systems and solar PV pumps were first implemented by the public funding in remote areas of the country. These were considered as the demonstration projects. Devices and systems were not very popular due to their high costs and low reliability. Thus, major constraint for using the solar devices at the beginning were cost and reliability. Major costs involved with solar cells and unreliability involved in control electronics and charging systems. The failure occurred mainly either in deep discharging in storage battery or uncharging the storage battery which is controlled by the electronic panel of the systems. Later on other devices like solar home lighting systems, garden lighting systems, solar lantern, solar cap and solar powered toys appeared in the market. To reduce the costs further, LED-based solar lantern with 2W LED panel coupled with $3W_P$ PV module put a big impact and emerged out as the most popular solar lighting systems. Large stand-alone systems like small PV power station for electrifying rural villages were set up to power the village for lighting. Looking at the peak, demand matching at present grid interactive systems have also been emerged out under the private public initiatives.

If the application of renewable energy is extended to rural community lacking the prospect of grid electrification, it should not just electrify villages, but should also bring measurable improvements in terms of addressing the poverty environment nexus identified as both a source and a symptom of underdevelopment in the rural areas of the developing world. Consistent with this approach, a step has been adopted in an Indian tribal village to plug them into the modern energy sources by means of community-based development and income generation activities, ultimately resulting to rise in their Human Development Index (HDI). Solar home lighting systems (SHL) were distributed among tribal families, thereby saving their kerosene consumption and reducing the kerosene-related indoor pollution. Both the SHL owners and local rice mill (moderator) deposited money for four years for maintenance of SHL and buying new sets to be distributed among newer families. These steps have brought impact on the economy, equity, social structure, empowerment and environment. The impact on lighting quality has been shown in Fig. 4.

Darkness using kerosene lamps   Brightness after installing SHL

Figure 4. Impact of solar home lighting system on the rural community.

To manage the PV systems, there is need for human capital — as a result need for human resources comes into picture very much. To develop human resources, courses on renewable energy system were introduced at the curriculum of both the undergraduate as well as post graduate level in the academic institutions. Even some of the universities started separate department for energy education. Thus, prospects on the applications solar energy is bright for the South Asian countries like India, Pakistan, Bangladesh, Nepal, Bhutan, Sri Lanka and Maldives. Thus, what we is required is to specify the benefits of the future generation. One such aspect is the need for international networking for dissemination of knowledge around the globe.

4. Views of the Author

The author of the present article is involved with PV research since 1978, the day he entered into the Jadavpur University as a research student to work in the solid state device group at the department of Electronics and Telecommunication Engineering. That day, initially, he was assigned to read two research papers and to find out the embedded philosophy within these. The first one was written by Paul Rapport based on the electron voltaic effect where the author's intention was to develop the solid state devices which will be able to convert the β radiation from nuclear wastes into electricity. The second one was by M. B. Prince where the author discussed about the behavior of a p-n junction solar cell under the influence of solar radiation and its quantum conversion capabilities. To this author, both these research papers were venturing for setting the path towards clean and safe energy technologies. In addition to these, the author conceived that PV power is very much complementary to the Indian monk Swami Vivakananda's famous deliberation at Chicago in the Parliament of World's Religion in1893. It appeared as: PV helps not fight, PV assimilate rather than destroy anything and PV maintains peace and harmony without making any dissention, what Vivekananda insisted in his famous religious deliberation.

Later, being a teacher to teach energy science and technology both in India and abroad, the author faced a number of challenges, many associated with the interdisciplinary nature of the subject particularly, when teaching energy sciences in master degree classes. As an experienced person in this field, the author had to cover the master topics as diverse in physics and astronomy, chemistry and material sciences, electrical and electronics engineering, even civil engineering and architecture, as energy is a key component which is very much associated with all the aspects of the above subject matters.

Solar PV power is now a reality and it is expected that by 2012 cost for photovoltaic power will have conventional grid electric parity. To reduce the price further, the research in the following areas are essentials. These are;

(a) Development of new recyclable materials and simple device structure for efficient conversion technologies.
(b) Efficient and cheap fabrication process for thin film modules for achieving high yield.
(c) Development of cheap and reliable electrodes for solar cells.
(d) Development of organic materials for efficient quantum conversion capabilities.
(e) Development of concentrating PV modules and its field performance evaluation for successful applications.
(f) Development of low powered reliable lighting systems capable of lighting individuals.

Therefore, global networking is essential that intends to develop connectivity between various actors with an intention that each individual actors can play comfortably and confidently in the domain of their actions and to integrate into the content.

The author narrated this story to the reader and hope there will be progress for the utilization of solar power to meet the projected demand not only at the rural remote place but also to involve socially marginalized section for the total development.

Biswajit Ghosh

Chapter 17

The Role of Research Institutes for the Promotion of PV: The Case of Fraunhofer ISE (Institute of Solar Energy Systems)

Prof. Adolf Goetzberger

Fraunhofer Institute for Solar Energy Systems, Germany

When I founded the Fraunhofer Institute for Solar Energy Systems in 1981, the world market for PV was about 4 MW. This is miniscule as compared to the Gigawatts produced today. At that time, solar energy was considered utopian and "experts", particularly from utilities predicted that it would never become a viable energy source. They pointed to the supposedly low solar radiation in Germany and did not even bother to look at the numbers. When randomly selected, people were asked their opinion about solar energy they answered that the Sahara had 100 times more solar energy than Germany. The generally held opinion was also that solar energy would never be economical.

Nevertheless it was clear to a small group of people, including myself that in the long run we would have to rely on renewable energy sources of which solar energy is the most important one. The Club of Rome had just published its impressive report and one oil crisis had already happened. It was obvious that planet earth had limited resources and particularly fossil energies were bound to be exhausted and would increase in price.

At that time the greenhouse effect was not yet a concern but very soon it became apparent that it provided a more immediate threat to our global future and again solar energy was the solution.

For starting a new institute I had a few advantages in this period. I was already head of a big institute, the Institute of Applied Solid State Physics which belongs to the Fraunhofer Gesellschaft, the largest institution for applied science in Germany. I had already established a modest working group for solar energy

Power for the World by W. Palz

Copyright © 2011 by Pan Stanford Publishing Pte Ltd
www.panstanford.com
978-981-4303-37-8

Figure 1.1.   Front view of Fraunhofer ISE.

within this institute. Thus a very dedicated group of people were already present to start a new venture. When I approached the management of Fraunhofer with the idea of founding a new institute for solar energy, I met with great scepticism but I was able to convince the president at that time, Dr. Keller, who gave me the go-ahead, overriding the opposition of his staff. From the beginning we had problems financing the fledgling institute. Almost exclusively the funds came from public sources since there was practically no solar industry. This created problems with our head office because Fraunhofer requires significant co-financing by industry. (In principle, a good idea because it assures close contact to industry.) There were a few companies producing small amounts of solar cells but they relied also on public support. On the thermal side, a collector market had already existed but the collectors were produced by very small companies, barely above the garage level. Consequently, many products were unreliable which gave solar energy a bad name for several years. Our administrators had to wait for a long time before the industry was able to fund R&D work on its own and it was considered to transfer the institute out of Fraunhofer into basic research. At one point even closing the institute was demanded by a fraction in the R&D ministry. Of course I fought for survival of the institute and was successful in this battle. What a difference it is today. Today, the ISE is not only the second largest institute of the Fraunhofer Gesellschaft with more than 800 employees but it also has a large measure of industrial contracts. It was also correct to keep it within Fraunhofer. The Fraunhofer framework provides numerous advantages for the institute because practically all imaginable technologies are represented within Fraunhofer and collaboration between institutes is encouraged.

There was not only opposition in the early stage. Citizens and also public media were always in favour of solar energy whereas government and political parties, with exception of the Greens were opposed. Without this public support the founding and further development of the institute could not have happened. It is also worth mentioning that the European Commission started very early an R&D application program lead by Wolfgang Palz for solar energy.

I should note that several crises helped the institute to survive. One was the accident of Chernobyl, the other one still another oil shortage.

In setting up the scientific program for the institute I formulated several guidelines that provided a good foundation for the future. One was that the entire range of solar energy technologies should be developed, including PV, thermal conversion, storage and systems technology. Concerning PV I decided on focussing on crystalline silicon although even then it was a widely held opinion that silicon would soon be replaced by another technology. This has not changed until today.

Noteworthy results of the institute in the field of PV during my term were:

- Development of the first transformerless inverter
- Setup of an internationally connected PV calibration lab for certification of cell efficiencies
- Systems technology for a non grid connected Black Forest inn (Rappenecker Hof)
- Design and construction of Germany's first self sufficient solar house

When I retired in 1993, my successor Joachim Luther took over at the helm of the institute. He led the institute very successfully until 2006 when his successor Eike Weber stepped in. Both of my successors were and are very successful in guiding the institute through good and not so good times, continually expanding its scope and staff.

We were not the only ones developing solar energy in Germany. Many institutes and universities are engaged in solar today. In 1990, the Forschungsverbund Sonnenenergie (Today Renewable Energy Research Association) was founded. It combined at that time four non-university research institutes. I represented ISE as a founding member. Today ten institutes are members of this association that coordinates research programs and exchanges results. Overall the strong R&D base in Germany has played a major role in the impressive growth of solar industry in our country. When industrialization took off, know-how and experienced personnel were available.

But R&D by itself can only provide the right starting conditions for industry. The next essential step is political action which happened just at the right time in Germany. It started with the 1000 roofs program. The real thrust for the explosive growth of PV in Germany was provided by the feed-in tariff. The rapid market development also demonstrated that development of new technologies advances much faster if it interacts with the market. It is not effective just to throw money at R&D with the pretext that more research is needed before the market can start. This argument usually comes from persons who do not want to see application of PV. On the contrary: The experience with the feed-in tariff has shown that feedback between development and production is essential for fast progress along the learning curve.

In summary, I am very proud today to see that the risk I took together with my co-workers in 1981 in founding the Fraunhofer Institute for Solar Energy Systems has led to such impressive results. If today's solar energy is considered to be the big hope for mankind, this is to some degree, due to the long and dedicated work of many visionary people with whom I had an opportunity to collaborate with in those years.

Adolf Goetzberger

A. Goetzberger received his Dr. rer. nat. degree in physics from the University of Munich in 1955. He spent 10 years in the USA - five years with the Shockley Transistor Laboratory , Palo Alto, CA. where he advanced semiconductor technology. Among his many contributions was discovery of the heavy doping gettering effect which is still used today in industry. Subsequently he spent five years with Bell Telephone Laboratories, Murray Hill, NJ where he published fundamental work about the $Si-SiO_2$–interface. He clarified the nature of surface states which are important for today's semiconductor devices. In 1968 he returned to Germany to accept a position of director of the Fraunhofer Institute for Applied Solid State Physics in Freiburg He modernized and greatly expanded this institute. In 1981 he founded the Fraunhofer Institute for Solar Energy Systems in Freiburg which grew into the largest solar energy laboratories in Europe and the second largest in the world. The institute was and is engaged in a broad spectrum of work in most aspects of solar energy conversion. Solar cell technology, solar materials research, thermal conversion, systems engineering and energy storage are the main activities. In 1993 he retired as a director of the institute but he carries on many publishing and advisory activities in the field of solar energy.

Although Goetzberger carried a heavy responsibility for management of the institute, he always maintained his scientific interests and guided the work of scientists and students. Scientific achievements: Fluorescent solar collectors, First theory of light trapping in thin silicon solar cells by diffuse reflectors, development of transparent insulation for buildings, planning and construction of the first self sufficient grid independent solar house in Germany in 1992. Many patents in photovoltaics, thermal solar energy, daylighting and systems.

He was President of the International Solar Energy Society from 1991 – 1993. He served on the Board of this society from 1987 to 1999 and since 2000 as an honorary director. From 1993 to 1997 he was President of the German Solar Energy Society DGS. He is now honorary president of this society. He is a fellow of IEEE. In 1995 he became doctor honoris causa of the Uppsala University in Sweden. In 2004 he was a member of the EU high level advisory board which prepared the PVTRAC vision report for 2030. He received the German cross of merit first class and the medal of merit of the state of Baden-Württemberg.

International awards:

1983	J. J. Ebers Award of the IEEE-Electron Devices Society
1995	Farrington Daniels Award of the International Solar Energy Society
May 1997	Karl W. Boer Solar Energy Medal of Merit Award
July 1997	Alexandre Edmond Becquerel Prize of the European Commission
September 1997	William R. Cherry Award of the IEEE
October 2005	Award for lifetime accomplishment of the Rhine Ruhr Int. Materials Conference
September 2006	Einstein Award of Solar World
December 2006	European Solar Award of EuroSolar
April 2009	Selected as European Inventor 2009 by European Patent Office and European Commmission

Chapter 18

Abandoning Nuclear in Favor of Renewable Energies

The Life Story of Giuliano Grassi — Florence, Italy

Giuliano Grassi

European Biomass Industry Association, Rue d'Arlon 63–65, 1040 Brussels, Belgium

Thanks to this publication — initiatives and efforts offered by my friend Wolfgang Palz — I shall briefly describe the history of my long diversified international life as engineer and scientist, essentially spent in the energy field, mostly in the renewable sector. In fact, out of 53 years of intensive activity, always stimulated by personal self confidence and enthusiasm for novelty, 20 years were dedicated to the nuclear sector and 33 years to the Renewable Energy (RE) sectors of photovoltaic (PV), wind and bio-energy, where I am still providing my contribution on potentially strategic areas.

Below I shall explain the motivations and contexts that, at a given moment, pushed me to change my professional activity, considering that the main desire of all human beings is the search of pleasure and satisfaction of diversified interests: science and innovation for me represents a vast area to stimulate my education and creativity.

During this long period, I have been involved in several exciting tasks: execution of studies, R&D and demonstration projects, responsible for the development of programs in support of advanced nuclear plants (fast breeders in France and advanced gas — cooled reactors in the UK), Management of RE R&D programs (EU) and elaboration of Industrial strategies for the European Biomass Industry Association, where I am still in charge as Secretary General.

However, I shall subdivide my professional life story in to three main periods:

- The first is in search of an enriching and satisfactory job, having identified the then emerging nuclear field as an attractive sector for a young engineer like me;

- A second period of 20 years with the establishment and consolidation of a comprehensive professional experience on advanced nuclear technology and atomic power plants, but with increasing doubts on the strategic validity of this option for the production of energy and the corresponding beginning of my self-education on Renewable Energy;
- A third period of 32 years up to now initiated by my decision to leave the Nuclear sector to join the Solar Energy Sector with the vision of a huge potential of diversified technologies and R&D needs.

1. First Period: Beginning of My Professional Activity as Engineer

For two years after my engineering degree at Pisa University I had three short employments in different sectors, namely:

- Magneti Marelli of Milan, for the development of pneumatically assisted servo-motors for trucks
- Lazzi of Florence, a public transport company
- Rafanelli of Florence, making good business on artistic wood works, but also interested on car pollution reduction where I was involved in experimental trials using a Volkswagen car equipped with a modified engine

In the mean time I had the chance to obtain an interview with a manager (Mr. Tams) of the English Electric Co. Ltd. in London. He offered me a job as research engineer at the Fundamental Development Department (Stafford Factory); I accepted it immediately for two main reasons: first, at that time (the 1950s), the UK was the most advanced industrial country in Europe due to the huge effort performed in connection with the World War II; second, the English Electric Co. was considered the most diversified Industrial conglomerate in the

Figure 1. English Electric Co. — Duston Hall (Stafford).

Figure 2. With family at Aix en Provence /France (near Cadarache).

UK, covering the aircraft and missile sectors, the atomic/oil/gas/steam/water/diesel power plants, the traction & transport sectors, the electrical/steam locomotives sector, the large civil works (Taylor Woodrow), the heavy electrical machinery, telecommunications (Marconi International), the marine equipment sector.

Thus a young engineer like myself, still without any professional experience but only with a sound theoretical education and attracted by research and innovation, saw in this offer an interesting opportunity to increase his level of professional knowledge.

Consequently I decided to leave my comfortable life in Florence and the traditional family country activity (wine making since the year 1620, from family records) in the beautiful Chianti valley located between Siena and Florence!

I left Italy for England on a cold winter day, with some disappointment of my father accompanying me to the railway station, with all my belongings enclosed in two voluminous bags and with a modest sum of 97 pounds sterling in my pocket.

Six months later I returned to Florence to get married and start a family life with Delia Ristori and the four children (Angela, Roberto, Susanna, Maddalena) that followed thereafter.

2. Second Period: Transition from Electro-Mechanical to Nuclear Activity

Some coincidences, a general interest and curiosity for the newly emerging sector of atomic power generation pushed me for searching of a job in this field. The main aspects that were claimed in those days attracted my interest.

The large estimated resource potential without economic constraints:

- Fission Thermal Reactors: 27 000 years;
- Fast Breeder Reactors: 1 100 000 years.

The extremely high energy content of nuclear fuel:

- 1 t of U-235 is energetically equivalent to 8700 t of oil
- 1 t of U-235 can produce 38.5 million kWh (electric)
- The high energy released at the fission of one atom of193 MeV (Mega electron Volt)

So the fission of all U-235 atoms available in 1 cm³ of natural uranium (0.72% of a total of 0.47×10^{23} atoms/ cm³) can produce the same amount of energy as 6 000 000 cm³ of oil.

The curiosity that the control of the chain reaction and power in an atomic plant is possible because a "Law of Nature" has established that a small portion (0.755%) of the total neutrons produced at the fission of U-235 atoms, during a chain reaction, is released with a delay of 56.6 seconds; this time span being sufficient for operating the control — bars (boron) to keep the chain reaction under control.

Here below are the facts that offered me the possibility of getting a job in the nuclear sector:

While working at the Stafford Factory of the English Electric Co. (now General Electric Co.) I started educating myself and collecting information on technology

Figure 3. Institute of Technology, Birmingham/UK.

problems related to atomic power generation from a friend (Vincenzo Cialella) working at the Nuclear Factory of English Electric in Leicester. Furthermore, I decided to follow a post-graduate course on Nuclear Metallurgy at The Institute of Technology of Birmingham (now Aston University) and on Nuclear Technology at the Technical Institute of Stafford with an old university friend, Alessandro Rutili.

But the real opportunity for reaching my desire to join the atomic power sector came from a casual visit to nuclear power plant at Windscale (now Sellafield) in Cumberland in 1958. This was the first commercial nuclear power plant in the world, officially opened by Queen Elizabeth II in 1956, shown on the figure below:

Figure 4.   Windscale Nuclear Plant/UK.

During this technical visit I met an Italian Engineer, Giorgio Ascione, a friend from Pisa University who was involved as project Engineer at "AGIP Nucleare" for the construction of the first Latina — Nuclear Power Plant in Italy. Enrico Mattei, President of the ENI Group, having the ambition to enter the business of nuclear electricity, founded in 1957 this large nuclear engineering company.

This casual meeting provided me with good contacts at AGIP-Nucleare, that offered me a job as R&D Engineer in Milan.

After a short period I was sent back again to England, the nuclear power group Parson & International Research Laboratory at Newcastle upon Tyne for training on uranium fuel-element manufacturing and nuclear technology.

At this R&D Center, I had the opportunity to acquire experience on several advanced technologies: construction and operation of a metal-plasma-deposition unit (protection of missile heads or the combustion chambers of supersonic planes like Concorde); superconductivity experiments for the manufacture of compact electric generators; large scale thermal simulation for the high temperature "Dragon Atomic Plant" under construction at Winfrith/Dorset.

Figure 5. Newcastle Upon Tyne: Int. Res. Laboratory.

Then I returned to Italy in March 1962 (Milan) where I spent two more years being involved in very diversified, activities, like:

- Construction and operation of an electron beam accelerator (300 kV)
- Design, construction and operation of a large thermal loop (for heat transfer tests on "nuclear fuel elements"
- Construction and operation of a small high-temperature gas reactor for magneto-hydrodynamic power generation trials
- Design, construction of a material irradiation facility for the experimental Nuclear Reactor at the "Military Center CAMEN" at PISA.

However, in the year 1962, several important events pushed me to envisage again the search for a new job.

On 27 October 1962, the President of ENI, Enrico Mattei, died in a flight accident. Thus disappeared a strong influential man, who wanted ENI to enter the nuclear electricity business.

In December 1962, there was a political decision by the Italian Government headed by A. Fanfani to nationalize the electricity sector, compelling ENI to abandon the nuclear power generation business. At the same time, strong attacks from government representative (G. Sargat) to the Italian Atomic Energy Commission (CNEN) obliged its president Prof. F. Ippolito to resign. On the 3rd of March 1964, he was also condemned and spent two years in prison.

All these facts created a very uncertain situation and perspectives for the entire nuclear activity in Italy convincing me to accelerate the search for a new job, but which job and where?

I considered that a very attractive solution would be to join EURATOM. Information on the very ambitious French program on fast nuclear projects and the establishment of the EURATOM — CEA (Commission for Atomic Energy) Association with envisaged participation of several EURATOM officials at the Nuclear Research Center of Cadarache naturally attracted my interest.

In fact my involvement in this program could have represented for me the realization of a dream!

Below was how I succeeded to join EURATOM. My Director, Prof. Carlo Tribuno was on leave from AGIP Nucleare to become the General Director at the Battelle Institute in Geneva.

As reliable and good references were considered essential, in a misty day of September 1962 (the day of the assassination of J. F. Kennedy in Dallas), I went to pay a visit to my director asking authorization to enclose his name as reference in my application form to EURATOM, EC-Brussels. Fortunately the requested favorable answer from EURATOM, implying of course the corresponding departure from AGIP, was addressed to my Director and not to the Director General! Finally I received a good offer for the wanted activity at the Cadarache Nuclear Center, starting 1 May 1964 where I stayed for more than six years.

My first activity was in the development of the conceptual design of large fast — breeder plants. In particular was the the design of "neutron — traps" to reduce the high flux of neutrons streaming from the reactor core. I was happy that the results of this study and my name were enclosed in the official CEA — Document presented at the last Geneva "UN Conference" on the "Peaceful Uses of Atomic Energy" (1964).

This was the first time that my name and my study appeared on a publication!

During my long stay at the Cadarache Center I was involved in activities related to the construction of the first two French breeder plants (Rapsodie at Cadarache and Phénix at Marcoule) in particular the following:

- Study and Design of the Hot-cell of the Rapsodie Reactor for the dismantling of irradiated components
- Heat transfer simulation testing on Rapsodie fuel elements
- Sodium testing of components and systems
- Design of the irradiated fuel-elements discharge machine and for its refueling (Phénix Reactor).
- Safety assessment during the discharge of irradiated fuel elements from the reactor. In case of malfunctioning of the discharging machine there would be an absolute need to guarantee the cooling of the fuel, avoiding the risk of plutonium contamination.

I was able to provide a positive contribution to this safety problem, by designing (CEA — patent) a sophisticated irradiated fuel transfer container equipped with thermo electric generators feeding magneto-hydrodynamic pumps to guarantee sufficient flow of liquid sodium through the irradiated fuel element for the evacuation of its thermal power. This complex system, utilizing a small part of the heat dissipated by the irradiated fuel and flowing through a compact constant thermoelectric bridge was able to generate a high current (250–300 Amp) that crossing the liquid sodium channel and the magnetic field was sufficient to cool

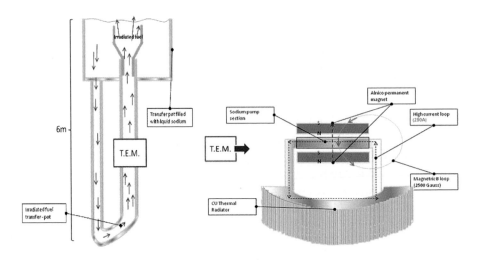

Figure 6. Thermoelectric generator feeding a magneto-hydrodynamic pump.

the irradiated fuel (see Fig. 6) maintaining, in any situation, its temperature below the critical level of 620°C.

General De Gaulle, the French President, was attracted during his visit in 1967 at the Rapsodie Reactor by this device and asked information on the simulation experiments under way. A second contribution was on the elaboration of a very challenging study to assess the emergency cooling of the reactor in case of a simultaneous shut down of the sodium circulating pumps and failure of the emergency electrical generator. In this situation there was a need to verify that the reactor core never reached a temperature level higher than 620°C. Under natural convection cooling, to avoid the risk of plutonium contamination, a large amount of plutonium (4800 kg) being stored in the reactor.

I remember that I was able to simulate the situation by 12 equations, but the problem could not be solved because the unknown parameters were 13. Fortunately, formulating the concept that "in nature the most probable event is the one that requires minimum energy demand", through huge computer calculations it has been possible to verify the validity of this formulation anticipating information on this critical safety aspect.

After 10 years of accrued experience on nuclear technologies in England and France, I accepted a new offer from AGIP Nucleare as Head of the Engineering Division for the Nuclear Fast Breeder Program that Italy had decided to implement in collaboration with the French CEA/EDF having as objectives:

- Construction of an experimental Reactor (PEC-Brasinone) for developing and testing nuclear fuel elements;
- Construction of a manufacturing plant for mixed plutonium-uranium fuel elements (AGIP N.);

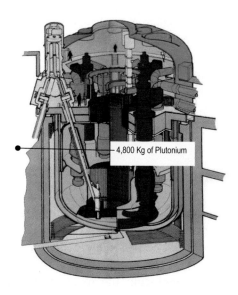

4,800 Kg of Plutonium

Figure 7. Superphénix reactor cross section (1240 MWe) at Creys Malville, EDF — ENEL — RWE.

- Participation in the construction and exploitation of the Superphénix power plant (1/3 of the investment being provided by Italy).

Therefore, during the following eight years (1970–1977), I was intensively involved in this program in particular,

- Development of major basic sodium components, measuring devices and electronic & electrical equipment needed for the envisaged experimental simulation program (electromagnetic pumps, heat exchangers, sodium purification systems, heat-pipe technology (from cryogenic to high temperature, 1000°C), etc.,
- Establishment of support laboratories for electrical, electronic measurement equipment for electrical, magnetic, temperature, vacuum, pressure measurements,
- Planning & execution of a wide program in support for the construction of Superphénix. At that time (1976) our division was in fact the only organization in Italy able to deal with the liquid sodium technology and to design, construct, operate large experimental facilities.

During the last one-and-a-half years (prior to my definite departure from Italy), I was assigned the responsibility of the experimental program in support of the Superphénix Reactor (1/3 of total planned program had to be implemented by the Italian partners: NIRA — ENEL — AGIP N.).

In 1977 AGIP N., decided to sell to Ansaldo of Genoa all the acquired know-how on sodium technology and to concentrate its business on the development and manufacture of nuclear fuels.

Nevertheless I refused to move to Genoa: having received five different employment offers (Dragon Project — UK, EURODIF — France, KARLSRHUE — Germany, BREDA-Italy, NIRA-Italy) I finally opted to go to Paris at Novatome KARLSRUHE — NIRA for the design of the new Superphénix II Nuclear Plant.

3. Third Period: Renewable Energies

Despite moving to Paris, I started to envisage leaving the nuclear sector, becoming conscious of its numerous weak points and risks. My wish was to join again the E.C. and possibly to be involved in the energy sector, hoping for a position in the new R.E. program!

The main motivations convincing me to abandon the nuclear sector are summarized below:

- Nuclear Power Generation is a very centralized method of production involving a small number of people in the decision, control and operation process.
- Risk of contamination (health hazard) is limited but possible, especially because plutonium, as nuclear fuel, is more and more utilized and available in large quantity from the dismantling of nuclear weapons.
- Plutonium is very dangerous not only as source of strong radiations (thousands of years) but also for its extremely high toxicity. A few kilogram could be sufficient, subdivided in deadly doses, for killing the entire world population!
- The risk of proliferation is limited but possible in a turbulent situation.
- Today, the volume of the fissile materials in circulation (production, refining, use, reprocessing, storage) could be sufficient for the production of about 100 000 atomic bombs!
- The production cost of nuclear electricity is not as competitive as it is being publicized, if the decommissioning, waste disposal, population insurance costs are enclosed.
- The amount of identified uranium resources (OECD-AIEA) is 4.75 million tons. Thus, considering that the annual consumption is around 70 000 t/y, the reserves of uranium could last only 65 years, therefore they are not very large and similar to the present world natural gas reserves. The introduction of Fast Breeders Reactors, considered essential for increasing by 50 to 100 times the amount of fissile nuclear fuel by converting inert uranium-238 into plutonium during reactor operation will be very, very expensive due to their sophistication and thus less competitive.
- The Nuclear power generation is not completely CO_2 free, as generally publicized. In fact for the refining of the mineral into U238 (uranium content:

100–200 gr/ton of mineral) its conversion into gas (UF6), its enrichment into U-235, its conversion into UO_2 and for the manufacture of suitable reactor fuel elements, a large amount of conventional energy is required causing CO_2 emissions as follows:

- 7-22 gr. of CO_2/kWhe produced (U.K. estimation),
- 60 gr. of CO_2/kWhe produced (EC-Ispra estimation).

My return to the EC (now EU Commission) was negotiated, during my permanence in Paris, with the EC services and facilitated by a new position opened in Brussels in 1978: the Renewable Energy R&D program. An additional asset proved to be my related technology experience previously acquired such as:

- Construction of solar-water heaters based on the utilization of heat pipes on parabolic concentration systems,
- Construction of rudimentary, miniature photovoltaic generators (using small silicon cells),
- Construction of a small inverter (50 Watt) for feeding a transportable oscilloscope with supply from a 12 V battery,
- Utilization of GaAs semiconductors (pulsed IR Lasers), material used also for high-temperature PV cells,
- Semiconductor — thermoelectric material for cooling (battery or PV cells),
- Construction of advanced "Liquid sodium — sulphur" battery of high capacity;
- Experience of agro — forestry activity in my family vineyard (45 ha of forest — 15 ha of agricultural land).

Figure 8. "I Fabbri" Grassi — since year 1620.

All these activities, although modest, had offered me some insight into the vast area of RE,

For nine years (1978–1987) I have been managing four sectors of EC research activities:

- Photovoltaic energy,
- Biomass energy,
- Wind energy,
- Photobiological, photochemical, photoelectrochemical processes.

This has been a lucky period of extremely broad diversified and professionally enriching activity, with challenging objectives but also important progress and achievements, and an immense pool of future activities.

During this period, I also had the opportunity to learn the art of management of wide research program from my friend and head of the RE program Dr W. Palz. Consequently, it proved very beneficial to me taking part into the management and implementation of program through contracts. In particular,

- Elaboration & definition of the sector activities with the assistance of experts,
- Preparation of programs to be approved by the internal EC structure, the European Parliament, the Council of Ministers,
- Selection and negotiation of projects,
- Preparation of contracts,
- Follow-up of research activities, their coordination & control,
- Preparation and execution of frequent contractor's meetings,
- Analysis & approval of final technical reports and related accountings.

Figure 9.   Visit to Japan

The programs were subject of continuous evolution, with the purpose of accelerating scientific-industrial cooperation among European countries on particularly sensitive promising sectors. The integration of the EC and national programs were considerably facilitated by the co-funding of projects in cooperation with consultative committees of EU member states.

All research results were the subject of a wide and comprehensive dissemination effort to the public through seminars, conferences, publications. Closer contacts and exchange of information with International Organizations of Third Countries (USA, Japan, Canada, Brazil, IEA) were also established.

Here I would like to underline some significant achievements obtained in the first 10 year of the R&D period: The 16 Photovoltaic pilot-projects, covering a total installed power well over 1 MW, with a total budget of €30 million.

This specific program proved successful not only in demonstrating the reliability of components and sub-systems, but also in verifying the feasibility of potential application of PV generators, like:

- Stand-alone power supply for remote villages/islands,
- Water pumping,
- Sea-water desalination,
- Production of hydrogen,
- Radio & TV broadcasting,
- Supply to airport electronic equipment,
- Supply to recreation centers.

Figure 10. PV Contractor's Meeting of the EU Program.

As I recall, other important R&D achievements were related to:

- Solar cell processing by cold junction formation,
- Continuous silicon crystallization growth at a high rate of 10 cm/min, to reduce silicon wafer costs,
- Identification of cost-effective resin materials for PV panels encapsulation,
- Amorphous Si deposition by magnetron sputtering,
- Cu2S-CdS solar cells, manufactured by spray technique (efficiency of ~ 6%),
- Hybrid energy concentration collectors,
- High efficiency systems.

Interesting results were obtained from an experimental program (suggested by our service) on support structures of wood for large PV generators (wood poles — steel wire), imitating similar configurations adopted in vineyards. This concept has anticipated good future perspectives of obtaining support structures of desirable low specific investment costs. Life-tests on 10 kW generators have shown good reliability over 20 years.

Figure 11. PV Conference Florence, Grassi third from the right, the EU Energy Commissioner second from the right.

Nowadays, after retiring from the EC, in the role of secretary general of EUBIA (European Biomass Industry Association), I still continue to follow-up the evolution and progress of the PV Sector.

Recently in 2007, I have installed in my country house (Casole di Greve in Chianti) a 3 kW PV generator (see Fig. 12):

Figure 12.   PV generator on Grassi house in Chianti, Italy.

EUBIA is also promoting the development of steam-engine generators (10 kWe to 1000 kW) fuelled by bio fuels that could represent, thanks to their characteristics and performances, an ideal hybrid PV-biomass combination. These bio-electricity generators could conveniently replace, in remote sites, the expensive battery-packages.

Another combined solution could be represented by PV systems associated with micro-gas-turbines generators, fuelled with low cost-low quality bio ethanol.

In Fig. 13 below, we present the figure of an interesting experiment: EC Program (Mercedes-ENEL) on ceramic-gas turbines, fuelled with fine biomass powder.

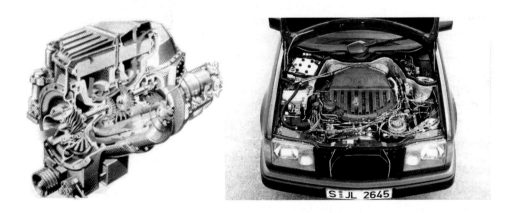

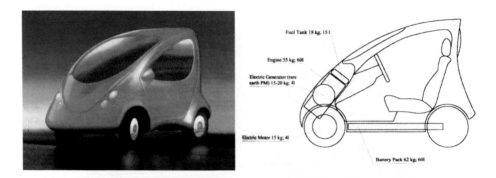

Figure 13. Advanced Italian car designs.

At medium term, the use of hybrid-vehicle (fuelled by bio fuels) in combination with PV generators could become an interesting application for houses. Below is the figure of a small city hybrid-car (ECO-FLY) for which I have in the past made a particular technological and coordination effort. It was unfortunately stopped with the death of the industrial promoter, Sign. Pasquali.

4. Concluding Remarks

Only well managed renewable energy resources in their so diversified forms, represent the most appropriate solutions to satisfy the different needs, always in evolution of mankind: vital needs, comfort needs, production activities needs.

Renewable electricity, in particular, is contributing more and more to the share of total energy consumption in individual countries (approaching now 40%; in future 60–70%) and will play fundamental role for modern society.

Photovoltaic progress in term of technological and cost improvements will thus be able to match the expected ambitious targets. For example, in continents like Africa, where only 15% of nearly 900 million inhabitants have access to electricity supply, photovoltaic is expected to play an essential role to satisfy the energy needs (perhaps the world's copper reserves would be sufficient for the implementation of a continental grid!) in a decentralised form, in combination with other renewable energies resources (hybrid systems) and through integrated forms of the coproduction of power, biofuels, food/feeds, industrial commodities.

Bioenergy, beyond offering good opportunities for combined photovoltaic–biomass hybrid power generation plants, will be able to satisfy with its huge potential other vital future needs of the world population. As an example, 70 000 commodities, derived actually from petroleum; jet-biofuels (requiring now over 340 000 m^3/y by 13 000 commercial jets in operation worldwide); the immense contribution to reduce the CO_2 emission level in coal power plants (biomass potential could now be only for the EU, equivalent to 200 nuclear power plants, in the future this could be 400 600 nuclear power plants equivalent contribution).

As for myself, believing strongly that everybody is in the position to provide some help for a generalised deployment of renewable energies, I want to my dedicate acquired knowledge and my effort (for the remaining lifetime that "Bon Dieu" will decide to great me) to the promotion of renewable energies and in particular hybrid PV-Biomass systems, seeing the immense benefits for the well being of population especially in the developing countries, the environment, health improvement and new job creations.

Chapter 19

Nonconventional Sensitized Mesoscopic (Grätzel) Solar Cells

Michael Grätzel

*Laboratory of Photonics and Interfaces, Institute of Chemical Science
and Engineering, Faculty of Basic Science,
Ecole Polytechnique Fédérale de Lausanne, Switzerland*

Following their inception in 1985,[1-6] dye sensitized mesoscopic solar cell (DSCs), also named Gräetzel cells after their inventor, have emerged over the past two decades as non-conventional contenders to p-n junction inorganic and photovoltaic (PV) devices. The DSC is the only photovoltaic device that uses molecules to absorb photons and convert them to electric charges without the need of excitonic transport. It is also the only solar cell that separates the two functions of light harvesting and charge carrier transport, whereas conventional and all of the other known organic solar cell devices perform both operations simultaneously. In this regard the DSC mimics the primary process in the photosynthetic conversion of solar energy by green plants where chlorophyll molecules absorb sunlight, generating positive and negative charge carriers after photo-excitation. The separation of light absorption and carrier transport achieved by the DSC greatly increases the options for the absorber and charge transport material allowing for attractive new PV cell embodiments of different color and transparency. The realization of light flexible forms is also possible.

The molecular sensitizer or semiconductor quantum dot is placed at the interface between an electron (n) and hole (p) conducting material. The former is typically a wide band semiconductor oxide, such as TiO_2, ZnO or SnO_2 while the latter is a redox electrolyte or a p-type semiconductor. Upon photo-excitation the sensitizer injects an electron in the conduction band of the oxide and is regenerated by hole injection in the electrolyte or p-type conductor. Alternatively, the sensitizer may be attached to a p-type oxide such as NiO,[7] which in this case reduces the excited dye, the latter being regenerated from its reduced form by electron transfer to an acceptor in the electrolyte. In both cases the role of the sensitizer is to absorb light and generate positive and negative charge carriers.

Note that minority carriers, i.e. electrons and positive charges (holes) in p- and n-doped semiconductors, respectively, play no role in the photovoltaic conversion process accomplished by the DSC. Because the sensitizer injects electrons in the n-type and holes in the p-type collector, only majority carriers are generated. These charges move in their respective transport medium to the front and back contacts of the photocell where they are collected as electric current. By contrast, in conventional p-n junction photovoltaic cells the photocurrent arises from minority carriers generated by photo-excitation of the semiconductor. The minority carrier must live long enough to reach the junction formed between the p and n doped semiconductor material before recombination with the majority carriers takes place. The electric field present in the vicinity of the junction separates the positive and negative charges generated under illumination attracting the electrons to the n-doped and the holes to the p-doped material. In order to impart a sufficiently long lifetime to the photo-generated electron-hole pairs, the use of very pure materials is required. The chemical purification of the semiconductor and/or production of thin films under high vacuum conditions entails a high cost for the photovoltaic converter. In the DSC the recombination of charge carriers occurs across the phase boundary separating the electron- from the hole conductor medium. This offers the prospective to fashion the interface in a judicious manner in order to retard the back electron transfer reaction

Figure 1(a) shows a typical band diagram of the DSC. Sunlight is harvested by the sensitizer that is attached to the surface of a large band gap semiconductor, typically a film constituted of titania nanoparticles. The electron microscopic

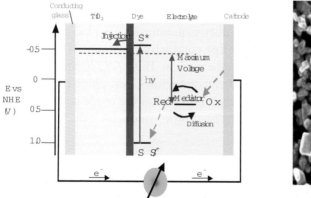
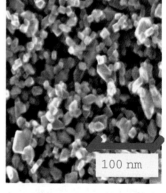

Figure 1 (a) Energy band diagram of a typical embodiment of the DSC. (b) Scanning electron microscopy picture of a nanocrystalline titania film.

image of such a mesosocpic titania film is shown in Fig. 1(b). Photo-excitation of the dye results in the injection of electrons into the conduction band of the oxide. The dye is regenerated by electron donation from a hole conductor or an electrolyte that is infiltrated into the porous films. The latter contains, most frequently, the iodide/triiodide couple as a redox shuttle although other have also been developed recently as an alternative to the I^-/I_3^- system. Reduction of S^+ by iodide regenerates the original form of the dye while producing triiodide ions. This prevents any significant built-up of S^+, which could recapture the conduction band electron at the surface. The iodide is regenerated in turn by the reduction of the triiodide ions at the counter-electrode, where the electrons are supplied *via* migration through the external load completing the cycle. Thus the device is generating electricity from light without any permanent chemical transformation. The voltage produced under illumination corresponds to the difference between the chemical potential (Fermi level), $\mu(e^-)$, that the electrons attain in the titania nanoparticles and the chemical potential $\mu(h^+)$ of the holes in the hole conductor. For redox electrolytes, the latter corresponds to the Nernst potential.

The energy levels in Fig. 1(a) are drawn to fit a frequently employed embodiment of the DSC based on the N719 ruthenium dye the iodide/triiodide as a redox couple and nanocrystalline anatase films as electron collector. The ground state standard redox potential of the N719 as well as that of several other ruthenium complexes and organic sensitizers, measured when the dye is adsorbed to the surface of the nanocrystalline anatase film, is around 1 V against the normal hydrogen electrode (NHE) while the Nernst potential of the triiodide/iodide based redox electrolyte is close to 0.4 V. Thus the regeneration consumes 0.6 eV, which constitutes the main loss channel in the presently employed DSC along with the lack of near IR response of todays' sensitizers. These two factors need to be improved in order to increase the efficiency of the DSC from currently 11–12 percent[8–11] to approach the theoretical limit of 32 percent for single junction solar cells. Improving the light harvesting in the 650–900 nm domain remains therefore one of the greatest challenges faced by present day research in the DSC field. Even moderate improvements in the photoresponse of the sensitizer in red and near IR wavelength region would greatly benefit the conversion efficiency.

Long term stability is a key requirement for all types of solar cell. A vast amount of tests have therefore been carried over the last 15 years to scrutinize the stability of the DSC both by academic and industrial institutions. Most of the earlier work has been reviewed.[12,13] Long term accelerated light soaking experiments performed over many thousand hours under full or even concentrated sunlight have confirmed the intrinsic stability of current DSC embodiments. Stable operation under high temperature stress at 80~85°C as well as under damp heat and temperature cycling has been achieved by judicious molecular engineering of the sensitizer, the use of a robust and non-volatile electrolytes such as ionic liquids.[14] and adequate sealing materials. In the early development stage of the DSC technology, the quality of device sealing was sometimes not appropriate in laboratory test cells, causing leakage of the volatile solvents. Most research groups with longer practical experience, including industrial enterprises, have overcome this by improving the sealing methods. Due to the direct relevance to the manufacturing of commercial products, little is published on these processing issues though. Good results on overall system endurance have been reported since several years demonstrating excellent stability under accelerated laboratory test conditions. These promising results are presently being confirmed under real outdoor conditions. From these extensive studies confidence has emerged that the DSC's can match the stability requirements needed to sustain outdoor operation for at least 20 years. This has paved the way for the recent worldwide surge in the industrial development and commercialization of the DSC. The commercial shipment of roll to roll manufactured mass produced flexible DSC modules has recently been announced (www.g24i.com).

Due to significant industrial up-scaling efforts, the conversion efficiency of DSC modules has been steadily rising over the last few years, the certified value measurer under AM 1.5 standard conditions reaching currently 9.5 percent.[15] Thus the DSC has now attained efficiencies on the laboratory as well as on the module scale which render it competitive to other thin film silicon solar cells. Their low cost and ease of production avoiding expensive high vacuum steps should benefit large-scale applications. Impressive stability both under long-term light soaking and high temperature stress has been reached fostering first industrial applications. These systems will promote the acceptance of renewable energy technologies, not least by setting new standards of convenience and economy.

References

(1) Desilvestro, J., Grätzel, M., Kavan, L., Moser, J. E. and Augustynski, J. Highly efficient sensitization of titanium dioxide. *J. Am. Chem. Soc.* 1985, *107*, 2988–2990.
(2) Vlachopoulos, N., Liska, P., Augustynski, J. and Grätzel, M. Very efficient visible light energy harvesting and conversion by spectral sensitization of high surface area polycrystalline titanium dioxide films. *J. Am. Chem. Soc.* 1988, 110, 1216–1220.
(3) O'Regan, B., Grätzel M. A Low-cost, high efficiency solar cell based on dye sensitized colloidal TiO_2 films. *Nature* 1991, *335*, 737–740.
(4) Nazeeruddin M. K., Kay A., Rodicio I., Humphrey-Baker, R. Müller, E. P. Liska, Vlachopoulos N and Grätzel, M. Conversion of Light to Electricity by cis-X_2bis (2,2'-bipyridyl-4,4'-dicarboxylate)ruthenium(II) charge transfer sensitizer (X = Cl-, Br-, I-, CN-, and SCN-) on nanocrystalline TiO_2 electrodes. *J. Am. Chem. Soc.* 1993, *115*, 6382–6390.
(5) Bach, U., Lupo, D., Comte, P., Moser, J. E., Weissörtel, F., Salbeck, J., Spreitzert, H. and Grätzel, M. Solid State Dye Sensitized Cell Showing High Photon to Current Conversion Efficiencies *Nature*, 395, 550 (1998).
(6) Grätzel, M., Photoelectrochemical Cells, *Nature* 414, 338–344 (2001).

(7)   Qin, P., Linder, M., Brinck, T., Boschloo, G., Hagfeldt, A. and Sun, L. High Incident Photon-to-Current Conversion Efficiency of p-Type Dye-Sensitized Solar Cells Based on NiO and Organic Chromophores. *Adv. Mat.* 2009, 21,1–4.

(8)   Chiba, Y., Islam, A., Watanabe, Y., Komiya, R., Koide, N. and Han L. Dye Sensitized Solar Cells with Conversion Efficiency of 11.1%. *Jap. J. Appl. Phys, Part 2*, 2006, *45*, 24–28.

(9)   Gao, F., Wang, Y., Shi, D., Zhang, J., Wang, M., Jing, X., Humphry-Baker, R., Wang, P., Zakeeruddin, S.-M. and Grätzel, M. Enhance the Optical Absorptivity of Nanocrystalline TiO$_2$ Film with High Molar Extinction Coefficient Ruthenium Sensitizers for High Performance Dye-Sensitized Solar Cells. *J. Am. Chem. Soc.* 2008, 130(32), 10720–10728.

(10) Cao, Y., Bai, Y., Yu, Q., Cheng, Y., Liu, S., Shi, D., Gao, F. and Wang, P. Dye-Sensitized Solar Cells with a High Absorptivity Ruthenium Sensitizer Featuring a 2-(Hexylthio)thiophene Conjugated Bipyridine, *J. Phys. Chem. C.* 2009, Article ASAP.

(11) Chen, C.-Y., Wang, Mi., Li, J.-Y., Pootrakulchote, N., Alibabaei, L., Cevey-Ha N.-L., Decoppet, J.-D., Tsai, J.-H., Grätzel, C., Wu, C.-G., Zakeeruddin, S. M. and Grätzel, M. Highly Efficient Light-Harvesting Ruthenium Sensitizer for Thin-Film Dye-Sensitized Solar Cells. ACS Nano 2009, 3(10), 3103–3109.

(12) Han, L., Fukui, A., Chiba, Y., Islam, A., Komiya, R., Fuke, N., Koide, N., Yamanaka, R. and Shimizu, M. Integrated dye-sensitized solar cell module with conversion efficiency of 8.2%. *Appl. Phys. Lett.* 2009, 94, 013305/1-013305/3.

(13) Lenzmann, F. O., Kroon, J. M. Recent advances in dye-sensitized solar cells. *Adv. Opto-Electr.* 2007 (Recent Advances in Solar Cells), 65073/1-65073/10.

(14) Grätzel, M., Recent Applications of Nanoscale Materials: Solar Cells, Chapter 1 *"Nanostructured Materials for Electrochemical Energy Production and Storage"*, Leite, R.E., ed. Springer New York (USA) 2008.

(15) K. Noda Sony, ESF Conference on Nanotechnology for Sustainable Energy, Obet 2010, AIST ralidated module efficiency.

Michael Grätzel

Chapter 20

The PV World Conference in Vienna

Wolfgang Hein

Eurosolar-Austrian Chapter, Vienna, Austria

It was at the beginning of July 1998 in the old imperial city of Vienna, which has become a remarkable wealthy town out of the ruins from the Great War. The Hofburg, residence of the Habsburg emperors for several hundred years, was the place for the Second World Conference on Photovoltaic Solar Energy and thus to start a new era for the power production system at least in Europe. Austria recently took over the European Union Presidency for the second half of that year. The first world conference of that kind took place four years ago in Hawaii and the initiative came now like before from Hermann Scheer, the prophet of the Solar Age.

I was at that time a director in the Austrian Federal Chancellors office and was in the preparatory process responsible for organising a national support committee and support from the government for the conference. The Conference Bureau from Munich and the Vienna Hofburg Conference Center prepared their part perfectly and the government departments and the city administration contributed as well. The world conference had about 1800 participants and the exposition some hundred exhibitors. It was a great event for the photovoltaics industry, which was less industrialised than it is now. The welcome should be given by the Federal Chancellor, with the Federal Presidents office there was no communication in the preparation for the conference.

The Federal Presidents commitment to the development of photovoltaic power production was quite weak, and was demonstrated by the following anecdote: After Thomas Klestil was elected for the Federal Presidency the first time in 1992 there was a lively discussion on Austrian TV called "Arguments" dealing with future energy supply. Panellists among others included the Austrian Minister for Economics Wolfgang Schüssel and the German Member of Parliament Hermann Scheer. From the invited guests in the studio came the question by solar activist Franz Niessler to the minister Schüssel, why does the Federal

Power for the World by W. Palz
Copyright © 2011 by Pan Stanford Publishing Pte Ltd
www.panstanford.com
978-981-4303-37-8

Building Administration not fulfil the President's wish to have solar panels on the roof of his residence. The President committed himself to that in his election campaign. Schüssel did not hesitate in the live discussion and expressed support for this project, if the President really wanted it. During the following weeks, the President was pressed and did not insist on the project, as his security people where negative because of the special attention tourists would give to the solar panels.

The Federal Chancellor Viktor Klima was in his first year within the position and before that was Minister of Finance and earlier, Minister of Science and Transport. Before he was asked into the government, he was the Financial Director of the Austrian oil and gas company OMV. He had insight into energy developments and as a long serving member of the government, he was not trapped in the energy business. As a minister, he did a lot for the implementation of the mobile telecommunication in Austria and for opening the telecommunication market. He was an insider estimating the importance of opening monopolised infrastructures to new technologies and new suppliers, if Europe wanted to stay competitive between other industrialised regions of the world. Viktor Klima could have used his welcome address for deliberations on analogies of telecommunication networks and power grids and as well as relative advantages of decentrally versus centrally organised structures, there were many options for a government chief with ambitions in technology policy. Finally he cancelled his speech. Maybe his origin from the oil industry or his cautious advisers prevented him from bringing official splendour to this outstanding conference in the historic halls of the Vienna Congress which followed the Napoleonic Wars. This drawback in the last moment could also come from the fact, that he was the new European Council President and therefore his presentation would have brought too much attention to the conference.

Instead of the Chancellor the minister for environment Martin Bartenstein gave the welcome address to the conference. Among other things he said:

"The overall capacity for photovoltaic (PV) power production grew in the last two years seven-fold. With growing rates of more than 20 percent annually photovoltaics help to bring electricity to more and more people in a sustainable way. There are still more than two billion people without electricity worldwide.

In 1997, the capacity of solar panels installed in Austria increased by 469 to reach 2208 kilowatts. So it grew 24 percent from 1996. The global development of the photovoltaics industry is comparable to the boom of the computer industry in the 70s. In 1995 there were 663 Megawatt of solar panels installed worldwide, 60 percent of that in Asia, Africa and Oceania.

In the recent EU council decision for integrating sustainability into all common policies there is great potential for renewable energies. The EU committed itself to double the share of renewables in its overall energy supply from 6 to 12 percent within the next 12 years. The program of investment includes the following goals: 10 Gigawatts both for wind energy and biomass installations, 100 communities or regions which rely totally on renewables as well as one million photovoltaic installations with an overall capacity of one Gigawatt and an overall size of 10 square kilometers. Half of these should be placed within the EU, the other half in developing countries. It is estimated, that 33 000–80 000 thousand working places will be created for these PV installations.

If you take the recent cost developments of PV power there is parity with conventional produced residential electricity expected within 20 years from now. The system costs decrease by 5 to 10 percent yearly. 1996 we had average production costs of about 11 Austrian shillings (about 80 Eurocents) per kilowatt-hour, in 2010 we can expect costs of about 3.7 Austrian shillings (about 27 Eurocents).

The integration of sustainability into all EU policies means as well a special task to national environment policies. Austria is the first EU presidency country confronted with this new common policy approach. Renewable energy use and energy efficiency will therefore be the main topics on the agenda of the informal environment ministers council from 17th to 19th July 1998 in Graz."

Martin Bartenstein was a tough Minister for the Environment and wanted to commit Austria at the Kyoto conference to a 20 percent emission reduction of greenhouse gases by 2010. The government (with three former oil company managers as members) accepted finally a 13 percent reduction. From 2000 onwards Mr Bartenstein was Minister for Economics (including energy) in the new right government and started a weak energy policy as compared with the ambitious Kyoto target. Later he resigned totally in approaching the target and became the greatest blocker within the government for measures enabling progress to the target.

Eighteen months after the world conference and with only three years in the position as a government chief Viktor Klima had to resign and became a manager in the car industry in South America. Wolfgang Schüssel took over that position after five years as Vice Chancellor and Minister for Foreign Affairs. Schüssel's conservative party came in the 1999 election for the first time only into the third place, but with the leader of the right wing party Jörg Haider, he reached an agreement against the strongest party, the Social Democrats under Klima. In the following changes to the government, I moved with my small department from the Chancellery to the Ministry for Transport, Innovation and Technology. My department was dismantled and I lost my involvement with the environment and energy questions.

Minister Bartenstein brought in 2002 a new Law for Renewable Power Production (Oekostromgesetz) through parliament, which made the installation of 700 megawatt wind power possible within the first two years and of almost 1000 megawatt in total within four years because of priority for renewable electricity from small installations and special feed-in tariffs. The law supported the installation of photovoltaics only marginally, that means until the neglectable overall capacity of 15 megawatt or 150 000 square meters. For about eight million inhabitants of Austria that was on average less than 2 Watts or 2 cells of 10 by 10 centimeters. The applications for PV installations according to that law became useless within the first day because of the small possible overall capacity. But not this blocking of PV installations was the reason for changing the law after less than three years, it was the great success the law had with wind power installations.

In the beginning of the new century the Austrian power production was based on hydropower for about 70 percent. The Oekostromgesetz brought about 3 percent wind power and 2 percent power from biomass. These two renewable power producers have greater potential in the winter half of the year whilst hydropower supplies more in the summer half, so wind and biomass power production fits well into the hydro power system. From the success of the Oekostromgesetz from 2002 onwards it was obvious, that Austria could phase out all coal and gas fired power plants within a few years. There are no nuclear plants and outside the oil refinery no oil fired power plants in Austria. So if the hundreds of biomass heating plants in Austria would produce also power for the grid and photovoltaics would no longer be blocked and available geothermal spots would be used for some power supply, Austria needed not any fossil or nuclear power within five years from now. The costs for power production would be a bit higher for some years, but they would be more stable and no more dependant from world market oil price changes. Within a couple of years electricity prices would be cheaper than in countries with electricity supply from nuclear or fossil power plants.

But after 2002 a united lobby of the pulp-, paper- and board-industry (because of competition with bio energy for timber) with the successful Austrian oil industry pressed strongly for a downsizing of the support for renewable electricity production. A new and very limited Oekostromgesetz was worked out and sent to the European Union for taking note while the old one came to an end. The new law was blocked by the EU because of exempting the great power demanders almost totally from contributing to the costs of renewable power support. The new law and EU's blocking resulted in a total halt for new wind power and biomass installations from 2007 on and brought many existing biomass plants close to insolvency. So since 2007, not only photovoltaic power installations are blocked but also new wind power and biomass power plants.

Wolfgang Hein

Born 1952 in Vienna

Diplom-Ingenieur for Technical Mathematics from the Vienna Technical University in 1975.

1976 Systems programmer at Siemens Austria

1976/77 Scholarship at the Institute for Advanced Studies in Vienna (Social Sciences)

1977–79 Technical Examinator in the Austrian Patent Office

1979–82 Department for Energy in the Federal Ministry for Trade: energy statistics, energy prognosis, energy report 1980

1982/83 Study on emissions from thermal power plants in Austria for the Federal Chamber of Labour and civil service

1984–1992 Institute for Economics and Environment of the Federal Chamber of Labour: clean air policy and legislation, water protection, policy on chemicals and waste, Environmental Fund, CO_2-commission, policy on transport and energy

1992–2001 Director in the Federal Chancellery, Department for policy-coordination: energy and environment affairs, Austrian Coordinator for the Environmental Program for the Danube River Basin, UN-Expert-committee for Renewable Energy sources and Energy for Development, participation in the UNESCO related World Solar Summit Process as personal representative of the Austrian Chancellor and leading officer in the Austrian Delegation to the World Solar Summit in Harare 1996. Representative of the Austrian Chancellor in the Austrian Delegation to the Climate Conference in Kyoto 1997.

1998 Head of the Austrian Support Committee for the 2nd World Conference on Photovoltaic Solar Energy in the Vienna Hofburg.

Since 2001 Federal Ministry for Transport, Innovation and Technology: support for research and technology development.

Founding President for the Austrian Chapter of Eurosolar 1989 to 1997, until 2006 Vice-president of Eurosolar.

Chapter 21

PV in Japan — Yesterday, Today and Tomorrow

Osamui Ikki and Izumi Kaizuka

RTS Corporation*

Japan has been one of the leading countries in dissemination of PV system and production of solar cells. Japan's efforts on PV power generation started in early 70s. While PV market in Japan experienced stagnation for these three years, it is expected that Japan's PV market will be revitalized again from 2009 as a new national target for PV capacity was set and new support framework for introduction of PV system has started. In July 2008, the cabinet announced the "Action Plan to Create a Low-carbon Society". This plan widely covers specific targets and measures regarding innovative technological development, dissemination of existing advanced technologies, and establishment of a framework to transform the entire nation into a low-carbon society. As for PV power generation, the national target was set; increasing introduction volume of PV power generation by 10 times (to 14 GW) by 2020 and by 40 times (to 53 GW) by 2030 from the current levels. Moreover, the government raised the target capacity to 28 GW by 2020 under the J-Recovery Plan formulated in April 2009. This paper reviews the history of PV deployment and industry in Japan.

1. History

The Sunshine Project, a national technological research and development (R&D) for new energy source, was started in 1974 under the initiative of The Ministry of International Trade and Industry (MITI), [currently The Ministry of Economy and

---

*RTS Corporation is a consulting firm specialized in photovoltaic (PV) power generation based in Tokyo, Japan. Founded in 1983, RTS has nearly 30 years of history and experiences.

Technology and Industry (METI)], triggered by the oil crisis in 1973. The Sunshine Project aimed to reduce dependency on petroleum and improve the vulnerable energy supply structure. Under the project, long-term R&D activities were conducted by cooperation of the industry, the government and academics, fore-seeing the year 2000, with the emphasis on (1) solar energy, (2) geothermal energy, (3) gasification and liquefaction of coal and (4) hydrogen energy. The photovoltaic power generation system (PV system) was categorized under the area of the solar energy, and fundamental and application R&D of solar cells and the PV system were started. In the early 1980s, New Energy and Industrial Technology Development Organization (NEDO) promoted R&D and demonstration (R&D, D) for practical application of the PV system, focusing on development of manufacturing technologies of the solar cell and application technologies of the PV system, aiming at cost reduction. A lot of researchers from the industry, universities, and national institutes working on R&D of the solar cell and the PV system were cultivated through the projects of NEDO in these years, and the foundation of today's framework of R&D and the market development were built. The R&D has been continued throughout the 2000s and the results from the R&D were commercialized into the PV products by the PV manufacturers in Japan.

In the early 1990s, basic technological issues on grid-connection of the PV system were solved because "Guideline of the Technical Requirements for PV Grid-connection with Reverse Flow" was established in 1993, and the preparation period of dissemination of PV system was started, while the prices of PV cell and modules were higher level. METI (MITI at the time) launched PV System Field Test Project for Public Facilities in 1992 and Residential PV System Monitor Project for individual houses in 1994, aiming at dissemination and cost reduction of the PV system. Thus, the markets of the residential PV system and PV system for public and industrial facilities were being formed. In order to promote installation of the PV system in the private houses more widely, METI converted the Residential PV System Monitor Program to the Residential PV System Dissemination Program in 1997. This program continued until the end of fiscal year 2005 and the residential PV system is still major application in Japan dominates more than 80% of the annual installed capacity in 2008.

2. Current Status of PV in Japan

With the support of the government, Japan's installed capacity has grown steadily. Accumulated installed capacity of the PV system in Japan in 2008 was 2 144 189 kW, exceeded 2 GW. However, after completion of the subsidy program for introduction of residential PV systems, Japan experienced the minus growth of the annual installed capacity in 2006 and in 2007. Total annual installed capacity of the PV system reached 225 295 kW in 2008, a 7% increase from 210 395 kW in 2007, which is the first increase in two years. Figure 1 shows the trends of PV installed capacity in Japan. Major application of the PV Market in Japan has been residential

PV system. Residential PV system accounts for almost 83% of the total domestic shipment amount of PV modules in 2008.

As was mentioned Residential, PV System Dissemination Program was terminated in FY 2005 and there have been no national incentives since 2006. The main drivers for the residential PV market in Japan have been the net billing program implemented by all the utilities and limited amount of the incentives from local governments. The net-billing program is a voluntary efforts of electric power companies and similar to net-metering, owners of the PV systems can sell the excess power from PV systems at the same price as the electricity charge, 24 Yen/kWh in case of typical contract for households.

In 2008, production volume of solar cells and PV modules in Japan reached 1224 MW in total, a 33% growth from the previous year as shown in Fig. 2.

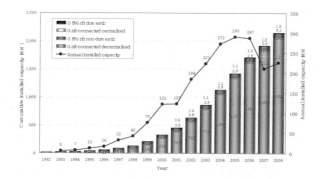

Figure 1.   Trends of PV installed capacity in Japan, (*Source* : IEA PVPS Task NSR of Japan.)

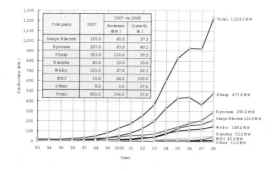

Figure 2.
Source : PV News, modified by RTS.

3. New Support Framework for PV

On June 9, 2008, the then Prime Minister Yasuo Fukuda unveiled the "Fukuda Vision", titled "In Pursuit of Japan as a Low-carbon Society". PV system is

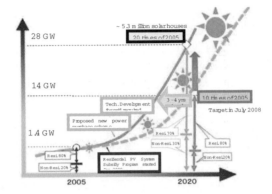

Figure 3. Prospects of PV power installation presented by the J-Recovery Plan.
Source : J-Recovery Plan.

Figure 4. Thin-film silicon PV system, Chidorigafuchi, Tokyo.

Figure 5. Residential PV systems in Ohta City, Gunma Prefecture.

Figure 6. Light-through PV systems in Motosumiyoshi Railway Station, Kawasaki City, Kanagawa Prefecture.

included as one of the key components of the Vision. The prime minister expressed his strong determination, with a target of "increasing PV system introduction volume ten times greater than the current levels by 2020 and 40 times by 2030 to realize "Solar Society".

In response to the "Fukuda Vision", the national government made a cabinet decision on "Action plan to create a low-carbon society" in July 2008. This plan widely covers specific targets and measures regarding innovative technological development, dissemination of existing advanced technologies, and establishment of a framework to transform the entire nation into a low-carbon society and so on. As for the PV power generation, a national target of increasing introduction volume of PV power generation by 10 times (to 14 GW, estimation) by 2020 and by 40 times (to 53 GW, estimation) by 2030 from the current levels was set.

In April 2009, the government formulated an economic stimulus measure named the "J-Recovery Plan". Under this plan, 'Low Carbon Revolution" is one of the three pillars, and measures for drastic acceleration of introduction of PV power generation are stipulated: drastic acceleration of introduction of PV power generation (public and agricultural facilities), promotion of PV power generation, expansion of programs to support introduction of new and renewable energy at private facilities, increase of subsidy for residential PV systems, promotion and support programs for environmental education for elementary/junior high schools installed with PV systems and other environment-friendly facilities, etc. Moreover, the plan set the new PV power target: 20 times of cumulative installed capacity, 28 GW by 2020. Figure 1 shows the prospects of PV power installation presented by the J-Recovery Plan. With this plan, the target for 2020 was raised from 14 GW to 28 GW.

In response to the series of these movements towards PV dissemination, Ministry of Economy, Trade and Industry (METI) started the new Residential

Figure 7. Dr. Palz, Osamu Ikki, CEO of RTS Corporation and their friends at PVSEC-18, Kolkata, India.

PV System Subsidy Program from January 2009, with the supplementary budget of FY 2008. The program provides subsidy of 70 000 Yen/kW with the cap of 700 000 Yen/kW for the introduction price of a PV system including the installation cost, amounting to 9 billion Yen in total. In addition to this, METI requested approximately 24 billion Yen for the subsidy program of residential PV systems in the FY 2009 budget request. METI expects that several tens of thousands of households in FY 2008 and approximately 100 000 households in FY 2009 will be supported by this program. METI also announced a new power purchase program for excess power generated from PV systems. This proposal was realized by the Act on the Promotion of the Use of Non-fossil Energy Sources and Effective Use of Fossil Energy Source Materials by Energy Suppliers that passed the diet in July 2009. Currently, each utility voluntarily provides net billing program and purchase excess energy generated by PV systems at the same rate as the electricity charge. Under the new law, excess PV power purchase is obliged to utilities and preferential buy-back price was set, approximately the double amount of the electricity charge. In case of residential PV systems, excess power will be purchased at 48 Yen/kWh. This price will be reduced year by year for new installation. It is expected this new scheme will shorten the cost recovery time of installation cost and promote energy conservation at households.

In R&D area, "international R&D centres for innovative solar cells" were selected to enhance R&D on ultra-high efficiency solar cell with over 40% of conversion efficiency. METI also established "Study group on low-carbon power supply systems" to identify challenges for expanded introduction of PV and other types of renewable energy, as well as actions for the challenges. METI also revised a roadmap for technological development of PV power generation, "PV2030" formulated in 2004 to drive cost reduction of PV systems.

4. Conclusion

In 2008 and 2009, the Japanese government has made a drastic turnaround in its energy policy to achieve a low-carbon society. As part of the turnaround, the government set a goal of installing 28 GW of PV systems by 2020, with a view to significantly expanding installation of PV systems. Installed capacity of PV systems in Japan in 2008 reached 225 MW. Annual installed capacity is expected to largely increase to 400–500 MW in 2009, through the following promotional measures and activities:

(1) enhancement of subsidies for residentia.l and other types of PV systems,
(2) a new program to purchase surplus PV power,
(3) support in taxation systems,
(4) action plans for promoting installation of PV systems with the initiative of the national government, and
(5) efforts by the PV industry to reduce costs.

With strengthened support and approaches for PV power generation, not only by the Ministry of Economy, Trade and Industry (METI) but also by other ministries, agencies and local governments, it is assumed that the installation of PV systems will grow in public and industrial markets in addition to the residential market, which is currently the major PV market in Japan. A series of ongoing promotional measures and activities to establish a framework for the dissemination of PV systems are expected to greatly contribute to reducing costs of PV systems. Consequently, the size of the Japanese PV market is estimated to grow to the level of 1–1.2 GW by 2012.

Osamui Ikki

Izumi Kaizuka

Chapter 22

PV in Europe, from 1974 to 2009: A Personal Experience

Helmut Kiess

Im Unteren Tollacher 11, CH-8162 Steinmaur, Switzerland

1. Insight Period: 1974–1988

Two or three years after the oil shock in 1974, a small group of Swiss scientists of various backgrounds and institutions (I was with Laboratories RCA in Zürich, a subsidiary of RCA corporation in the US) gathered privately in order to discuss the question of how to scope with the energy problem, in particular also whether solar photovoltaics in Switzerland would be a route to go. However, first a-Si cells, patented in 1977, had efficiencies in the range of less than 1%, obviously too small to contribute in a sensible way to the supply of electrical power. This would be certainly true for countries, where solar inputs are in the range of 1200 kWh/year, like in Switzerland, or less. The fabrication of the first crystalline Si cells and research goes back to the years between 1950 and 1960 induced by the emerging satellite technology. Best small-sized laboratory cells reached in the 1960s efficiencies of about 11%, which could be improved to about 17% in the 1970s. However, cells could be only produced at high cost. Cost was not a real problem for satellites, but it would be crucial for terrestrial large area solar photovoltaics. So some of us, including me, had doubts whether cell technology could be improved and cost reduced to a degree that it might substantially contribute to electrical energy supply. During these years I had many discussions with A. Rose, at that time one of the world's leading scientist in photovoltaics and photoconductivity. He was convinced that human ingenuity would overcome these problems and that the abundant radiation of the sun to the earth could be tapped efficiently and at reasonable cost. Would this become reality?

Power for the World by W. Palz
Copyright © 2011 by Pan Stanford Publishing Pte Ltd
www.panstanford.com
978-981-4303-37-8

In 1986, the Chernobyl accident increased doubts in nuclear energy. In addition, the public became more aware of the reasons for global warming, all of which favoured public funding for an increasing commitment to research in renewable energies.

2. Innovation Period during the Decade 1988–1998

In the meantime Laboratories RCA were taken over in 1988 by the Swiss Paul Scherrer Institut. After the takeover, a reorientation of the laboratory's programs was indicated. Our group considered solar photovoltaics being a promising subject of research and we had projects in our minds, aiming at an improvement of the efficiency of crystalline Si photovoltaic solar cells. In 1988, I attended for the first time a PV-conference, namely the 8th European Photovoltaic Conference in Florence, which confronted me with the state-of-the-art in 1988 in solar photovoltaic energy. A broad approach had been given to this field by an international scientific community. Generally the essence of the papers presented in Florence can be summarised as follows: Dominant were exploratory research and engineering in cell technology regarding processing, optics, and investigation of various materials as candidates for solar cells. Equally important were investigations on modules. However, also systems and even the storage of electrical energy were the topic of some papers, surprisingly for the early stage of terrestrial solar PV in 1988.

2.1 *State of the Art at 1988, Some Details*

(1) Crystalline Si solar cells: Efficienies of single crystalline silicon solar cells were 20.4% at AM 1.5, and for multicrystalline Si cells 16.6% at AM 1.5.
(2) Thin film cells: a-Si:H 8–9% at AM 1.5, $CuIn(Se)_2$ 12%, CdTe 12–13%. The area of the cells was often below 1 cm^2, upscaling to larger areas ended up in significantly smaller efficiencies. Further materials evaluated were CdSe, WSe_2, InSe. Thin film materials offer a large variety of deposition technologies which may end up in different quality, reproducibility, need for post deposition treatment of the films. A large number of techniques were explored, however, the best one could not yet be identified.
(3) Cells for concentrator systems: Single crystalline Si cells: 28.4% efficiency at concentration of 100 suns, 28% with GaAs single junction at 500 ×, 24.5% for multijunction AlGaAs-on-GaAs cell. The prediction at that time was, that 35% efficient of multijunction cells might be reached within 5–10 years of continuous research.
(4) Optics and light trapping: Light trapping and concentration were important topics. Discussed were also fluorescent concentrators.

(5) Last not least modules, systems and projected costs of solar photovoltaic energy intensively were discussed and investigated.

2.2 The Decade Between 1988 and 1998

There had been, indeed, an impressive progress during the years between 1988 and 1998. After various tentative proposals, our group decided at the end of the 80s that a program might be sensible dealing with light trapping in solar cells, in particular of crystalline Si cells. The intention was, in the long term, to allow the fabrication of thin crystalline Si cells avoiding significant light loss and achieving cells with more than 20% efficiency. At the end of this program in 1996, our light trapping scheme with gratings was applied to 50 μm thin crystalline Si cells ending up with a short circuit current equivalent to that of 100 μm thick cells. The best cells we fabricated were 21% efficient, though they were 250 μm thick. Thin cells suffered from insufficient surface passivation.

The trends for the years to follow 1988 could be envisaged at that time by increasing cell efficiencies using improved cell technology, by concentration of the sun light and investigations of tandem and triple cells in order to get beyond the limitation of the efficiency of cells fabricated from a single material. Indeed, by 1998, the efficiency of crystalline Si had been pushed to 24.4%. Concentrators were not really in the focus of discussions and tandem and triple cells remained essentially the domain of space cells.

Further trends would be: investigations of various materials for thin film solar cells, improving their efficiency, test their reliabilty. In the long run, thin film solar cells might offer a technology advantage over that of crystalline Si technology due to high speed of cell and module fabrication. Significant progress was made. Three types of materials were found to remain candidates for thin films, CdTe, CIS ($CuInSe_2$) or ($CuInGaSe_2$), a-Si:H and tandem structures (micromorph Si and a-Si:H/a-SiGe tandem structures).

Starting in 1986, a series of workshops were devoted in the years to follow to fundamental research on quantum solar energy conversion. After an internal discussion during the workshop 1994 concerning the topics, organisation etc. we concluded that it would be advantageous to form a society. At the 10th International Conference on Photochemical Conversion and Storage of Solar Energy at Interlaken, I presented this idea to a broad scientific community. Support for the foundation of a "European Society for Quantum Solar Energy Conversion" was obtained. The formation of this society could be realized. It should promote fundamental research on photochemical conversion and storage of solar energy as well as research on frontiers on excitation and recombination processes in solids. Unconventional proposals and basic questions in solar energy conversion had now a platform for interdisciplinar discussions between chemists, electrochemists, physicists, was open for students, and visible to a wider range of scientists.

In order to push solar photovoltaic, the proof should be given, that results obtained in laboratories and on test fields could be validated on a large scale. Thus in 1990, the 1000 roof program was launched in Germany. This was, indeed, a step required to prove that solar photovoltaics can significantly contribute to the electrical energy supply in Europe and that funding of research and technology would be continued. A mental barrier could be overcome because the results showed that solar photovoltaics works on a larger scale, that it can contribute to the electrical power supply, that solar panels operated well in northern latitudes of Germany with insolation of 900 kWh/year, that dc to ac conversion of electrical power posed no serious problems, and finally that large installations on roofs, along roads etc. are possible. However, despite the success of this program, industrial production capacities remained low at a modest level.

2.3 *State of the Art 1998, Some Details*

(1) Crystalline Si solar cells: Efficiencies of single crystalline silicon solar cells were 24.4% at AM 1.5, and for multicrystalline Si cells 19.8% at AM 1.5. Module 30×30 cm^2 with shingle brickwork arrangement of the cells: Efficiency 23%
(2) Thin film solar cells: a-Si:H mass production achieved, stabilized efficiencies 5 to 6%, near term goal: 8% stabilized efficiency. Triple junction cell a-Si — a-SiGe$_x$ — a-SiGe$_y$: 13% stabilized; tandem cell (socalled micromorph cell) a-Si — μc-Si: 12% efficiency
(3) Cu(In,Ga)Se$_2$ 17.7% small cell size, in pilot fabrication 11.2% for modul size of 3800 cm^2
(4) CdTe 16% small cell size, in pilot fabrication about 10% for modul size of 4500 cm^2
(5) Industrial production capacities: Crystalline Si: 215 MW worldwide, 55 MW Europe
(6) Industrial production capacities: CdTe, CIS and CIGS: 20 MW worldwide, 0.2 MW Europe
(7) Industrial production capacities: a-Si:H: 49 MW worldwide, 2 MW Europe

3. Innovation and Industrial Production During the Decade Between 1998 and 2008

After my retirement in the year 1996 from the laboratories and the end of my activity as lecturer in 1998, I acted essentially as an observer of the solar photovoltaic scene and the task remaining was to consult various companies and writing reports to the Swiss Energy Department (Bundesamt für Energie) on progress in crystalline Si solar cells and new developments.

The great push to solar photovoltaic industry in Europe was induced by the political decision in Germany to stimulate solar photovoltaics. This ended up in a law on the feed in tariff (the so-called Einspeisegesetz), a law which affected the solar industry worldwide because a huge solar energy market developed in Germany and, by a similar stimulation programs, in the second half of the decade in Spain and in France. In response to the demand the production capacity of solar cells in Europe increased from 57.2 MW/year in 1998 to 1.9 GW/year in 2008 by a factor of 33, and worldwide from 284 MW/year to 4.3 GW/year by a factor of 15. Thin film solar PV modules reached a market share of about 13%, whereas concentrating photovoltaics was still an emerging market which is marked by an installed capacity of only 17 MW in 2008.

Converting the solar energy market into a real business had a tremendous impact on research and development. The competition between crystalline silicon and thin film and concentrator systems on the market gave rise to efforts in the research to improve existing systems or to investigate new concepts for solar cells. Thus single crystalline silicon PV modules with 20% efficiency were developed and are on the market. These modules deliver the same electrical output on half the area as compared to 10% efficient modules. Is this the answer to the 10% efficient CdTe thin film technology which is at present the cheapest cell technology and with which grid parity can be achieved, depending, of course, on the solar irradiance at the particular solar area? Are area related cost for solar PV systems small enough and is area for solar abundantly available in Europe? Already these few questions indicate, that research in cell technology was and will be an issue even beyond 2008. Hence it could be observed, that investigations were enforced on multijunction cells and on unconventional ideas which would allow to get beyond the efficiency of cells fabricated from a single material (with a single gap).

It is well established that a photovoltaic device can achieve in theory an efficiency well above 50% if it consists of a monolithic stack of cells made of various, apropriately selected materials. At the conference in Milan in 2007, such devices were presented with efficiencies between 40 and 41%, at 100 to 200 fold concentration of the sun light. A different device, using clever optics and arranging the different cells side a side, a module was fabricated with an efficiency of close to 43% at 20 fold concentration. Indeed, a significant progress: devices without huge machinery or moving parts convert sunlight that efficient into electrical power! In order to reduce the complexity of the device fabrication, physicists want to design cells prepared of a material which fulfils in one the properties of three stacked materials. Efficiencies are predicted to reach 62% at highest concentration of the sun light. Other routes as well for highly efficient cells have been proposed, which are promising but are still in an early stage of experimentation.

3.1 *State of the Art 2008, Some Details*

(1) Crystalline Si solar cells: Module efficiency commercially available 20%. Multicrystalline Si cells: Module efficiency 18%
(2) Thin film solar cells: a-Si:H module, stabilized efficiencies 5 to 6%
(3) Micromorph a-Si (tandem cell) module 10% efficiency
(4) CIS ($CuInSe_2$) and CIGS ($Cu(In,Ga)Se_2$) module 11% efficiency

(5) CdTe modules 9–10% efficiency
(6) Triple junction cell on the basis of GaAs. 35% without concentration
(7) Triple junction cell at a concentration of 240 suns: 40.7% efficiency
(8) Module of five different cells, not stacked, concentration of 20 suns: 42.8% efficiency

4. Epilogue

The optimism and prediction of A. Rose in the 70s of the 20th century proved to be true. An incredible progess in solar photovoltaics has been achieved. Even in these days, solar photovoltaics is an exciting field due to new developments. Indeed, it is by no means evident which route the conversion of solar irradiation into electricity should go: Huge solarthermal installations (Desertec) are being planned in North Africa and near the eastern sunbelt which are to supply electrical energy to Europe through submarine cables in the Mediterranean Sea. Or else, would it not be wise to use instead of solarthermal, the highly efficient solar photovoltaic devices of low maintenance cost and of say 30% to 40% efficiency? Of course, those cells might require another 10 to 20 years before being ready for industrial production. Anyway, I am convinced that solar photovoltaics will be a dominant technology of the future for the conversion of sunlight into electrical energy as A. Rose had already predicted 40 years ago.

For this short article I did not mention the names of the many scientists, engineers and politicians who were crucial for the success of solar photovoltaic energy conversion. I hope for their understanding. The list would be too long.

Helmut Kiess

Chapter 23

PV in Berlin — How It All Began: The Story of Solon, Q-Cells, PV in Brazil

Stefan Krauter

Photovoltaik Institut Berlin, Wrangelstr. 100, D-10997 Berlin, Germany

Sometime around 1994/1995 we had the 25th meeting of our association (aim: the development of the International Solar Centre — ISC), the goal was noble (the salvation of humanity?), the participants were pretty much the solar scene in Berlin (four professors, five solar NGO representatives, a former spokesperson for the Green Party, a solar artist, two solar freelancers), but somehow the implementation was taking a while. The centre was supposed to be financed through the renting of rooms to companies operating in related industries, but as a result of the "reunification", there were not only rooms but also whole streets available in Berlin for almost nothing. As well as that, nobody was brave enough to try to talk the individual participants out of their beloved favourite projects (a solar boat hire, a solar crèche, a solar school, a giant solar art exhibition) with the aim of drastically reducing costs. We had received support from the Berlin Senate, but this was just enough to turn the association into a GmbH (limited company) and to finance Jo Leinen (a former anti-nuclear activist who made a career through the State, Federal, and European Parliament).

In any case, most of the time was spent drafting really very balanced press releases and brochures to publicise this wonderful project. Interestingly enough, in doing this the age of our association was played down (in contrast to most institutions, which like to accredit themselves with a somewhat longer "tradition" than is actually the case), like it would be for a diva past her prime and still awaiting for her big break-through. One day during our Charlottenburg debating club period, a completely non-academic group from Kreuzberg appeared: Birgit Flore, the spokesperson (boss of the "Economista" business start-up for women, successful in the importation of Nicaraguan coffee), Reiner Lemoine [co-founder of the socialist engineers collective "Wuseltronik"[1]) and Alex Voigt (he had founded a mini-company for solar heating installation, and had already installed a Wuseltronik solar charge controller in a shooting box belonging to some Frankfurt bankers, and knew a couple of potential financiers from that scene. He indirectly excused himself for this by sometimes mentioning "coincidentally" that he was in the front-line at the demonstrations against the construction of "Startbahn West" (*"Runway West" for the Frankfurt airport which destroyed a lot of woods in the region*)].

This colourful bunch had, in any case, read the Greenpeace study on the feasibility of efficient solar factories, were also familiar with our dazzling ISC brochure, and wanted to immediately get cracking with the factory in our solar centre. We had to cautiously admit that the opening time was not really in the foreseeable future, and, on top of that, a factory wouldn't really fit in with the ISC (although otherwise, or in the beginning at least, every absurdity which contained the word "solar" found a place there). Somewhat disappointed by the stonewalling on the part of my association colleagues, I asked some of the potential political PV factory founders whether they could use someone who knew something about PV.[2]

I was summoned a couple of days later to the "Economista" barracks, and cross-examined in an interview there before it was finally adjudged that I could participate. Extremely proud of my membership of this illustrious circle, I told my friend and neighbour Hartmut Duwe, who was immediately keen to participate. We also thought of Paul Grunow, who I knew from Rio de Janeiro and Hartmut from its dissertation time at the Hahn-Meitner Institute (nowadays, Melonholz Gesellschaft), as a potential candidate. He was going through a rough time after a disappointing year in Brazil (a long story) and needed a new challenge.

As I had already "got acceptance" on the committee for the huge need for PV know-how for the building of the factory, there were soon eight of us; along with the ISC visiting committee, my friends Stefan Fütterer (inverter specialist at Wuseltronik, soon to be boyfriend of Birgit) and Saleh El-Khatib (one of the first PV grid feed suppliers) were also on board.

---

[1] I had the impression he was less of a political macho (as he was described by others), and was more like the Love Parade founder Dr. Motte, something which on first glance could be described more as "well meaning" or "naïve".

[2] As it turned out much later, what was at the time a completely insignificant event (it frequently happened that crackpots would come to visit and disturb the peace of our ISC circle) turned out to be the turning point in my development and perhaps in the development of PV, or possibly this development was an inevitability: Who knows why it all didn't happen then 10 years earlier; from a purely technological standpoint, there was nothing available at this point in time which hadn't been there 10 years previously. Or was it all down to the feed-in act after all?

The discussion style definitely took some time to get used to, let's just say you invariably had to show what stuff you were made of, and in comparison the ISC seemed like a holiday camp. Hartmut was most often in the thick of the action; as a result of his active membership of the Green Party and as an East German he had an extraordinary ability to discuss in depth even the most insignificant items on the agenda. He also displayed a certain reticence when it came to "cut to the chase", thus increasing the latent uncertainty of the project (which was never spoken of, as this would have been akin to "blasphemy"), which resulted in an overcompensation that manifested itself in the above described discussion style.

Clemens Triebel wasn't exactly a direct co-founder, but rather a motor and passionate main agitator; he would certainly have made an excellent actor, and indeed, he was on the frontline for many discussions. The same is true of Johannes Grosse Boymann, who, although not there so often (though when he was, he really was), without much ado put up 25 000 German Marks towards the founding (at the time this was not an insignificant amount for Kreuzberg "Sponti"[3] circles).

During the founding discussions a young notary by the name of Thomas von Aubel was in attendance, and became immediately involved in the discussion on the legal form of the company. The old political guard around Birgit, Clemens and Reiner were in favour of a non-profit GmbH (Ltd.) or some form of cooperative. That was a bit too much secret society for me, and more than anything it was clear to me that we should make participation as simple, quick and unbureaucratic as possible, just to get near the scale of the means required. For me, the simplest and most effective form of participation was company shareholding, arch-capitalistic of course, but nevertheless effective. (A couple of years previously, I had bought some GEA shares for dust filter technology, the only "environmental shares" available at the time on the German market. Solar shares were first available with us.) I wouldn't have been able to come through alone with my opinion, but the notary became starry eyed, saying that he thought it an excellent idea, and that should a public limited company be founded, he would definitely be buying a shareholding (which he then did so a number of times in the future, in particular with the founding of Q-cells). Eventually the others were also convinced and Solon AG, the first European solar public limited company, was founded. (The Solarfabrik Freiburg AG almost beat us to this; after I mentioned our planned legal form at the PV-Forum in Bad Staffelstein in 1995, the idea was then adopted in double quick time by the Freiburgers.)

Alex then brought the Wella heir Immo Ströhr along, who unobtrusively but attentively observed proceedings, and then decided to come on board in the form of a generous development contract for a solar thermal system for the Spanish market. At the founding of the public limited company, he became the majority shareholder and later stood by the company at critical moments, meaning that

---

[3] Sponti: member of a leftist political activist movement in the 1970s and 80s propagating, amongst other things, "being spontaneous".

Solon AG survived much better in the long term than the southern German companies which at that time were the dominant presence in the field (PV in East Germany only came a couple of years later).

At the time at which the name was determined, I was at a workshop in Brazil; I found the name less than ideal, mainly because the domain www.solon.com already belonged to an American lawyers association, though I eventually got used to it.

At the workshop in Brazil I was asked if I would be interested in a five-year tenure as a professor at the Federal University of Rio de Janeiro. During a restless night, I went over the alternatives: In the long term Brazil offered, from a purely technical point of view, the better opportunities for PV; in comparison to Central Europe it has double the annual solar radiation (and on a typical winter's day four times the daily radiation), meaning that the solar electricity production costs would only be half as high. At the same time, there was no electricity infrastructure in rural areas, so PV electrification would be perfectly suited to development there. With the enormous hydro power stations with their gigantic dam storage it would be relatively easy to realise an energy supply based 100% on renewable energy. Moreover, my mentor on site, Prof. Richard Stephan, fired my enthusiasm for the country (he often organised mini-excursions when getting a lift with him to the university, located 25 km from the city centre and the residential areas). My Ph.D supervisor, Prof. Hanitsch, considered our Solon endeavour to be a suicide mission, and thus advised me go for Brazil. The new Brazilian currency, the Real, was at the time worth 10% more than the US$ (and 30% more than the Euro), meaning that relatively speaking the salary was also attractive. And not forgetting that I had also made the acquaintance of an exotic woman on the bus ….

At Solon, of course, nobody was amused by my decision, but it was received with a certain degree of understanding, given the offer. The first instance after my departure, it looked like my decision was more right than ever: The shares were at times 80% below their original value, it would have been better if the Italian laminator had come together with its own Italian mechatronic engineer (an analogy to Italian sports cars?), and a profitable feed-in act for PV was still a long way away. Lars Podlowski was often asked about laminator problems, later his engineering company was taken over, and then at some stage he also became CTO at Solon.

Q-cells: At the time of the founding of Solon in 1996, the establishment of a cell production plant at a later stage was planned. However, in the beginning, nobody dared to take the first step. Even the "established authorities" in the industry at the time warned against it (however, they also had done the same regarding the founding of Solon). The project was then put on ice for the foreseeable future, before certain "centrifugal forces" accelerated the initiative to establish a cell production plant: on one hand the aforementioned discussions, and on the other the loss of ideals, e.g. resulting from the appointment of Mr Hasenbein, the chairman of the board "suggested" or rather installed by our bank, Berliner Sparkasse, the desire for glory on the part of certain shareholders with close ties to the banks, etc. Actually, Q-cells would have been established in Berlin: In any case a proposal for the founding of a cell production plant was soon presented to Berlin Partner GmbH and the Berlin promotion bank IBB, which, however, apparently due to lack of credibility, was not processed. At this moment, Manfred Kessing, the Mayor of Thalheim near Bitterfeld appeared. He showed a strong interest in the project and took care of a grant and the participation of the state of Saxony-Anhalt, as well as a building permit. Of course the "location" was by no means the dream of the founders; all came from Berlin, most had family there, and couldn't really warm to the industrial wasteland of Bitterfeld, but the invitation was clear and the funding had been secured for the most part. There are a number of reasons for Q-cells subsequent meteoric rise:

First and foremost the uncapped feed-in act for PV electricity, without that, the planning, financing and operation of large industry-scale PV factories would not have been possible (previously there had only been what could be at best described as manufacturing workshops). The almost unlimited expansion possibilities at the location, and the uncompromising decision for a later entry to the stock market, the increasing experience of the founders, and the philosophy that, if you can't do without a "finance nerd" then be sure to get a good one in at the beginning (before they buy themselves a slice of the action at a later stage): thus Anton Milner, a top-class McKinsey adviser from England who had previously worked in the oil business, became the first employee of Q-cells.) Definitely not a snooty aristocrat, but rather a good mate who gets on well with everyone, can cut

Figure 1. Preparing the foundation of Solon SE (Stefan Fütterer, Birgit Flore, Stefan Krauter, Reiner Lemoine, Hartmut Duwe, Alexander Voigt, Paul Grunow).

to the chase, and prefers flat, efficient hierarchies. Moreover, he had a trace of those "delusions of grandeur" which are necessary to be able to think really big and outside the established structures. There were still no manufacturers for production lines for solar cells; rather the individual suppliers first had to be brought around to conceiving the existence of such a thing, let alone manufacturing and supplying it. Paul and Reiner were in their element. The marketing department also made a significant contribution to this success: The large Q in bright blue, trendy, straight talking, lively one liners, the best parties at trade fairs and congresses, and the relaxed manner of the relatively young team attracted a lot of attention. Previously the PV business was more of the appendix of the large, inflexible, traditional electricity and mineral oil corporations: Siemens, AEG, Kyocera, Sharp, Arco, BP, Shell, Total — who from the younger generation would really want to invest their money with them or buy solar cells from them? The Q-cells sales and marketing team approached the module manufacturers directly, talked (and listened) to them, and Paul solved their problems — as a result the cells always sold well, although they were not always the cheapest (nowadays, with the competition having learned from this, and the Chinese cell manufacturers having become so cheap, Q-cells no longer has it so easy).

But I was in Rio: of course we stayed in contact, Paul even came twice to my Rio congress with which I attempted to keep the "spirit" of the huge Rio World Summit in 1992 alive, and aspired to further the idea of climate protection through renewable energy (the most recent: www.rio9.com, next www.Rio.12.com). Running parallel to the congress, soon there was also a small trade fair, the LAREF (the "Latin America Renewable Energy Fair"), at which Solon and Q-cells were represented a number of times. But in reality I was a part from the developments at Solon and the establishment of Q-cells. I learned Portuguese, established a small PV laboratory at the university, and attempted to teach my students something about solar energy and sustainable development. Together with a former student I then established a small planning and installation company for PV called "Rio Solar Ltda". The exotic woman from the bus moved in with me and together we had a wonderful roof terrace apartment on the Copacabana. But after eight years (following my time in Rio I accepted a guest professorship in Fortaleza, in the north east of Brazil), I discovered that I was not genetically predisposed to abstinence from the business of efficiency. I telephoned regularly with Paul, for whom Q-cells had become soulless and "contaminated with management consultants", and who because of this had decided to take six months out with his family to Hawaii (Maui — to surf). We authored a scientific publication together and considered what our options for the future were. I couldn't stop thinking about the phenomenal development of PV in Germany, and I was annoyed with the ignorant Brazilians who held fast to the development of nuclear power (and still do to this day), and on top of that built lots of new overpriced fossil fuel power stations (an overreaction to the energy crisis in 2002), leaving PV out in the cold. Seeing as in either Brazil or Germany (or indeed anywhere else), there was hardly any training available in the area, in 2004 I wanted to establish the International University for Renewable Energy (IURE) in Brazil, with applied research and a permanent exhibition, as well as courses for everybody (probably just a throwback to the ethos of the ISC). We even received support from UNESCO, but on site we had to struggle against national sensitivities: the language used in class was supposed to be Portuguese in any case, and the whole thing was to be inspected by the Brazilian decision board, CAPES. The German ambassador Prot von Kunow was very taken with our affair, supported us, attended our conference together with all the German ambassadors of the Mercosul, but then cautioned me against setting up the IURE in Brazil as CAPES was controlled by the established federal universities and they would hardly allow anything like this — that had been his experience a couple of years previously with the founding of an international university for economics.

All signs indicated that Paul was correct in his pessimistic reading of the situation, and this convinced me in 2005 to establish a scientific institute for R&D of PV module technology in Berlin, which was exactly my area of expertise. Initially, it was supposed to be a Q-Cells Institute at the TU-Berlin together with Reiner, but Reiner developed a brain tumour and retired from Q-cells, though he supported us up until his death in the establishment of a photovoltaic institute.

The Photovoltaik Institut Berlin now has 32 employees, and is now less involved with science than with testing and certifying of PV modules. In December 2009, owing to a lack of space, the institute moved from the TU-Berlin in Charlottenburg to Kreuzberg, back to where it all began. As coincidence would have it, one of its near neighbours is the wind measurement company "Ammonit", a "capitalist" spin-off of the socialist engineers collective "Wuseltronik".

(a)

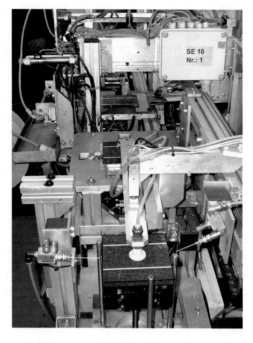

(b)

Figure 2(a) and (b). Self-made equipment for the preparation of cell-stringing at Solon ten years ago.

Stefan Krauter

EDUCATION

1994–1998	Habilitation (qualification as university lecturer) at University of Technology Berlin (TUB), Germany.
1995–1996	Post-doc studies at Prof. Dr. Richard Stephan at Federal University of Rio de Janeiro (UFRJ/COPPE/EE), Brazil.
1994	Post-doc studies at Prof. Martin Green and Prof. Stuart Wenham at University of New South Wales (UNSW), Sydney, Australia. 1989–1993 Ph.D. on PV-modules at University of Technology Berlin (TUB) at Prof. Rolf Hanitsch
1982–1988	Master in Electrical Engineering at University of Technology Munich (TUM).

EMPLOYMENT HISTORY

since 09/2010	Professor for Sustainable Energy Concepts at Paderborn University.
since 10/2006	Director at Photovoltaic Institute Berlin (PI-Berlin AG).
2008–2010	Professor for Energy Systems at Biberach University of Applied Science, Germany.
since 2005	Professor for PV Energy Systems at University of Technology Berlin (TUB).
2003–2005	Visiting Professor at the State University of Ceará (UECE), Brazil.
2003–2005	General Chairman of RIO 3 & RIO 5 & RIO 6 & RIO 9 congresses and Latin America Renewable Energy Fairs (LAREF 2003/05/06/09).
since 2000	Director of Rio Solar Ltd. (planning and set-up of PV power plants), Rio, Brazil.
1998–2003	Visiting Professor at Federal University of Rio de Janeiro (UFRJ/COPPE), Brazil.
1996–1998	Foundation and set-up of the joint-stock PV company SOLON SE, Berlin.
1994–1998	Lectureship "Solar energy conversion" at University of Technology Berlin (TUB).
1989–1994	Scientific assistant, Department of Electrical Engineering at University of Technology Berlin (TUB): Set-up of laboratory "Practical for Photovoltaic Energy Systems".
Scientific output:	105 publications, 4 books
Awards:	Green Price of the Americas 2006, IEEE Poster Award 2006, Berlin Solar Award 1995.

Stefan Christof Werner Krauter
Prof. Dr.-Ing. habil
www.stefankrauter.com
E-Mail: info@stefankrauter.com

Chapter 24

Three Steps to a Solar System — 1–40% and 100%

Harry Lehmann

UBA, Dessau, Germany

Since the advent of humankind, energy supply has had a decisive influence on its social development, since it sets the social and economic framework within which people live. In terms of energy use, human history to date can be roughly divided into three periods: the age of the hunters and gatherers, the "pre-industrial solar age", and the "fossil/nuclear industrial age". We are currently moving towards a fourth period: the industrial solar age.

The characteristics of a sustainable society have been outlined and discussed many times. They include, among others:

- a resource-optimised service oriented economy (factor 10 increase of productivity and absolute decrease of the consumption of resources),[1]
- a switch to renewable resources/energies and the efficient use of these resources,
- a conserving use of land, and, ultimately,
- a critical review and re-assessment of prosperity and of how much of it is sufficient.

The energy industry must be quick in completing the switch to renewable resources, since the changing climate requires fast action. But this is not the only reason for a switch to solar energy. Solar energy, geothermal energy, biomass, wind, hydro and tidal energy are the Earth's only real energy income. The exploitation of its material reserves of coal, oil and natural gas, which formed in

---

[1] See also: Schmidt-Bleek and Factor 10 Club — 1994 — several publications.

Power for the World by W. Palz
Copyright © 2011 by Pan Stanford Publishing Pte Ltd
www.panstanford.com
978-981-4303-37-8

the past over a very long period of time, is simply a huge-scale redistribution at the expense of future generations. Oil will be the first fossil energy carrier which will have reached its mid-depletion point.[2]

The problem is compounded by those people who do not yet even have a share of the energy so abundantly available to the rich part of the world. One third of the world population is not connected to an electricity grid, and this figure is rising. Given the decentralised nature of this supply problem and the depleting fossil energy reserves, it will take renewable energy technologies to supply these people with electricity and thereby with information, light, cooling and other basic services. For all of these reasons, renewable energies are far more than just alternative, more environmentally friendly energy technologies — they are the energy basis for a more sustainable civilisation model.

A sustainable energy supply will need to be based on three pillars: firstly, on renewable energies; secondly, on an efficient use of available resources; and thirdly, on a conscious decision about the limits to consumption, i.e. on sufficiency. This "great transformation" requires paradigm changes, requires several steps, shifting thinking in people and societies.

1. Equal Treatment

1986: In Wackersdorf thousands protested against nuclear energy but just anarchists — no, ordinary citizens from all walks of life as well. At the same time, the first German PV symposium with some 70 participants took place in Banz Abbey in Bad Staffelstein. A few weeks after that the nuclear power plant in Chernobyl was on fire. In Germany, children's playgrounds were closed and food taken off the shelves due to radioactive contamination. From that time on to this day, a huge part of the German population no longer view nuclear energy as an energy source for the future.

The weeks and months following these events saw much interest in alternatives, ideas, scenarios and visions for a future energy supply without nuclear energy and, in the long term, without fossil fuels. Energy policy moved to the centre of the debate and the motto "nuclear phase-out" could be heard throughout the country. Both the Green and Social-Democratic Parties took up this debate and called for the shut-down of nuclear power plants, which stood for a fifth of electricity production. In addition, Germany had to find a solution for "forest dieback", caused among other things by electricity production based on fossil fuels. The issue of climate change emerged increasingly in the debate.[3]

---

[2] See also: Energy Watch Group: "Crude Oil — The supply outlook" (http://www.energywatch-group.org) This study shows that "peak oil", i.e. the global maximum oil production, has already been in 2006. Other studies of the group show that global coal and uranium resources will also become scarce in this century.

[3] Shortly after Chernobyl, Hermann Scheer founded EUROSOLAR in Bonn, the "Aachener Solar Förder Verein" like many other local solar initiatives were founded.

Some scientists wanted to solve the problem with an efficiency initiative while there were others, including myself, who spoke of renewable energy sources as being one of the pillars of a future energy supply. Looking back, our calculations that by 2030 renewable energies could cover about 30% of electricity supply in Germany were probably quite bold. People still remembered the failure of "Growian", a 3 MW wind turbine (dismantled 1987). Photovoltaic systems had not yet moved on from the "tinkering" stage.

At that time Germany's electricity supply was dominated by monopolies, some of which were partly state-owned. The energy legislation consisted of laws which dated back as far as the time before World War II, and focussed on supply security. These monopolies had to have their prices approved, the condition being that supply security in the regions and a cost-effective infrastructure had to be ensured. Environmental costs could not be included in the prices. In 1989, the legislation had to be changed, however, to allow the cost of flue gas treatment to be charged. The amended legislation also allowed energy suppliers to purchase electricity from private producers at a higher price — on a voluntary basis.

But by how much could prices increase and what was an adequate price to be paid to private producers of electricity from renewable energy sources? And how to find a strategy to reduce the costs for PV, wind and other renewable technologies? Different ideas where discussed at that time, from "fair cost compensation" of the costs of installing a renewable energy system, to "fair cost compensation" through payment for produced energy, to tax incentives and last but not least research.

The governing Christian Democratic Party was criticized for not taking the renewable energies serious. They responded with a "1000 PV Roofs" and a similar wind energy program — a first step of market penetration. Due to the failure of Growian, where the developers shut down the windmill after a few days, already having swiped the subsidies, German policy was not willing to pay for installations alone. Grid connection and paying for electricity was the solution. In 1990, the German parliament unanimously passed the "Act on the Sale of Electricity to the Grid" (Stromeinspeisegesetz — StrEG/Electricity Feed-in Law) with support from all parties. It forced utilities to allow grid connection and guaranteed producers of renewable energies up to 90% of the retail price for electricity. In that time the German utilities didn't take serious the renewable energies — so they made no objections against this first step into a feed in tariff.

This allowed what is known as the "Aachener Model". STAWAG, the local utility in Aachen, opposed the installation of renewable energy systems. It needed several years until all legal actions and all initiatives against this model where defeated. From today's perspective — there arguments slowed down the process, but helped to identify the weak points of the concept.

In 1991, in Burgdorf (Switzerland), the director of the local utility proposed that every PV System installer will get 1 Swiss franc for each kilowatt-hour supplied to the grid. This covered part of the costs and induced an installation of 30 solar system per year. In Germany, we discussed that the real price of PV and wind energy, had to be calculated for the private producer equal to the calculation of prices in utilities ("equal treatment"). The price utilities could charge for electricity was

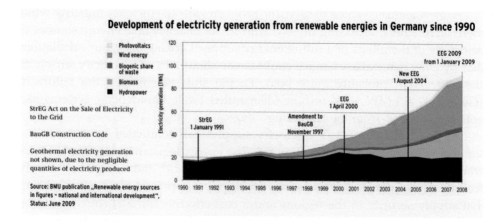

Figure 1. Development of electricity generation from renewable energies in Germany since 1990. (*Source*: Federal Ministry of the Environment, Germany, BMU, 2009.)

controlled at that time by the state governments. So it was logically that after all court ruling the state government of North Rhine-Westphalia had to set up a commission, a roundtable to generate a calculation method for electricity from renewable energy sources. This commission was comprised of representatives of the consumers, renewable energies (Eurosolar), utilities and the state government.

We worked some months before we published a methodology for the calculation of prices of kilowatt-hours produced by renewable energy suppliers. It was different for wind and PV, it included system cost, maintenance, insurance, other costs and what was most important the amount of interest to be paid on the investment. With the last part it was clear that the private producer was allowed to create profit from his investment into renewables. Equal Treatment! This calculation method was adopted by other utilities and formed the base for the later calculation of the feed in tariff.

In 1990, the umbrella association of German electricity utilities paid an advertisement of the size of a full page in mayor German newspapers with the following message: "We congratulate Denmark for their share of 1.7% electricity from wind energy! In Germany due to physical barriers it is not possible to produce more than 1% from wind and PV!" What a mistake.

2. A Further Step — Coming out of the Niche

The Enquête Commission[4] with the somewhat bulky name "Sustainable energy supply in the context of globalisation and liberalisation" was convened by the

---

[4] The Enquête Commission consists of members of the German Parliament and an equal number of representatives of academia and society. Its mandate is to prepare a report on a complex issue, if possible by consensus. Extensive minority rights in the work of the Commission ensure broad discussion. Wolfgang Palz and I have been members of this Enquête Commission.

German Bundestag in March 2000 to explore development paths towards a sustainable energy supply up to 2050 and derive recommendations for policy action, with a view to the World Summit in Johannesburg in August 2002.

This first gave rise to the question: What is a sustainable energy supply? The result of the first part of the work was a system of economic, ecological and social indicators which can be used to "measure" the sustainability of the state and development of an energy system. The necessary reduction of greenhouse gas emissions (with CO_2 as key indicator) by 80% by 2050 for Germany was at the heart of this system. Already at this stage it became clear that despite a lot of common ground between the parties one core conflict could not be eliminated: Should or should not nuclear energy be regarded as sustainable? Many (particularly Greens and Social Democrats) believed that it is unsustainable, for example because of the waste problem which will be a legacy for many generations to come and because of the risk of damage which (while small) would be of a huge scale. Representatives of the Free Democratic Party and parts of the Conservatives did not share this opinion.

As background information it is important to know that this enquête commission was established at the suggestion of the conservatives and the liberals. The strategic goal was to demonstrate to the Social Democrats-led Federal Government (the Coalition) that a nuclear phase-out is inappropriate from the viewpoint of climate protection. Therefore, there was still extensive agreement about the climate problem, and the Third Assessment Report of the Intergovernmental Panel on Climate Change (IPCC) gave cause for a serious precautionary policy.

Other key topics were the availability of fossil energy sources[5] and the associated potential for geopolitical conflicts, the impact of globalisation and liberalisation on energy markets and the role of (state) institutions at national and international level in orientating the energy system towards sustainable development objectives. There was some common ground in respect of all of these issues. The coalition saw the problematic aspects as demanding political action whilst the Opposition wanted to leave a lot solely to "the markets". All agreed, however, that the present energy system is unsustainable and that the internalisation of external costs is a key element of a more sustainable system.

Especially the "scenarios" were the subject of heated debates, for they were the centrepiece of the work of the study commission. The commissioning of two institutes, which worked with two different, competing, (computer) models, the establishment of three "scenario philosophies" (transformation efficiency, REG/REN offensive and fossil-nuclear mix), the failure to reach agreement on key dataset components, all this shows how prone to conflict the work was. In addition to a number of demographic and economic parameters, the requirement for all scenarios was that they had to achieve an 80% reduction in CO_2 emissions by 2050.

---

[5] In this commission I raised the problem of the Mid-depletion point of oil — but only a minority of commission members followed the analysis.

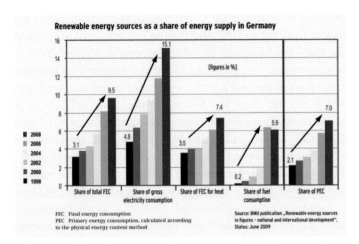

Figure 2. Renewable energy sources as a share of energy supply in Germany. (*Source*: Federal Ministry of the Environment, Germany, BMU, 2009.)

The "transformation efficiency" scenario was to do so through a massive increase in efficiency in the use of fossil energy sources and with the option of CO_2 capture and storage in geological formations.

The "REG/REN offensive" scenario was to achieve the objective through a massive efficiency increase on both the supply and demand side and by expanding renewable energies significantly at the same time so that they would account for 50% of primary energy consumption by 2050.

The "fossil-nuclear energy mix" scenario allowed new construction of nuclear power plants to resume after 2010 and exploited efficiency potential and renewable energies according to strictly economic criteria.

One investigation initiated into full solar-based supply by 2050, and the modelling of a fast nuclear phase-out brought numerous additional scenario variants.

The Free Democrats and the Conservatives considered the continued use of nuclear energy to be essential for cost reasons and justified this with a single analysis which, based on certain assumptions, arrived at lower costs compared with the other scenarios (although the scenario was the most costly when including "external costs"). They did not, however, embrace this fossil/nuclear scenario, since it foresaw new construction of up to 100 nuclear power plants in Germany, it therefore seemed unimaginable for it to gain social and political acceptance, and was unfavourable in times of election campaigns.

Despite or maybe even because of these differences, the final report is very interesting to read for all those who are interested in energy issues. Especially the results are of interest, despite the complexity of the matter:

- A sustainable energy supply based on energy saving, efficiency and renewable energies can be realised in Germany by 2050,

- The economic costs of the various development paths differ only marginally,
- In the view of the majority of the enquête commission, only the "REG/REN offensive" scenario (with elements from scenario 1) can be regarded as sustainable.

This was the first time that it was stated in a report to a government, to a parliament, that renewable energies are not a niche technology but capable of guaranteeing full supply (100%). At the time this report was published the share of renewable electricity in Germany was 6%.[6]

Germany takes a lead in climate protection and supports a 30% reduction of Greenhouse Gases by the EU with its own commitment of a 40% reduction by 2020 compared to 1990 levels. In the logic of these scenarios several others followed. One important was a scenario, developed by the German Federal Environment Agency (UBA), showing the feasibility of achieving this target.[7]

Energy-related CO_2 emissions account for over 80% of German GHG emissions. The UBA assumes that the 40% target is achievable if this largest share of total emissions is reduced accordingly. For energy-related CO_2 emissions, the 40% target translates into an annual CO_2 emission of not more than 571 million t in 2020. This corresponds to a CO_2 emission reduction of 224 million t CO_2 compared to 2005.

Technical development has led to strong cost reductions in the area of renewable energies, and this will continue in the future. Thus, these technologies can be

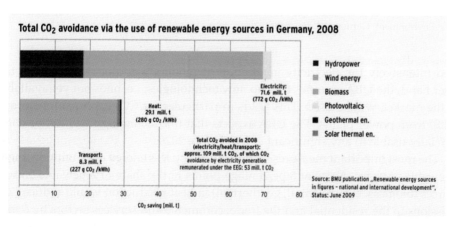

Figure 3. Total CO_2 avoidance via the use of renewable energy sources in Germany. (*Source:* Federal Ministry of the Environment, Germany, BMU, 2009.)

[6] See German Parliament's Enquête Commission. 2002. *Sustainable Energy Supply Against the Background of Globalisation and Liberalisation*. Deutscher Bundestag. Referat Öffentlichkeitsarbeit. Available in book: ISBN 3-930341-62-x or via internet: www.bundestag.de/gremien/ener/index.html.

[7] "A Climate Protection Strategy for Germany — 40% reduction of CO2 Emissions by 2020"; UBA Dessau — Erdmenger *et al.*; 20th World Energy Congress "The Energy Future in an Interdependent World", World Energy Council, 2007.

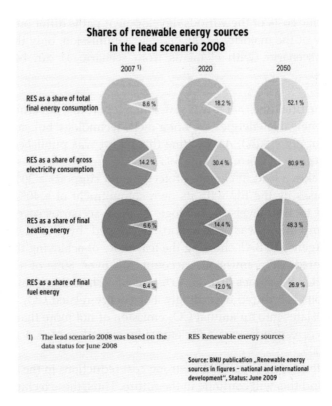

Figure 4. Shares of renewable energy sources in the lead scenario of 2008. (*Source*: Federal Ministry of the Environment, Germany, BMU, 2009.)

used intensively for electricity generation at moderate additional costs. On the other hand, the UBA assumes that no new technologies, i.e. ones not yet available on the market, will be used. This holds in particular for CO_2 capture and storage (CCS) from power plants. The UBA expects that this technology will not be commercially usable to any significant extent before 2020.[8]

The most important measures: higher efficiency, less energy consumption, more renewable energies. A rise of 6 percentage points in the share of renewable energy sources (biomass, solar thermal, geothermal) in heat production would reduce CO_2 emissions in the residential and the trade, commerce and services sectors by 6 million tonnes and those in industry by almost 4 million tonnes. Support via new legislation, analogue to the Renewable Energy Sources Act (EEG), is necessary.

The increase from 70 to 140 TWh/year proposed in this scenario, which is aligned with several studies, is borne primarily by Multi-Mega-Watt wind turbines, both onshore and offshore, and electricity generation from biomass.

---

[8] See: UBA 2006, Technische Abscheidung und Speicherung von CO_2 — nur eine Übergangslösung, Dessau August 2006.

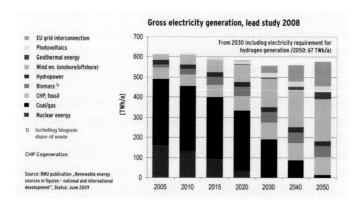

Figure 5. Gross electricity generation till 2050 in the lead scenario of 2008. (*Source*: Federal Ministry of the Environment, Germany, BMU, 2009.)

This will generate emission reductions in electricity generation of 31 million tonnes/year. The Renewable Energy Sources Act (EEG) is the single, most successful instrument for promoting renewable energies in Germany. Its basic principle — guaranteed compensation (fixed price) for electricity fed to the grid — has proven to be cheaper, more efficient and more effective than other instruments used in this field, not only in Germany but also in all of Europe. 40% is one step — the next goal is 100% Solar System.

3. Full Solar Supply or the "Great Transformation"

Today, at the advent of the solar industrial age, tried and tested basic technologies[9] as well as a number of experimental technologies are already available. These technologies can make energy from renewable resources available on a scale many times that of current global energy consumption, and can continue to do so for millions of years. The only limit will be the availability of land area, water and resources (e.g. scarce metals). To harvest the necessary energy from renewable energy sources for a full solar supply a great transformation of our local, national and global energy system is needed.

Any solar energy supply system that provides secure and year-round supply consistently utilises locally available resources. It does so with plants (large or small, networked or stand-alone) that are adapted to local potential and circumstances. Exchange of energy within a region or between regions, and storage of "surpluses" are characteristic features of a solar energy supply system.

---

[9] Biomass, hydropower and pump storage, wind energy, solar thermal collectors, photovoltaic systems, centralised solar thermal power plants, tidal power plants, solar architecture and geothermal energy.

In such a system, the various technologies for the utilisation of renewable energies, and the potentials of the various regions complement one another with their individual strengths and weaknesses to make energy provision function all year round. An approach of this kind makes it possible, in particular, to offset temporary fluctuations in energy provision as experienced with some renewable energy technologies. If there is no wind in a region, one would first turn to other locally available sources. If these are insufficient, energy would be supplied by plants from other regions.

This system needs to be controlled much more "intelligently" than today, starting with energy production planning using weather forecast models all the way to the consumers, who need to adapt their energy consumption to the supply. Foresighted management guarantees stable energy supplies for consumer through local compensation, interregional exchange, and storage of energy generated both with technologies whose energy production is weather- and season-dependent and ones whose energy is available all the time. This is possible with today's communication technologies. These also facilitate the networking of very small, decentralised and scattered production units to form a "virtual power plant".

The solar industrial age will be characterised by global availability of renewable energy sources and by technology which can be used both on a small, decentralised and a large, centralised scale. To limit the costs, these technologies will be mass produced on an industrial scale.

4. Scenarios — A Look into the Present and the Future

Scenarios are today's "crystal ball." Even scientists use them to look very far into the future. Full solar supply was already discussed in the 70s — e.g. "Le Groupe de Bellevue" in Paris (1978).[10]

One study investigating the possibility of such an energy system is the "Long-Term Integration of Renewable Energies into the European Energy System" (LTI).[11] The LTI project worked on "extreme" scenarios with very different but ambitious economic, social and ecological goals over the next decades.

Based on two simplified archetypes of behaviour — exhibited by those who are motivated to protect the environment ("sustainable" scenario) and those who are interested in consuming (fair market scenario) — two scenarios were developed which result in an 80% reduction of CO_2 by 2050. Because of the extremely varying assumptions, the examined scenarios represent two extremes of a

---

[10] More information at can be found on: www.solarmissionpossible.info.

[11] I had the chance to be part of this research group financed by the European Commission "APAS" programme investigating Europe's possible energy futures. Source: LTI Research Team. 1998. Long-Term Integration of Renewable Energy Sources into the European Energy System. ISBN 3-7908-1104-1, Heidelberg, Germany.

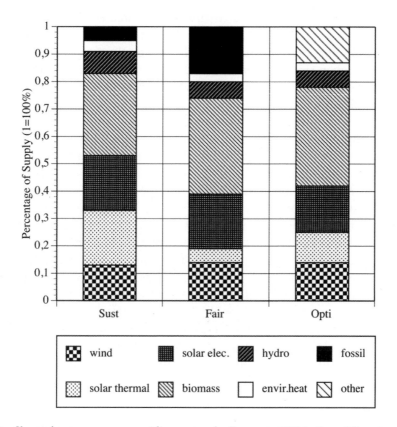

Figure 6. Share of energy sources providing energy for Europe in 2050 in three different scenarios. Energy efficiency and savings have lowered demand by 38 (fair) to 62 (sust) percent. Geothermal energy, imported solar-generated hydrogen and others are not included in the sustainable and fair market scenarios and are together in the "opti" scenario under "other", the opti scenario is a full solar supply scenario. (*Source*: H. Lehmann *et al.* and LTI Research Team 1996/1998.)

possible development and are not meant as prognosis for the future. Rather they were designed to learn as much as possible about supplying solar energy to Europe.

Reality will be a mixture of different trends and incorporate aspects of both scenarios. A third optimised version of the supply system is shown in the "opti" scenario — a 100% renewable energy scenario.[12] This last scenario was much better in terms of availability of electricity over the year.

The LTI project showed that the European energy system can be changed until 2050 to use energy in a sustainable way. There are no fundamental technical or financial hurdles that inhibit an exclusively solar/renewable energy supply system for Europe.

---

[12] Source: H. Lehmann, Wuppertal Institute, 1998.

Since supply security in the electricity sector is unalterable — production and consumption must match at all times — some of the energy systems were reproduced in computer-based simulations. This showed that a system based fully on solar energy, when suitably designed and controlled, functions reliably all year round. As an example of such studies, we will analyse "Solar Catalunia" a little bit more in detail.[13]

The objective of initiating the Catalonia study[14] was to show Catalonia's ability to supply its own needs for energy from renewable sources and, thus, to provide a fact-based vision of a future energy supply as an alternative to the present fossil/nuclear system. The study is focused on Catalonia's electric energy demand — and how it can be reduced — and the design of a reliable renewable electricity supply system.

The scenarios highlight a development towards halving electricity intensity in the three most important sectors of electricity consumption until 2050, which — of course — is a great challenge, but feasible from a technological point of view.

The future development of generating capacities for all renewable technologies was calculated using so called "logistic growth functions", showing the typical s-curved shape for growth with saturation effects in the later stage of development. This reflects the underlying assumption that growth cannot be unlimited if any of the resources growth depends on is limited. This approach also required incorporating assumptions regarding the future development of technology specific investment costs. While wind energy, geothermal power plants, biomass

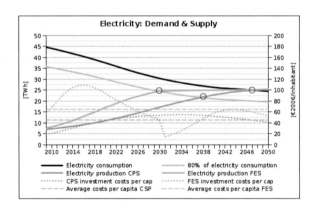

Figure 7. Figure: Development of electricity demand and supply (Scenario "Fast Exit" FES and "Climate Protection" CPS) until 2050. Cost of both scenarios. (*Source*: S. Peter *i.*, "Solar Catalonia — A Pathway to a 100% Renewable Energy System"; ISuSI 2007 and 2009, www.isusi.de.)

---

[13] Another study is also the "Energy Rich Japan" Study — which has also an animation of the dynamics of the supply and demand of electricity.
[14] S. Peter *et al.*, "Solar Catalonia — A Pathway to a 100% Renewable Energy System for Catalonia"; 2007 and revised version 2009.

plants and solar thermal power plants were expected to show half of today's specific investment costs by 2050, the specific costs of photovoltaic were expected to fall to one third of today's costs; hydropower was assumed to remain on current cost levels.

The "Fast Exit Scenario" shows an increase of renewable generating capacities from less than 200 MW in 2006 to about 6.400 MW in 2030 and further to almost 12 000 MW in 2050. Wind energy contributes most to the total renewable capacity — almost three fourths (2030) respectively two thirds (2050) of the total renewable capacity consists of wind energy. Photovoltaic shows a dynamic extension too, resulting in an 11% share of total renewable capacity in 2030 and 23% in 2050. The shares of biomass, geothermal, solar thermal plants and additional hydropower are substantially lower if compared to wind energy or photovoltaic.

Validation of supply security was based on dynamic computer simulation of electricity supply for all the four seasons of the year. The simulation showed no indication for undersupply at any time.

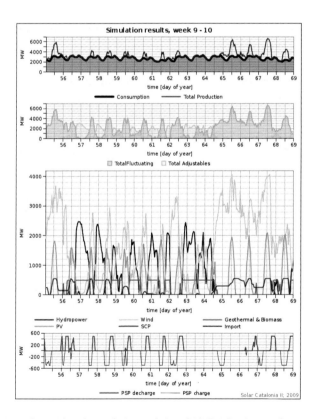

Figure 8. Electricity demand and supply in week 9 and 10; Production peaks occur around midday, driven by good solar radiation and wind conditions. At no time demand exceeds production (SCP Solar Central, PSP Pumped Hydro Storage). (*Source*: S. Peter i., "Solar Catalonia — A Pathway to a 100% Renewable Energy System II"; ISuSI 2009, www.isusi.de.)

The two scenarios of the Catalonia study show the feasibility to achieve a fully renewable electricity supply, one until 2035 (Fast Exit Scenario), the other until 2045 (Climate Protection Scenario). The realisation of these goals is not a matter of potentials, but it is a matter of setting and pursuing ambitious targets, encouraging policy and people and — of course — the financial investments Catalonia and it's people are willing to take. The scenarios show that the financial aspect is not that big obstacle that one might expect. With an annual investment into renewable capacities peaking at 104€$_{2006}$ per inhabitant in the "Fast Exit Scenario" (2050) and 85€$_{2006}$/cap in the "Climate Protection Scenario".

Compared to the Catalonian Gross Domestic Product (181,029 million € in 2005) the annual costs of the scenarios are 0.2% of the GDP for the "Climate Protection Scenario" and 0.3% for the "Fast Exit Scenario" on average.

Such scenarios have been published for different regions/countries. All show that a 100% renewable energy system is possible.

5. To Sum up I can Say: 100% Solar System is Possible!

The analysis of scenarios that consistently go as far as a 100% supply from renewable energy shows the following:

(1) A reduction of greenhouse gas emissions by at least 50% worldwide and by 80%–100% in industrialised countries by 2050 (compared to 1990 levels) is technically and economically feasible from today's perspective. The path towards a full renewables-based and efficient energy system, is a realistic future option and no dead-end. There is no need for nuclear power, neither fission nor fusion.

(2) The supply and demand systems described in the scenarios offer possible target ranges for restructuring the energy sector. Any restructuring towards renewable energies will not need to be limited to the ideas described in those reports. Other systems, other combinations of technologies are possible as well.

(3) Since supply security in the electricity sector is crucial — production and consumption must match at all times — some of the energy systems were reproduced in computer-based simulations. The simulations showed that a system totally based on solar energy, with suitable design and control, works reliably all-the-year. This is also shown by experiment.[15]

(4) Although investments will initially be necessary to stimulate development towards a sustainable energy system, the restructured system will not result in higher costs than the present system in the medium and long term. All this scenarios are economically feasible.

---

[15] A number of firms and associations have demonstrated this with their "regenerative combined power plant", which connects and controls 36 wind, solar and hydropower installations across Germany. Source: www.kombikraftwerk.de

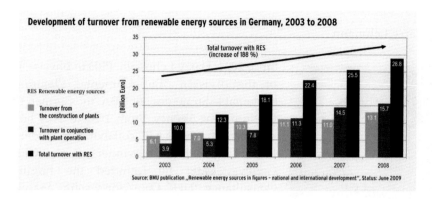

Figure 9. Development of turnover from renewable energy sources in Germany. (*Source*: Federal Ministry of the Environment, Germany, BMU, 2009.)

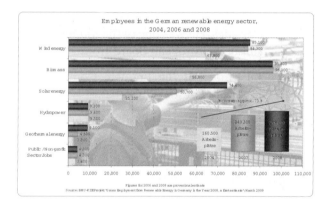

Figure 10. Employees in the German renewable energy sector. (*Source*: Federal Ministry of the Environment, Germany, BMU, 2009.)

Analysing older scenarios, it is obvious that market introduction and expansion of renewables have beaten even the expectations of optimistic scientists in such regions that implemented suitable support mechanisms. "First-mover" countries (such as Germany) are in a highly favourable win-win situation, due to the creation of jobs and export opportunities. This should make us confident that these scenarios can in fact be implemented.

Today's society must take action to implement a renewable strategy. The most important step is to start right now, since every day that passes by without enforcing a renewable energy strategy only increases and complicates the problem — because energy consumption is increasing, money is still being invested in fossil/nuclear systems and finding ways to solve the problem of climate change merely gets postponed.

Dr. Harry Lehmann

Since 2004, Harry Lehmann, PhD in physics, is head of the division "Environmental Planning and Sustainable Strategies" of UBA, the German Federal Environment Agency. From 1980 to 1984 he was a member of the "UA1" collaboration under Carlo Rubbia (Nobel Prize winner) at the CERN in Geneva. Later he founded the Engineering Consultancy "UHL Data" for systems analysis and simulation in the field of energy and environment. He was also head of the Systems Analysis Group at the 'Wuppertal Institute for Climate, Environment and Energy' after it was newly created in 1991 under the direction of Ernst Ulrich von Weizsäcker. Since 1985, Harry Lehmann has been teacher at the University of Luneburg and other universities. He was one of the founders of Eurosolar in 1988 and was the Vice President of its European Board from 2000 to 2006.

Harry Lehmann was a member of the Enquête Commission on Sustainable Energy Supply of the German Bundestag (2000/02).

He also was a member of the Commission "With new energy technologies in the next century" of the Parliament of Bavaria.

He was a Director of Greenpeace International and is still a member of the World Renewable Energy Council, the Energy Watch Group as well as a member of the "Factor X Club" for resource productivity and sustainable use of natural resources.

Chapter 25

France Did Not Want to Look for the Sun...

Alain Liébard and Yves-Bruno Civel

*Observatoire des Energies Renouvelables, Fondation Energies pour le Monde,
146, rue de l'Université, F-75007 Paris, France
observ.er@energies-renouvelables.org, www.energies-renouvelables.org*

Back in 1980, some French pioneers, among them Robert Lion, Serge Antoine, and
Joël de Rosnay founded the "Comité d'Action pour le Solaire". Their first task was
to draft and publish the "Manifesto for a Solar France". That visionary text anticipated that by the end of the 20th century, 20 to 25% of all energy needs could be
met with the solar, heat, light, wind, water and bio-energies. The backlash of the
oil price shock and the choice of nuclear electricity have rapidly buried the ambitions of this text. France refused to look for the Sun and its eclipse was going to
last for quite a long time. Then in 1984, the two undersigned of the present text
take over the job at the 'Comité d'Action pour le Solaire' and keep fighting to keep
alive the fantastic idea of the renewable energies. The first is an architect and
became the president. The second is a journalist and became the director general
of the association and editor in chief of the journal "Action Solaire". Since those
days, we acted like a durable twosome and carried out, in a quasi general indif-
ference, the research of proves of the pertinence and the effectiveness of the
"renewables" and the big challenge of information.

First of all, we are convinced that the best ally of the renewable energies is
Europe and that the policy of the European Union is more than just the addition
of the EU Member States' policies. For numerous European workshops and
symposia we drafted the "daily newsletters", in particular for some of the EU PV
Conferences and the EU Solar Architecture Conferences. We drafted charts,
chaired panels and discussions, and contributed our ideas. We were also the
executive arm of the "Club de Paris for RE" in the 1990s. We have edited and

published for the EU Commission "Solar Europe" and the "Renewable Energy Journal". We have created EurObserv'ER, the European barometer of the renewable energies. Our strategy: to demonstrate that the renewable energy challenge, the way it was addressed in Europe, the United States, and in Japan, could also become a chance for France.

Outside the strict "EU frame" we have developed numerous activities in France. Aiming at the defence of the overall values, we have implemented a lobbying that was more cultural than economic; a highlight was and still is the Journal "Systèmes Solaire": it has imposed itself as a general reference in France. In 2010 it can celebrate in Paris the publication of its nr 200 issue. In parallel, we have set up the foundation "Energies pour le Monde". By now it has achieved the electrification of up to 10 000 villages in the developing countries with the renewable energies and in particular photovoltaics.

Implementing in practice what others have developed as a remarkable concept in "Power for the World", we are addressing by now new large-scale programmes to provide the access to energy to millions of people. Via the competition "Habitat Solaire, Habitat d'aujourd'hui" (Solar Living, Today's Living), we have promoted solar architecture and "low energy" buildings. More recently we have initiated a European system of accounting and certification of renewable energy production and contribution to ensure honest and sincere transactions and exchanges.

At a time when in France the subject of the renewable energies was virtually taboo, we had to be patient and inventive. We had to convince and seduce. Our prime objective was to bring the renewable energies into the mainstream of France's society and economy; we understood early on that we could not achieve it "against" the society but "with" it. The path was long and full of obstacles.

Twenty-eight years after the publication of "Manifeste pour la France Solaire", one of us, Alain, has been appointed by the government, president of the operational committee for the renewable energies, a co-operative structure of brainstorming as part of the French President's initiative "Grenelle del' Environnement". As an experienced manager of an NGO, Alain has chaired that endeavour. It was very original and very "French", consisting of addressing all French stakeholders of the renewable energies to establish the ways and means of achieving the objective of 23% RE in France's energy mix by 2020, an objective defined by the EU Directive in 2008. That fantastic democratic exercise did produce a remarkable report, the "Development Plan of the Renewable Energies of High Environmental Value". Its conclusions have by now become French law.

One of our greatest victories and satisfactions consisted doubtlessly to get the French Government to recognise the immense promises of the solar photovoltaic electricity, that this "poor" of the RE family, totally despised by the "nomenklatura" had one day found its right place among the big players in the energy field. The vision of photovoltaics for the development of countries of the "Solar Belt" is ours since 1972. The other one, making PV a "must" on all façades and building roofs and leading eventually to the "building of positive energy", is ours since 1980. Grid-connected solar electricity and stand-alone solar electricity are the two faces of the same coin: we are firmly convinced that the great development of the PV industry is going to demonstrate rapidly its positive effects on the social and economic development of the poorest countries. We witness with satisfaction that France is committing itself and starts realising that the "green growth" is not a growth "as usual"; it is a metamorphosis coming along with a technological and cultural revolution, aiming at a new balance. Not to say a "modus vivendi", the right way for living durably together!

The job is not yet finished. We got to keep our vigilance and spirit of "anticipation" allowing us to be always one or two ideas ahead. We are happy to have been part of that fight. We always thought it was worth for our planet and the little men who are living there!

Yves-Bruno Civel and Alain Liébard

Chapter 26

On the International Call for Photovoltaics of 2008

Daniel Lincot

CNRS, Institute of Research and Development of Photovoltaic Energy (IRDEP)
(Joint between CNRS-Chimie Paristech) and EDF, 6 Quai Watier,
78401 Chatou, France

There was no doubt that solar energy is immense and that it represents thousands time the whole energy consumption by humankind, that only one hour of solar energy would be able to cover one year of the world's energy demand, but in the years 2000, when the discussions about climate change and fossil fuel exhaustion started to become more and more important, it was still considered as a dream or as an option for the next century. In the beginning of 2008, after a G8 meeting about world energy issues, the solar energy option was not even mentioned in the press releases among the solutions, which could be considered to face this challenge, involving mainly the concept of "clean" coal, nuclear and large wind mill conversion for renewable energies. *Solar energy, and photovoltaic conversion especially, were simply ignored as a short-term strategic priority at the high international decision maker levels.* The fundamental reason was very simple, photovoltaic achievements and prospects, were not "visible" at this level of decision. What represents 5 GW installed, worldwide in 2007, in front of hundreds of GW installed just of electricity supply (about four times more for the whole energy supply including fossil fuels) when decisions have to be taken for the next 10 to 20 years? Almost nothing! Even impressive growth rates of 40% per year are not impressive at such a low level of market penetration. Also, looking to the roadmaps made by specialists at these times — solar energy endeavor onset at a significant level was only expected in 2030 — would have also discouraged more motivated decision makers to engage strong fighting for short term scenarios were photovoltaics would take a significant share: *if specialists themselves are not expecting significant*

Power for the World by W. Palz
Copyright © 2011 by Pan Stanford Publishing Pte Ltd
www.panstanford.com
978-981-4303-37-8

contributions why non specialists should be more credible! This was a strange situation, with on one side an extraordinary development of photovoltaics (PV) at the scientific, technological and industrial levels, but on the other side a poor consideration at the political level, due to a lack of visibility and ambition, and reinforced by a systematic opposition of many established and powerful energy conservative lobbies!

Throughout the 80s and 90s, the key word for photovoltaics was economic competitiveness and *growth only governed by the market law* like for *a standard consumer product*. In contrast to accompanying policies for the established energy sectors (coal, petrol, nuclear) where considerable public support (justified by the strategic importance of the energy sectors for society) have been engaged, renewable energies development had to rely only on market driven policies. As the sector was not competitive, the result was evident: the growth was kept slow! The same would have taken place for the other energy industries, especially nuclear. PV would have stayed at a low level much longer, if public support policies would not have been implemented in pioneering countries like Germany and Japan. Since 2000 feed-in tariffs have catalyzed the endeavor of a real PV industry, and revolutionized the photovoltaic paradigm, bringing the PV domain under the growth assisted initial conditions. *This was starting to make the "only market assisted" growth scenarios and roadmaps obsolete;* but in 2007–2008, old schemes were prevailing again, and the voices to support photovoltaics were very weak and even becoming weaker in the context of emergency created by the acceleration of climate change considerations. Back to coal (but "clean") and nuclear was in the dominant position. *The danger was present to face the same situation as about 40 years before, in the 70s-80s,* after the oil crisis, when solar energy, after being considered a credible alternative for energy supply, similar to nuclear, was suddenly dropped in many countries, especially in France. After the victory of the nuclear option, solar energy has been systematically denigrated and blocked. The winners wanted a complete victory over their competitors!

In 2007, I had the feeling that, despite the success of feed-in tariffs in Germany and the unexpected surprise of their introduction in France in 2006, followed by the remarkable process of "Grenelle de l'Environnement" initiated by the environmentalist Nicolas Hulot, the situation was again fragile and that the pressure for the efforts to be dropped again, *even in Germany*, was increasing. I had a clear feeling that a new match point was in preparation and that if solar energy development was not able to reach a sufficient critical mass rapidly, systematic attacks about higher costs, intermittency, dilution, "non-competitiveness" may again condemn for years and years its endeavor.

These were my worried thoughts when, in autumn 2007, came the immense surprise and honor to be chosen as general chairman of the 23rd European Photovoltaic Solar Energy Conference, which was held in September 2008 in Valencia (Spain)! This was 30 years after my decision, walking in the "Jardin du Luxembourg" in Paris on a sunny day, to engage myself in solar energy research. In 1979, I was presenting a communication at the 2nd EUPVSEC, in Berlin. I would never have imagined such situation to occur one day. France had become weaker and weaker along the years, and the only previous French general chairman was Ionel Solomon, 20 years ago, in Florence. Rapidly came the question what do I have to do in this new situation that would contribute best to the progress of photovoltaics? I analyzed from the impressive growth of the PV industry and the dynamism of the research sector, that the situation was excellent. But this was an internal view point from an actor of the PV community. The situation from the outside as discussed before was in fact not so excellent, and the visibility and image of photovoltaics at the political and decision makers levels was not good (even if the public opinion warmly supported solar). This brought me to the conclusion that my specific action and contribution should be to convey a strong message from the PV community to the rest of the society, that it is a credible short term alternative for renewable energy supply at the world level. The other point was to use this exceptional opportunity to work at the French level for improving the situation in the country.

In fact it was initially very difficult to get in touch with the high level French representatives: to my letters were not answerd. In parallel, I received an excellent support and help from French colleagues, especially Yves Bruno Civel, Alain Liebard and Jean Louis Bal, who are all pioneers and eminent actors in the PV sector. No real progress was made up to the beginning of 2008. Then I started to make a parallel between the Kyoto protocol, the IPCC declaration and the PV situation, and the idea came about launching an international call for the *acceleration* of the deployment of photovoltaics, supported by scientists. In the committee of preparation of the conference, in March in Brussels, I submitted this proposal to Antonio Luque and Wolfgang Palz, who were enthusiastic about it and shared the analysis. The writing of the initial text was the affair of a few days and exchanges with Antonio, which then made improvements following the suggestions of the colleagues. The idea was to recall the interest of photovoltaics, and the importance of joining scientific, industrial, educational and political efforts for its acceleration. Then mail after mail started the process of discussion with colleagues. It was not so easy, and for many of them it was not evident to take such an initiative, some of them were considering that the statements were too evident, for others to much emphasis was given to science with respect to industry, while for others it was the opposite with a too high industrial commitment. This was a proof that in fact the text was sufficiently balanced to promote *both research and industrial developments together*, which was the objective. It was sometimes discouraging, but the enthusiastic supports from initial signatories (A. Goetzberger, A. Shah, M. Konagai, I. Solomon, J. E. Bourée, among many others) was pushing to go ahead. Then the call started to spread towards 10, 20, 50, in April and May, many countries were represented with renowned researchers. In May, I organized a press conference in Paris about the situation of photovoltaics, in relation to the Valencia conference but also with the announcement of the PV Call. It was at the library of "Chimie-Paristech" Institute. The attendance was important, with about 20 journalists, with Wolfgang Palz, Jean Louis Bal, Yves Bruno Civel and Alain Liebard. I discovered at this occasion the importance of explaining what solar energy represents quantitatively, the equivalence between one year of solar irradiation and one to two barrels of petrol was then reported in most of the papers, and has probably a decisive impact on the public opinion! Presenting the international call was not easy at the beginning, and the skepticism was initially present within the audience, which suddenly turned into support when making the parallel with the Kyoto protocol, also based on scientist opinion first, but in that case to point out a solution instead of a danger. The following days the call was very positively reported in the French and related international media (AFP). This was a great success! In parallel, the situation with respect to the contact with high level French representatives was improving. France was at this time at the presidency of the European Union, and my objective was to invite a personality of the government to attend the Valencia conference to help increase its political impact. I will remember my first meeting, in April, attention with Helène Pelosse, who was deputy director of the cabinet of Jean Louis Borloo, Minister of Ecology and Environment. She received me with a great attention and our first discussion was about solar energy resources and photovoltaic conversion principles and technologies! I was expecting a more general discussion and, again, as for the press conference, rigorous information was the sesame for convincing. Photovoltaics and solar energy are not sufficiently known at high political levels! Helene Pelosse, who was in parallel participating in the difficult negotiations on the "Energy-Climate package" for the European Union became a strong supporter for the photovoltaic development, which may also had a decisive impact on the minister himself. Later in July, the Mediterranean Union project was discussed including the Solar Mediterranean plan which was placed as a priority. In Valencia, Helene Pelosse represented the European Union Presidency at the PV conference in Valencia. She became then head of the newly created International Agency for Renewable Energy (IRENA), an ideal place to support PV deployment!

September 2008: the opening session of the conference. A few days before, I was still wondering about the decision to present or not the call in my welcome message, it is so unusual. At the same time were the running of the US elections, and the battle was between MacCain "Drill Baby Drill" and "Yes You Can" Obama's, and the result was not written: I decided to present the call at the opening session! It came out in the international press along with the great success of the conference. On Friday the 5th of September, the French main newspaper "le Monde" was devoting in the middle of its front page a paper on photovoltaics, with a cartoon entitled "Solar Energy is setting itself as the energy of the future". This was like a revolution! At the same time, during the conference, the European Photovoltaic Industry Association (EPIA) launched its accelerated scenario for photovoltaics indicating a possibility to raise up to 12% of solar electricity in Europe by 2020. Yes we can!

January 2010: about one year has passed from the Valencia call. Many advances had taken place: Barack Obama has been elected and the USA are becoming a solar leader, PV industry is growing even more after the financial crisis. However, the sky is not blue, clouds are again accumulating. In France the campaign against "excessive" support to PV is increasing again, aggressive papers are published against photovoltaic "profiteers". This must motivate us to maintain our strong determination for making solar energy a reality, as soon as possible, keeping the sense of general interest and humanism which govern our action.

Daniel Lincot

Daniel Lincot started his research in the field of photovoltaics in 1978 with a PhD in the field of cadmium telluride solar cells at the Solid State Physics laboratory of CNRS. After his PhD, he was engaged in 1980 as permanent researcher at CNRS, in the laboratory of electrochemistry and Analytical Chemistry of Ecole Nationale Superieure de chimie de paris (Chimie-Paristech) in the field of semiconductor photoelectron chemistry. He discovered there the interest of chemical and electrochemical processes for the synthesis of thin films from solutions, and became strongly involved in the development of thin film solar cells based on copper indium diselenide (CIS), with results which contributed markedly to the advances in the field, at the highest international level. He became director of the Laboratory of Electrochemistry in 2001 while creating in parallel, as vice director, with EDF (Electricity of France) the institute of photovoltaics, IRDEP, in 2005. The first project was o develop an electrode position technology for industrial production of CIS solar cells. He became director of IRDEP in 2009. He received in 2004 the silver medal of CNRS. His activities are presently devoted to the acceleration of the development of photovoltaics in France and at the international level, both from research contributions and communication towards the public. He was general Chairman in 2008 of the 23rd European Photovoltaic Solar Energy Conference in Valencia (Spain), which was the world's largest conference in the photovoltaic area.

*International Scientists for
Photovoltaic Solar Energy*

International Call of Scientists for an Accelerated Worldwide Deployment of Photovoltaic Conversion of Solar Energy

Presented at the 23rd European PV Solar Energy Conference in Valencia 2008

Humanity is facing a dramatic challenge created by the progressive exhaustion of fossil fuels and the onset of climate change most probably due to CO_2 pollution. The utilization of renewable energies, as a major component of the global action, is becoming an absolute and urgent necessity.

Presentation of the Valencia call for Photovoltaics at the opening session of the 23rd European Photovoltaic Solar Energy Conference. From left to right : Hélène Pelosse (now Director general of the newly created International Renewable Energy Agency IRENA), M. Florez-Lanuza, A. Gonzales-Finat, E. Macias, W. Palz, M. T. Eckhart, H. J. Fell, D. Lincot (General Chairman). Photo : E. A. Gunther

Photovoltaics, which is the direct production of electricity from solar energy, is a key solution for universal sustainable energy production. It has unique characteristics:

- It is safe, clean, robust, efficient and highly scalable technology, making its implementation very easy all over the world. It can be used both for developed countries and for developing countries.
- It is already associated with a fast growing and dynamic industry with a growth rate of about 40% per year.
- Photovoltaics electricity costs are becoming more and more competitive. However stronger efforts for R&D in Photovoltaics will allow accelerating this evolution and saving precious time.
- Photovoltaics is the child of modern solid state/electronics science and technology. The prospects of large improvements of conversion efficiencies from scientific and technological progress are very high. R&D in Photovoltaics will benefit from novel developments in Physics, Chemistry, Materials Research and Biology.
- Photovoltaics has practically unlimited potential. It will benefit all populations and future generations.

Therefore, we, as scientists from various domains, consider that PV solar electricity has to be a key part of the response that is needed NOW to solve crucial energy, environmental and climate concerns. On the basis of the remarkable results of developing Photovoltaics by public action, via feed-in laws, we ask for the launch of a strong concerted action at the international level to accelerate worldwide photovoltaic solar energy production and its enrichment through large research, development, cooperation, and education programs.

*International Scientists for
Photovoltaic Solar Energy*

List of Supports

International Call of Scientists for an Accelerated Worldwide Deployment of Photovoltaic Conversion of Solar Energy

J. AbuShama (manager R&S Solopower, USA), D. Abou-Ras (Helmholtz Center Berlin for Materials and Energy, Berlin, Germany), **M. Acciarri** (Università degli Studi di Milano — Bicocca, Italy), **C. Algora** (Instituto de Energia Solar, Spain), **Zh. Alferov** (Nobel Laureate, Vice President of the Russian Academy of Sciences, Russia), **N. Allsop** (Helmholtz Zentrum Berlin for Materials and Energy, Berlin, Germany), **G. Almonacid** (University of Jaén. Spain), **V. M. Andreev** (Head of Photovoltaics Lab.,Ioffe Physico-Technical Institute, Russia), **J. Anta** (President of the PV Spanish Industry Association (ASIF), **P. Alpuim** (Universidade do Minho, Portugal), **H. Arakawa** (Tokyo University of Science, Japan), **T. Anderson** (University of Florida, USA), **K. Baert** (IMEC, Belgium), **C. Ballif** (Institute of Microtechnology IMT, Switzerland), **P. Banda** (Director General. Institute for Concentration Photovoltaics Systems — ISFOC, Spain), **K. Barnham** (Imperial College London, UK), **A.K. Barua** (General Chairman PVSEC18), **B. Basol** (CTO SoloPower, USA), **G. Bauer** (University of Oldenburg, Germany), **A.A. Bayod-Rújula** (Research for Energy Resources and Consumption, CIRCE, University of Zaragoza, Spain), **M. Belhamel** (Director, Renewable Energy Development Centre, Algeria), **V. Bermudez** (Institute of R&D for Photovoltaic Energy IRDEP, France), **J.C. Bernede** (University of Nantes, Frances), **J. Bertomeu** (Universitat de Barcelona, Spain), **R. Bhattacharya** (National Renewable Energy Laboratory NREL, USA), **B. Bibek** (Director Solar Energy Centre New Delhi, Vice Chair of 18th PV Conference-2009), **A. Blakers** (Australian National University, Australia), **A. Bouazzi** (University of Tunis El-Manar, Tunisia), **J.E. Bourée** (Ecole Polytechnique, France), **C. Brabec** (Konarka, USA), **H. Branz** (NREL, USA), **G. Bremond** (Institut des Nanotechnologies de Lyon, France), **F. Briones** (Institute of Microelectronics of Madrid, CNM-CSIC, Spain), **L. Brohan** (Institut des Matériaux Jean Rouxel, Nantes, France), **B. Richards** (Deputy Director Joint Research Institute in Energy, Managing Director of Progress in Photovoltaics, Edinburg, UK), **M. Burgelman** (Gent University Belgium), **J. Byrne**

(Intergovernmental Panel on Climate Change, Nobel Prize Laureate 2007, Director of Center for Energy and Environmental Policy, University of Delaware, USA), **D. Cahen** (Weizmann Institute of Sciences, Israel), **M. Camenzind** (Air Liquide, USA), **J. Carabe** (Acciona Solar, Spain), **M. Cassir** (Head of Fuel Cell Group, Ecole Nationale Supérieure de Chimie de Paris, France), **K. Shane-Ching** (Laboratory of Coordination Chemistry, Toulouse, France), **E. Chassaing** (IRDEP, France), **Frederic Chandezon** (CEA Grenoble, France), **P. Chatterjee** (Head Energy Research Unit, Indian Association for the Cultivation of Science, Kolkata, India), **N. Chaudron** (AGFPE, France), **P. Chauduri** (Head Energy Research Unit, Indian Association for the Cultivation of Science, Kolkata, India), **N. Cherradi** (Yingli Green Energy Holding co., Ltd, China), **Xiaohong Chen** (SAFC, USA), **N. Ciro** (SUMCO, Italy), **PK Chiang** (Vice President Arima PV, taiwan), **Y.B. Civel** (Direteur Observatoire des energies renouvelables, France), **P. Clancy** (Air Liquide USA), **J. Cochard** (NTNU, Trondheim, Norway), **D. Cohen** (Director, Oregon Sunrise, USA), **Al. Compaan** (University of Toledo, USA), **M. Contreras** (NREL, USA), **A. Cravino** (University of Angers, France), **A. B. Cristóbal López** (Project manager, Instituto de Energia Solar, Spain), **A. Cuevas** (Australian National University, Australia), **M. Dachraoui** (Tunis University, Tunisia), **M. R. de Aguiar** (University of Campinas, Brazil), **A. De Lillo** (Resp. Fonti Rinnovabili, ENEA, Italy), **E. Dalchiele** (Institute of Physics, Montevideo, Uruguay), **A. Darga** (Ecole Supérieure d'Electricité, France), **J. C. Denis** (Ecole Supérieure d'Electricité), **P. Destruel** (Institut Laplace, Toulouse), **B. Dimmler** (Managing Director Würth Solar, Germany), **N. Dhere** (Program Director, Florida Solar Energy Center, USA), **Z. Djebbour** (Ecole Supérieure d'Electricité, France), **C. Dragon-Lartigau** (University of Pau, France), **S. Dumoulin** (SINTEF, Norway), **J. Dupin** (CERN, Switzerland), **Y. Dobriansky** (University of Warmia and Mazury, Poland), **K. Durose** (Director, Durham Centre for Renewable Energy, UK), **M. Eckhart** (President of the American Council on Renewable Energy, ACORE, Washington DC, USA), **A. Ennaoui** (Helmholtz-Zentrum Berlin für Materialien und Energie GmbH, Germany), **L. Escoubas** (IM2NP — Paul Cézanne University, France), **G. Esposito** (University of Virginia, USA), **A. Etcheberry** (University of Versailles, France), **A. Fahrenbruch** (University of Stanford, USA), **D. Faiman** (Director, Ben-Gurion University's National Solar Energy Center), **C. Ferekides** (University of South Florida), **G. Férey** (French Academy of Sciences, Vice President of the Chemical Society France), **F. Finger** (Forschungszentrum Jülich, Germany), **A. Freundlich** (University of Houston, USA), **V. Fthenakis** (BNL and Columbia University, USA, co-initiator of the Grand Solar Plan), **D. Fuertes Marron** (Instituto de Energía Solar-ETSIT, Spain), **K. Fukui** (Manager PV development department, Kyocera, Japan), **T. Fuyuki** (Nara Institute of Science and Technology, Japan), **R. Gamboa** (Instituto Politécnico de Leiria, Portugal), **J.M. Girard**

(Air Liquide, France), **T. Glenn** (NREL, USA), **G. Goaer** (Photowatt, France), **A. Goetzberger** (Fraunhofer Institute for Solar Energy, President of the Becquerel PV Committee, Germany)X.G. Gong, Professor of Physics, Dean of Research and Director of center for computational sciences and engineering , Fudan University, China), **R. G. Gordon** (Harvard University, USA) **M. Green** (Alternative Nobel Prize, Director of the Centre for Third Generation Photovoltaics, New South Wales University, Australia), **J.F. Guillemoles** (Head of New concepts project, IRDEP, France), **E. A. Gunther** (PV Bloger , USA), **T. Gutiérrez (**Head of Photovoltaic Polycrystalline Materials, Energy Department, CIEMAT, Spain)**, A. Hamidat (**Head of Photovoltaic applications, Centre for the Development of Renewable Energy, Algeria), **S. Hayase** (Kyushu Institute of Technology , Japan), **P. Helm** (WIP Renewable Enegies, Germany), **C. Henry** (Columbia University, USA), **F. Hergert** (Johanna Solar Technology GmbH, Brandeburg, Germany), **K. Herz** (ZSW, Germany), **C. Heske** (University of Nevada Las Vegas, USA), **J. Herrero** (CIEMAT, Spain), **L. Hirsch** (University of Bordeaux, France), **W. Hoffmann** (Vice president of European Photovoltaic Industry Association (EPIA), CTO Energy and Environmental Solutions at Applied Materials, Germany), **J. Hüpkes** (Research Centre Jülich, Germany), **J.C. Hummelen** (University of Groningen, The Netherlands), **H. L. Hwang** (Taiwan), **Y. Hyashi** (University of Tsukuba), **A. Ihlal** (Laboratory of materials and Renewable Energy, Agadir, Marocco), **H. Iida** (Namics, Japan), **O. Inganäs** (Linköping University, Sweden), **A. Jaeger-Waldau** (European Commission Joint Research Center, Italy), **R. Janssen** (Technical University Eindhoven, Netherlands), **J. Jimeno** (Director of the Technological Institute of Microelectronics, UPV/EHU, Bilbao, Spain), **M. Kane** (Director of the Laboratory of Semiconductors and Solar Energy, University Cheikh Anta DIOP , Senegal), **V.K Kapur** (President CEO ISET, USA), **E. Katz** (Dept. of Solar Energy and Environmental Physics, Ben-Gurion University of the Negev, Israel), **S.G. Katsafouros** (Imel, Greece), **L. Kazmerski** (NREL, USA), **L. Kerr** (Miami University, USA), **J. Kessler** (University of Nantes, France), **R. Kishore** (National Physical Laboratory, India), **J.P. Kleider (**Director of Research CNRS, LGEP-Ecole Supérieure d'Electricité, France), **E. Klugmann-Radziemsk** (Gdansk University, Poland), **H. Koinuma** (Special Advisor to National Inst. for Materials Science, Japan), **M. Konagai** (Tokyo Institute of Technology, Japan), **F. Krebs** (Denmark), **M. Krunks** (Tallin University of Technology, Estonia), **J. Krustok** (Tallin University of Technology, Estonia), **S. Kumar** (Indian Institute of Technology Kanpur, India), **K. Kushiya** (Deputy General Manager, Showa Shell Sekiyu K.K, Japan), **A. Labeyrie** (Astrophysics, French Academy of Sciences), **B. Lambert** (Director Worldwatch Francophonie), **C. W. Lan** (Chairman of Taiwan Photovoltaic Industrial Association (TPVIA), Taiwan), **M. Leduc** (President of the

International Scientists for
Photovoltaic Solar Energy

French Physical Society, France), **M. Lemiti** (Head of PV Group, Institute of Nanosciences of Lyon, France), **D. Levi** (NREL, USA), **N. Le Quan** (Head of Research and Development, Photowatt, France), **A. Liébard (**Président de l' Observatoire des Energies Renouvelables et de la Fondation Energies pour le Monde, France), **D. Lincot** (Chairman of the 23rd European Photovoltaic Solar Energy Conference, Valencia, Institute of Research and Development of Photovoltaic Energy –IRDEP, France), **A. Luque** (Becquerel PV Prize, Director of the Solar Energy Institute, Spain), **I. Luque-Heredia** (CTO — Solfocus Europe), **M. Lux-Steiner** (Helmholtz-Zentrum Berlin für Materialien und Energie GmbH, Berlin), **E. Macias** (President of the European Photovoltaic Industry Association (EPIA), President of the Alliance for Rural Electrification (ARE)), **A. Madan** (MVSystems Inc., USA), **Y. Maigne** (Director of the Foundation Energies pour le Monde, Member of the French Academy of Technologies, France), **G. Malliaras** (Cornell University, USA), **X. Mathew** (Centro de Investigacion en Energia-UNAM, Mexico), **P. Malbranche** (Institut National de l'Energie Solaire, France), **D. Marsacq** (Director of New Energies Technologies Department –CEA, France), **S. Martinuzzi** (Université Paul Cézanne, Aix-Marseille III, France **), D. Mayer** (President of the European Renewable Energy Centres Agency –EUREC, Ecole des Mines Paris, France), **B. McNelis** (Managing Director, ITPower, UK), **E. Mellikov** (Tallin University of Technology, Estonia), **D. Mencaraglia** (Ecole Supérieure d'Electricité, France), **S. Menezes** (Interphase Solar, USA), **J. Méot** (Director SOLEMS, France), **D. Mercier** (Head of Xergies, Energies Renouvelables, France), **R. Mosseri** (Research Director at CNRS, Paris), **J. C. Muller** (INESS, Strasbourg, France), **T. Nakada** (Aoyama Gakuin University, Japan), **J. Nelson** (Imperial College London, UK), **S. Niki** (AIST, Japan), **R. Noufi** (Vice President for Research, Solopower, USA), **A. Nozik** (Director, Colorado Center for Revolutionary Solar Photoconversion, USA), **S. Palomares-Sánchez** (Facultad de Ciencias, UASLP, Mexico), **W.Palz (**Chairman World Council Renewable Energy, Brussels/Paris), **J. Pierce** (Queen's University, cananda), **N. Pearsall** (Vice-President of European Renewable Energy Centres Agency –EUREC, University of Northumbria, UK), **A. Penzias** (Nobel Prize, New Enterprise Associates), **E. Perezagua** (President of EU-PV Technology Platform, Isofoton, Spain), **C. Perkins** (NREL, USA), **V. Perraki** (University of Patras, Greece), **S. M. Pietruszko** (Head of Centre for Photovoltaics, Warsaw, Poland), **D. Mataras** (University of Patras, Greece), **T. Peterson** (Solar Power Consulting, USA), **Physical Society France, S. Pizzini** (University of Milano Bicocca and Chairman of Nedsilicon spa, Italy), **L. Pirozzi** (ENEA, Roma, Italy), **J. Poortmans** (IMEC, Belgium), **M. Powalla** (ZSW, Germany), **Haiyan Qin** (China General Certification Center, Beijing, China), **F. Rabago Bernal** (Instituto de Fisica, UASLP, Mexico), **J.K. Rath** (Utrecht University, Netherlands), **U. Rau**

International Scientists for
Photovoltaic Solar Energy

(Helmholtz Research Center, Germany), **S. Ray** (Program Chair of PVSEC18, Energy Research Unit, Indian Association for the Cultivation of Science, Kolkata, India), **B. Rech** (Hahn Meitner Institute, Germany), **A. Ricaud** (University of Savoie, France), **P. Roca I Cabarrocas** (Ecole Polytechnique, France), **A. Rockett** (University of Illinois, USA), **A. Romeo** (University of Verona, Italy), **M. Rosso** (Ecole Polytechnique, France), **T. Saitoh** (Tokyo University of Agriculture and Technology, Japan), **E. Sader** (Physics Department, University of Birzeit, Palestine), **R. Saez** (Helmholtz Zentrum Berlin, Germany), **G. Sala** (Universidad Politecnica de Madrid, Instituto de Energia Solar**), W. Sampath** (Colorado State University, USA), **H. Scheer** (Alternative Nobel Prize, General Chairman of the World Council for Renewable Energy, Germany), **R. Scheer** (Hahn Meitner Institute, Germany), **M. Schmidt** (IBM, Mainz, Germany),**H. Schock** (Helmholtz Research Center, Berlin, Germany), **R. E.I. Schropp** (Section Leader of Nanophotonics, Utrecht University, The Netherlands), **C. Sentein** (CEA SAclay, France), **Dinghuan Shi** (Chairman, China Renewable Energy Society, Beijing, China), **Zhengrong Shi** (Chairman and CEO Suntech Power Holdings Co., Ltd., China), **P.K. Singh** (head of PV Group, National Physical Laboratory, New Dehli, India), **S.N. Singh** (National Physical Laboratory, New Dehli, India), **W.C. Sinke** (ECN, The Netherlands, and EU PV Technology Platform), **J. Sites** (Colorado State University, USA), **A. Shah** (PV Becquerel Prize, Institute of Microtechnologies, Switzerland), **G. Sissoko** (Université Cheikh Anta DIOP, Dakar, Senegal), **A. Slaoui** (President of the European Materials Research Society, France), **I. Solomon** (Ecole Polytechnique, French Academy of Sciences, France). **J. Song** (Korea Institute of Energy Research, Korea), **B.J. Stanbery** (CEO, HelioVolt Corporation, USA), **R. Tena-Zeara** (Institute of Materials, Microelectronics and Nanosciences of Paris East, France), **I. Tiginyanu** (Vice President of the Academy of Sciences of Moldova, Moldova), **A. Tiwari** (ETH Zurich, Switzerland), **M. Topic** (University of Ljubiana, Slovenia), **P. Torchio** (Institute of Materiasl, Microelectronics and nanosciences of Marseille, France), **Chuang Chuang Tsai** (National Chiao Tung University, Hsinchu, Taiwan), **M. Umeno** (Chubu University, Japan), **N. Usami** (Tohoku University, Japan), **M. Ushijima** (Tokyo Electron LTD, Japan), **A. Vallêra** (Vice- Rector of the University of Lisboa, Portugal), **J. Van Duren** (Nanosolar, USA), **M. Vanecek** (Head, Dept. Optical Crystals, Academy of Sciences of the Czech Republic), **W. van Sark** (Copernicus Institute, Netherland), **P.Varadi** (Co-founder of Solarex, Founder of PV GAP, Geneva, Switzerland), **AC Varonides** (Univ. Scranton, USA), **P. Vasseur** (Past Director of the Research Centre of Ecole Polytechnique, France), **J. Vedel** (CNRS, France), **P. Vivo** (Tampere University of Technology, Finland), **T. Wada** (Ryukoku University, Japan), **S. Wakao** (Waseda University, Japan), **E. Weber** (Director of the Fraunhofer Institute for Solar Energy ISE, Freiburg, Germany), **S. Wei** (NREL, USA), **R.W. Welker** (University of

Maryland, USA), **S. Wiedeman** (Global Solar, USA), **S. Williams** (Plextronics, Inc., USA), **C. Wronski** (Pennsylvania State University, USA), **N. Wyrsch** (Institute of Microtechnology IMT, University of Neuchatel, Switzerland), **M. Yamaguchi** (Becquerel Prize of the EU-PVSEC and the 2008 winner of the William Cherry Award from IEEE, Director of the Super High-Efficiency Photovoltaics Research Center, Toyota Technological Institute, Japan), **T. Yamaguchi** (Wakayama National College of Technology, Japan), **A. Yamamoto** (University of Fukui, Japan), **S. Yanagida** (Osaka University, Japan), **I. Youm** (University Cheikh Anta DIOP, Dakar, Senegal**)**, **T. Yoshida** (Gifu University, Japan), **S. Yoshikawa** (Kyoto University, Japan), **K. Yoshino** (University of Miyazaki, Japan), **A. Zaban** (Bar-Ilan University, Israel), **G. Zangari (**University of Virginia, USA), **Z. Zapalowicz** (Poland), **M. Zeman** (Head of Solar Cell group, Delft University of Technology, The Netherlands), **M. Zazoui** (President of the Association for the Development of Renewable Energies of Marocco, Hassan II Mohammedia University), **Meifang Zhu** (Graduate School, Chinese Academy of Sciences (GUCAS), Beijing, China).

Contact : PVCall@chimie-paristech.fr

Le solaire s'impose comme l'énergie du futur

Electricité Le secteur photovoltaïque croît au rythme de 40 % par an

*As a result of the 23rd European Photovoltaic Solar Energy Conference in Valencia, the front page of the important French journal "**Le Monde**", was dedicated to solar energy with a full paper entitled "**Solar Energy is setting itself as the energy of the future**" and illustrated by a splendid cartoon.*

Chapter 27

High Efficiency Photovoltaics for a Sustainable World

Antonio Luque

Instituto de Energía Solar, Universidad Politécnica de Madrid, Spain

1. Introduction

The energy consumption of Mankind, with its consequential waste, has to increase enormously when in few decades the consumption patterns of more than two billion inhabitants (China, India, etc.) adopt the consumption patterns of the developed world. The only solution can be the use of renewable energy.

From the Sun the Earth receives a power of about 10 000 times higher than the present consumption of Mankind. It is the only renewable source of energy big enough to support human activity. But it has to be converted into usable energy (electricity) with high efficiency. Efficiencies below 0.01% would not permit the supply of the present needs of Mankind, not to mention its expansion. Biomass and wind energy are related, to certain extent, to this low efficiency problem because both forms of renewable energy are derived from solar energy through the natural processes of conversion.

In contrast, photovoltaics (PV) provides such high efficiency. An efficiency of about 10% for a surface normal to the sun (that is the one receiving the 10 000 times the present consumption) is today possible with two-axis tracking panels implying that only 1/1000 of the Earth surface can provide the present Human needs, or about 1/250 of the Emerged Lands. This is incomparably less than the space required for food and means that there is not a problem of space, despite the frequent claim to the contrary.

Power for the World by W. Palz
Copyright © 2011 by Pan Stanford Publishing Pte Ltd
www.panstanford.com
978-981-4303-37-8

Photovoltaics has often been considered as ideal for a very distributed electricity generation pattern. Power supply for remote electronic equipment, essential electricity for rural areas in developing countries, electricity for isolated homes, and, at most, homes and buildings that inject electricity into the electric grid. The idea of PV central power plants has indeed been contemplated but, at least in Europe and Japan, as a remote possibility, not the most desirable.

2. The 2008 Spanish Boom

However what happened in 2008 in Spain was extraordinary. Under the stimulus of a, perhaps too high, feeding tariff, 2.6 GW were installed in the country in only one year, able to produce about the same electricity as a (not too big) nuclear power plant. But installed in one year! Not in the one decade that a nuclear plant requires for its construction or the several years needed for almost any other power technology.

In Spain, of the 3.4 GW that form its PV park, 98% are built on the ground; 44% is above 5 MW. Of the 50 biggest power plants on the ground in the World by the end of 2008, 40 were in Spain and the biggest one, of 60 MW, was also in Spain. In addition 40% had some kind of tracking.

The lessons learned are important:

- Business men (mostly Spanish but also German as well as from other nations) realized that good climates are better than poorer ones for doing business with solar (although the high tariff was a determinant)
- Banks have accepted the financing of PV (often up to 90%) with the only guarantee of the PV plant.
- Bigger plants are cheaper: the management costs are much less; furthermore they are more bankable because the banks have few experts and cannot waste their time on small projects (again a reduction in cost).
- Tracking provides more energy and is often preferred. Any reluctance about tracker reliability has virtually disappeared (although examples of malfunctioning trackers have been found, but are usually fixed quickly) further stressing again the need for businessmen to collect as much energy as possible. This is opposite to the case of homes where architectural integration sets (not too big) limitations on the optimization of the energy collection.

In general it can be said that the response of businessmen is that of favoring high efficiency as was expected from the general statements in the introduction of this paper. We present in Fig. 1 a big PV plant with trackers that shows a gentle integration with the land. Cattle can be raised around the panels and crops can be grown. This picture is not an exception. There are several large plants taking advantage of these characteristics.

Figure 1. 10 MW 2-axis tracking flat panel PV power plant at Jerez de los Caballeros (Spain).

But, of course our plea for efficiency is not blind. Efficiency often requires a higher cost and a judicious choice must be made. For instance, highly efficient silicon solar cells are deemed in most analyses to be the best way of reducing their costs, but one company, producing high efficiency cells has stopped its production because the extra cost for this high efficiency was not paid for by the extra efficiency. Likewise, if flat modules become very cheap, as is happening today, once Chinese production becomes more predominant, it might happen that the tracking option will become less cost effective.

3. A Market Forecast Model

In 2001 we published a market forecast model[1] based on a set of differential equations that coupled the PV module learning curve with demand elasticity. To date, the model has accurately predicted the market evolution. Notice that it has even predicted the acceleration of the annual market growth rate in recent years.

Actually the paper stated in its conclusions in 2001:

"First, we predict several years of explosive market growth...

But this period cannot last too long, not more than one decade. If it does, the capital involved will become excessive. Powerful voices will cease to consider PV as a curiosity and will question its cost-effectiveness. Other voices not less powerful will support PV. The equilibrium will determine subsequent growth. This will induce a slower market

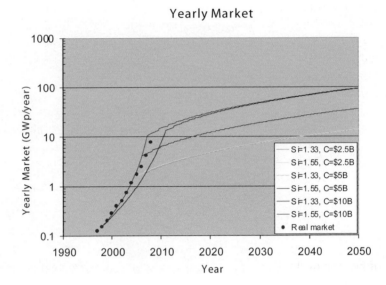

Figure 2. Real yearly market (after Photon International, successive years) and predicted market for several combinations of the model parameters (see Reference 1). The model was drawn up using data from up to 1998 and published in 2001.

growth but at levels no longer negligible, at least in terms of business volume but not enough for pollution abatement. Price decrease will continue but slowly. For the new half century prices will not be competitive with conventional electricity unless some of the following events take place:

- *Electricity prices rise,*
- *Commercial schemes substantially reduce market costs,*
- *New inventions reduce the initial costs or offer more cost-reducing potential".*

The preceding words seem to describe accurately what has happened in Spain by the end of 2008. The feeding tariff has been greatly reduced but in addition, the administrative procedures that had been trimmed for more than one decade, have changed and virtually no new module was installed in 2009. But indeed the world continues and PV still grows, mainly in Germany (the steady motor of the PV deployment) and good perspectives seem to appear in other countries. We have to wait until next year to see what happened globally in 2009 but maybe we shall realize that we are actually starting the phase of slower growth as predicted by the model.

But the model predicts that to permit PV to become a major pollution abatement factor, one condition was to find cheaper schemes for its commercialization. Maybe this is what has happened with the establishment of big plants where the assumed (in the model) scattered nature of the PV market has disappeared.

The model also predicts that technologies with cheaper initial costs or more cost-reducing potential might also lead to PV becoming a major pollution-abatement factor. We are witnessing the development of thin film technologies, probably of lower initial cost. We also want to present recent efforts to produce a technological breakthrough, defined in this context as more cost-reducing potential, that is, a faster learning curve.

Before doing so, I want to appeal to your indulgence if you, dear reader, feel indignant by my business-oriented approach. I can assure you that I share the feelings of intergenerational solidarity that probably moves you. But I think that History shows that the noblest human endeavors cannot come to fruition if less noble egoistic motivations are not present too.

4. The FULLSPECTRUM Project and the ISFOC

On the suggestion of the European Commission (EC), in the Spring of 2002, we conveyed to my university's Conventions Centre amid the Cercedilla forest, three dozen of world class scientists to elucidate why the PV learning curve, of around a 20% price reduction every time that the cumulated production is doubled, cannot be faster. It was agreed that one important factor is the inability of prevalent solar cells to convert the full solar spectrum properly. Actually only the photons with energy close to the bandgap of the semiconductor of which the solar cell is made are effectively converted. The photons with less energy are lost and for those of higher energy only the bandgap energy, at most, is converted. The efficiency limit of solar cells working in this way was calculated in 1961 by Shockley and Queisser[2] and is of about 40% at most. Several solutions were presented to overcome this limitation and a book was published[3] with them.

Then, a proposal for the FULLSPECTRUM integrated project was presented to the EC that was granted for the November 2003–October 2008 period. The proposal involved 19 European Research Centers[4] and was perhaps the first organized attempt to find a breakthrough in PV with a research in a network. The proposal included medium-term and longer-term solutions that should assure a lasting flow of inventions to feed a fast learning curve (this approach is rather unique in R&D projects).

Its motivation is that if a learning factor as fast as that of semiconductor memories is introduced into the model in Ref. 1 the cost of PV electricity would become, in few years, as low as that of the wholesale electricity generated by the current power technologies, and not only as the retail price on the users side, which is the dream of the current PV community. Thus utilities would be inclined to buy PV electricity rather than any other alternative. In this way the use of PV electricity would reach the level of penetration that is possible to manage for intermittent power generation (solar and wind) with the dispatching mechanisms used in the electric networks that according to reference 5 is about 1/3 of the consumption. In Spain the penetration of intermittent energies (mostly wind) in its electrical network is already over 14% of the demand with no major difficulties.

One of the medium-term approaches was the development of multijunction (MJ) solar cells. These are monolithic stacks of solar cells made of different semiconductors. The more energetic photons are converted in the top cell letting less energetic photons pass to the following cell where the most energetic of the transmitted photons are absorbed and the rest are transmitted and so on. These cells are currently made of thin films of GaInP, GaInAs, and Ge, the latter material being a wafer for the support of the stack. Within FULLSPECTRUM, FhG/ISE achieved a efficiency world record of over 41% that had formerly been held by the American firm Spectrolab and has recently been recovered by this firm. In any case, we in Europe are already playing in the first league in this area that formerly had been practically the monopoly of the US, geared by the powerful space programs of this nation.

But while the MJ cells are available, the way of using them for terrestrial applications is in its infancy. FULLSPECTRUM also devoted resources to developing concentrators to bring these cells to land. As they are very expensive, their use can only be afforded by casting a big amount of luminous energy on them by using optical elements that concentrate the light. The concentration factor must be high, a minimum of 500 but preferably 1000, to pay for the expensive cells.

As a complement of the FULLSPECTRUM actions, the Institute for Photovoltaic Concentrator Systems (ISFOC) was created by the Regional Government of Castilla La Mancha (under our design) and two consecutive international calls for tenders were issued with the condition that any tender has to propose at least 200 kW of concentrators operating with MJ solar cells. Three Spanish companies, two from the USA, one from Germany and one from Taiwan were selected. Currently about one MW has already been operating for about a year, presenting the expected good performance. One MW has just built and the remaining one MW is in different construction and approval phases. When everything is finished, the plant will contain 3 MW of concentrators with MJ solar cells on three different sites.

Figure 3 shows the 800 kW of MJC concentrator panels from three different companies installed at the central site of the ISFOC institute.

The reason why several companies were invited to participate in the ISFOC program is that when starting a new technological endeavor, the odds of technical and commercial failure are high. Therefore redundancy has to be provided so that success is that at least some of the selected companies can find a substantial place in the PV market.

Longer-term concepts were also researched in FULLSPECTRUM in order to permit new innovations to be incorporated into the PV sector and maintain a quick learning in the learning curve for long time. Among these concepts the intermediate band (IB) solar cell is one of them. In this concept[6] a band of energy is permitted within the bandgap so that not only high energy photons can pump

Figure 3. 800 kW MJC CPV panels installed at the central Puertollano site of ISFOC. Modules of three different companies are installed there.

electrons from the valence band to the conduction one but also lower energy photons can pump them by using the IB as a relay. The Fermi level must split into three quasi Fermi levels, for the valence, intermediate and conduction band in order to permit high voltage. This cell is a sort of spatially integrated MJ cell stack and has similar efficiency potential. More precisely, it is 63% under the same assumptions in the Shockley and Queisser paper.[2] It was implemented for the first time in 2004[7] using the confined levels of InAs QD in a GaAs host crystal as the IB. Although high efficiency has not yet been achieved, the basic principles have been experimentally demonstrated.[8,9] Today more than 30 centers worldwide have published ISI registered papers on this topic.

An alternative way of preparing IB solar cells is by developing alloys with such an IB. Solar cells clearly based on this principle have been prepared in O-doped InTe[10] that while presenting still low efficiency show clearly their operation as IB solar cells.

In principle, IB solar cells based in nanotechnology would be suitable to replace ordinary cells in a MJ solar cell scheme to be used in concentrators for still higher efficiency. Alternatively, alloy-based IB cells could be prepared as thin film modules with improved efficiency. But as said before this research will be long term.

Other new concepts, able to exceed theoretically the Shockley and Queisser efficiency limit, and therefore able to lead, in our view, to a fast learning curve, are the multiple exciton generation solar cells[11–13] and the hot carrier solar cells.[14,15]

All these new concepts, including MJ solar cells are the object of sensible high interest worldwide. Three large network-based projects are now in development in Japan that confess to be inspired by FULLSPECTRUM (one of them is also called FULLSPECTRUM) and up to four large network projects are under development in the USA.

5. Summary

In summary, we think that likely two of the conditions we considered necessary for PV to become ecologically relevant have possibly been met:

On one hand the fast development of PV in 2008 in Spain shows the way towards a substantial reduction in the costs of PV commercialization. It is is only fair to give Germany the credit for having invented the basic mechanisms of the feed-in tariff, of which the Spanish explosion is a consequence. The trend towards large ground-based plants of low BOS and commercialization costs will probably continue worldwide.

On the other hand there is an increasing activity in the World for research in concepts of high efficiency potential. Cost constrains in the manufacture of these concepts might be eliminated if the CPV industry develops healthily and the creation of the ISFOC in Castilla La Mancha might be an initial step for this. These novel cell concepts can also be of interest for the industry of thin film modules or related concepts by increasing their efficiency potential. All this should give rise to a faster learning curve.

References

[1] A. Luque, "Photovoltaic markets and costs forecast based on a demand elasticity model," *Progress in Photovoltaics: Res. Appl.,* vol. 9, pp. 303–312, 2001.

[2] W. Shockley and H. J. Queisser, "Detailed balance limit of efficiency of p-n junction solar cells," *Journal of Applied Physics,* vol. 32, pp. 510–519, 1961.

[3] A. Martí and A. Luque, "Next generation photovoltaics: High efficiency through full spectrum utilization," Bristol: Institute of Physics Publishing, 2004.

[4] A. Luque, A. Martí, A. Bett, V. Andreev, C. Jaussaud, J. A. M. v. Roosmalen, J. Alonso, A. Räuber, G. Strobl, W. Stolzi, C. Algora, B. Bitnar, A. Gombert, C. Stanley, P. Wahnon, J. C. Conesa, W. G. J. H. M. v. Sark, A. Meijerink, G. P. M. v. Klinko, K. Barnham, R. Danz, T. Meyer, I. Luque-Heredia, R. Kenny, C. Christofides, G. Sala, and P. Benítez, "FULLSPECTRUM: A new PV wave making more efficient use of the solar spectrum," *Solar Energy Materials and Solar Cells,* vol. 87, pp. 467–479, 2005.

[5] T. B. Johansson, H. Kelly, A. K. N. Reddy, R. H. Williams, and L. Burnham, *Renewable Energy Sources for Fuel and Electricity.* Washington DC: Island Press, 1993.

[6] A. Luque and A. Martí, "Increasing the efficiency of ideal solar cells by photon induced transitions at intermediate levels," *Physical Review Letters,* vol. 78, pp. 5014–5017, 1997.

[7] A. Luque, A. Martí, C. Stanley, N. López, L. Cuadra, D. Zhou, and A. Mc-Kee, "General equivalent circuit for intermediate band devices: potentials, currents and electroluminescence," *Journal of Applied Physics,* vol. 96, pp. 903–909, 2004.

[8]   A. Luque, A. Marti, N. Lopez, E. Antolin, E. Canovas, C. Stanley, C. Farmer, L. J. Caballero, L. Cuadra, and J. L. Balenzategui, "Experimental analysis of the quasi-Fermi level split in quantum dot intermediate-band solar cells," *Applied Physics Letters,* vol. 87, pp. 083505–3, 2005.

[9]   A. Marti, E. Antolin, C. R. Stanley, C. D. Farmer, N. Lopez, P. Diaz, E. Canovas, P. G. Linares, and A. Luque, "Production of Photocurrent due to Intermediate-to-Conduction-Band Transitions: A Demonstration of a Key Operating Principle of the Intermediate-Band Solar Cell," *Physical Review Letters,* vol. 97, pp. 247701–4, 2006.

[10] W. Wang, A. S. Lin, and J. D. Phillips, "Intermediate-Band Photovoltaic Solar Cell Based on ZnTe:O," *Applied Physics Letters,* vol. 95, p. 011103, 2009.

[11] J. H. Werner, S. Kolodinski, and H. J. Queisser, "Novel optimization principles and efficiency limits for semiconductor solar cells," *Physical Review Letters,* vol. 72, p. 3851, 1994.

[12] A. Luque, A. Marti, and A. J. Nozik, "Solar cells based on quantum dots: Multiple exciton generation and intermediate bands," *MRS Bulletin,* vol. 32, pp. 236–241, Mar 2007.

[13] A. Luque and A. Martí, "Entropy production in photovoltaic conversion," *Physical Review B,* vol. 55, p. 6994, 1997.

[14] R. T. Ross and A. J. Nozik, "Efficiency of hot-carrier solar energy converters," *Journal of Applied Physics,* vol. 53, pp. 3813–3818, 1982.

[15] P. Wurfel, "Solar energy conversion with hot electrons from impact ionisation," *Solar Energy Materials and Solar Cells,* vol. 46, pp. 43–52, 1997.

Antonio Luque

Antonio Luque is a Full Professor at the Universidad Politécnica de Madrid since 1970, he serves at the Instituto de Energía Solar he founded in 1979. He invented the bifacial cell in 1976 and, to make them, in 1981 he founded Isofotón, a solar cell company with a turnover of 300 million € in 2007. In 1997 he proposed the intermediate band solar cell. Today, more than two dozen of centres worldwide have published on this topic in (ISI) registered journals. He is the recipient of several important prizes and distinctions, including the membership to the Real Academia de Ingeniería of Spain, the honour membership of the Ioffe Institute in St. Petersbourg and the Becquerel Prize of the EC.

Chapter 28

Promoting PV in Developing Countries

Bernard McNelis

IT Power, UK

I have written this chapter as a very personal account of my own small efforts in promoting PV in Developing Countries. Over the past 30 something years I have been privileged to travel and work in 50 odd countries of the developing world. This has been at times exhilarating, at times depressing, even infuriating. I write about a small selection of these experiences. I apologise to any individuals, countries or institutions I might offend.

1. Looking at Solar

My first real job was in electrochemistry research. I thought I was going to work on fuel cells, but in between my interview and starting the job in 1971 in the Chloride Research Centre in Manchester (R&D for the UK's biggest battery manufacturer), they had given up on fuel-cells. So at first I worked on the fundamentals of lead-sulphuric acid (not changed since the 19th Century), and later, because I was keen and pushy for something more exciting, on more interesting things, like using air-breathing (fuel-cell) electrodes to measure oxygen levels underground. But I wanted something still more exciting. In 1972 I read an article in New Scientist, on photoelectrochemical cells — "solar-charged" fuel cells. Now that was exciting, so I suggested to my boss that we should look at this area of research. He replied that fuel cells were difficult enough, so don't be stupid! But I read more and more, and became more and more interested in all aspects of solar energy. I read about photovoltaic cells, and immediately thought that when these got cheap, it would be a big new market for batteries, which was Chloride's business. So I ran to tell my boss the exciting news. He just laughed. I didn't give

Power for the World by W. Palz
Copyright © 2011 by Pan Stanford Publishing Pte Ltd
www.panstanford.com
978-981-4303-37-8

up, so eventually I was authorised to do a very small internal study and report on solar energy. I spent every hour of every day on this. I read Farrington Daniels "Direct Use of the Sun's Energy". I contacted the handfull of people working on solar in the UK, notably Brian Brinkworth at Cardiff University, John Page at Sheffield University, Peter Dunn at Reading University and Fred Treble at the Royal Aircraft Establishment in Farnborough. I heard about The *Sun in the Service of Mankind* Conference in Paris, in July 1973, and asked if I could attend. "Not a chance", was the boss's reply! Then I found out that a young lady called Mary Archer was organising a UK section of the International Solar Energy Society (ISES). I joined and I went to the Inaugural Meeting at the Royal Institution in London (where Mary worked on photochemistry with Nobel Laureate Sir George Porter) in January 1974, and I have been an avid supporter of ISES ever since.

My solar study went far beyond its supposed scope, time and budget. But my own enthusiasm paid off at Chloride. They sent me to Southampton University for Martin Fleichmann's Electrochemistry Summer School, and offered me a big promotion to work in a new venture on sodium-sulphur batteries. But in this time I had also met Solar Power Corporation (SPC), for which Elliot Berman had secured backing from Exxon to start manufacture of PV modules which were very cheap at the time, in Braintree, Massachusetts.

2. Into PV

SPC offered me a job, even though my knowledge of PV, or the practical side of batteries, was pretty rudimentary. I jumped and accepted with excitement. Elliot not only had a vision for PV, he had a passion to bring electricity to the developing countries; I was inspired! My friends at Chloride and the Electrochemistry Group of the Royal Institute of Chemistry thought I was completely mad not to accept the promotion and to leave for the Solar (dream) world.

Elliot had submitted an abstract to the Photovoltaic Power Generation Conference to be held in Hamburg in September 1974. (I call this "Zeroth" EPVSEC). My dear boss then kindly gave me the job to prepare the paper and present it. I had never presented in international conference before. So I went to Hamburg as a real novice. Almost all the papers were on space PV. There was just one session on "Terrestrial PV" including me. I prepared my presentation, but I hadn't written the paper. The chairman of my session was someone from the French Space Research Centre, CNES. This rough and tough guy took no pity on the young, inexperienced English guy. Banging the table he shouted at me, "Where's your manuscript?!". But the presentation went ok; I handed two 5 Wp modules to the audience to examine (no one else did such an act). Both were returned to me, but some smart person had realised the weakness — and scrapped off the silicone rubber back sealant!

When with SPC, I visited the Exxon R&D Centre in Linden, New Jersey to see research on "Next generation" PV, I thought this was a great company. But, to cut a long story short, Exxon was the first oil company to pull out of, as well as into, PV. So in 1975 Elliot, I and the rest of SPC had to find alternative employment. Exxon are still the bad guys, promoting denial of the science of global warming, just like the tobacco companies denied smoking damages health.

3.   Into All Things Solar

I never considered leaving solar and I managed to work on solar architecture, solar thermal collectors, long-term heat energy storage, with Dominic Michaelis in the Solar Energy Developments (SED) Partnership. Dominic was a brilliant passive solar architect, but he was also an inventor with clever designs for tracking systems, low cost solar-thermal collectors, solar cookers, even wave power devices. He invented and demonstrated the solar balloon. Dominic also wanted to include PV in his house designs.

I also worked for General Technology Systems (GTS), on a range of solar projects. In 1977 we won a contract from the Barbados Government, with EC funds, to design a solar air-conditioning system for a new building of the Ministry of Agriculture, the Central Argonomic Research Unit (CARU). The European Development Fund was actually paying for three new buildings. Construction of the first two buildings was already well underway and their design included flat plate collectors and Lithium Bromide/water absorption chillers, to be supplied by Tadiran of Israel. The ministry wanted similar for CARU, which had been designed by architects Robertson Ward Associates. We reviewed the design, which was for a courtyard building with natural ventilation, good shading etc., indeed the architects had cleverly designed the building not to need air conditioning! The building would by warmer inside than an air-conditioned building, which I consider more comfortable. But the ministry was adamant that they wanted the building actively cooled, not just ventilated, so there was no compromise on having air-conditioning added, and we were told to get on with the solar design. I took the opportunity to suggest we consider all the solar (and wind) options not just copy the existing "conventional" solar designs. This way we were able to get to consider PV with a vapour compression chiller. I even took a 30 Wp PV module (ARCO) with me on my second trip to show to the people at the ministry people. This was of course valuable and well packaged. The Bridgetown Customs Official insisted it to be completely unpacked, and I tried hard to explain to him what it was. He got increasingly angry and eventually said: "Shut up, just show me where the hot water comes out?". I did the comparative study. None of the options were economic, and my calculations suggested the PV/compressor would cost about 20% more than the solar collector/absorption "conventional" option. I then argued that as PV was expected to fall dramatically in cost, in the longer term PV air conditioning would more likely become economic and commercial, than the "conventional" solar-thermodynamic, so why not look to the future? My argument was accepted and we were given an extension to the contract to design the PV System. This was quite exciting, and a lot of effort went into it. However, there was another problem. The costs of the other buildings escalated and there was no longer sufficient European Aid funds to pay for the air-conditioning for CARU. Even the building construction became doubtful. Assistance was at hand though, from the British High Commission. A UK aid grant to buy Leyland busses for Barbados had been underspent and they were happy to divert the surplus to pay for the 10 kWp PV system. But this had to be done within the Financial Year, or the funds would be lost. The BHC also required, naturally that the PV System had to be "British Made". There was not much chance of this, as there was no PV manufacture in the UK. We managed to compromise to "British Designed and Built", and ordered the (American) Solarex modules from Solapak and the (American) Carrier compressor chiller from their UK distributor. We tested the compressor, and our clever design of varying the number of cylinders operating with the current available (i.e. simulating the PV output changes with irradiance) worked exactly as predicted. The kit was shipped and paid for inside the Financial Year, but I have no idea what happened to it, as the CARU building was not built, and I have not been back to Barbados since.

We also won a contract from the EC to be co-ordinator of the Eurelios project which built the world's first 1 MW central receiver solar electric plant at Adrano in Sicily. There had not been a plan to have a coordinator, but Albert Strub in Brussels was concerned about the arguments between the competing contractors in France, Germany and Italy so brought in Brits as being neutral and with no commercial interest in the technology. In the same timescale, the "tough guy" I met in Hamburg moved to the European Commission in Brussels, and managed some of the contracts I was working on.

4. Into Intermediate Technology

E. Fritz Schumacher's book "Small is Beautiful" was published in 1973. He was the Chief Economist at the British National Coal Board, but he was interested in more than coal. Influenced by Gandhi, he argued for, what were, new ideas at the time on ecology, environment, decentralisation…. He said particularly that giving "high-tech" equipment like huge combine harvesters to try and transform agriculture in developing countries was flawed. Instead, improvements on "low-tech" equipment would be more appropriate — Appropriate Technology. He launched the charity Intermediate Technology Development Group (ITDG) to put his ideas into action. This was the start of the appropriate technology movement. Schumacher traveled and lectured all over the world, and ITDG was inundated with enquiries from people in developing countries, like "how can I improve my donkey-drawn plough", to more relevant to this subject "can I build a solar water heater". ITDG did not have the resources to deal with the volume of enquiries, so it enlisted the support of volunteers. A series of advisory "panels" were set up, with un-paid Chairpersons and experts as members. Examples were the Transport Panel, the Health Panel, and of most relevance, the Power Panel, chaired by Peter Dunn (an old pal of Schumacher). This should have more accurately been called the Energy Panel, but even today there is confusion between power and energy. ITDG also hired a part-time engineer as the Power Projects Officer, Peter Fraenkel. The Power Panel was further divided into Sub-Panels including the Wind Panel (which included members Garrad and Hassan) and which "evolved" into the British Wind Energy Association (BWEA). I joined the Solar Panel (a silly name) in 1975, and later became Chairman. There was a sub-sub-panel on Solar Cookers who met and dreamt up all sorts of crazy designs. One was a mini power tower cooker for refugee camps. I was almost expelled when I (seriously) asked had anyone done a technoeconomic comparison of their designs with a PV/battery/microwave oven. The Solar Panel was the "breeding group" for IT Power, as discussed later. ITDG's "Power Projects" included work on wind pumps, small wind generators, solar stills, micro-hydro, with practical work being done at Reading University. My aim with the "Solar Panel" was to actually work in developing countries, with PV as a very appropriate technology.

5. Into Africa

My big break, and one of the most important in my life, was in the summer of 1978. Wolfgang Palz called GTS from the EC and invited me to join him on a trip to Francophone Africa, as part of the EC's Solar Energy for Development Initiative. Actually, he really wanted my French-speaking boss, but I conned him and when he called, I didn't even know where Bamako was — I covered the phone and shouted to a colleague — "Get me an atlas!!"

In Mali, we saw the world's first commercial PV pumps in action and met the legendary White Father, Père Bernard Vesperien. He had started Mali Aqua Viva, with the aim of providing clean water to villages in the San region. He installed mostly manual pumps, but he was very quick to see the tremendous advantage of a mechanised water delivery system, powered by solar. We visited PV pumps, including the one just installed at Koni, which had been partially paid for using EU aid. We also visited villages without clean water. The Mali story is covered elsewhere in the book. When I saw at first hand the living conditions of the poor people in Mali, and the tremendous benefits brought by clean water pumped by PV, I at once decided that my life's ambition would be to make a contribution, however small, to using solar energy to change the lives of people like those I met in Mali. Figure. 1 illustrates one of many examples of happy faces thanks to clean, PV-pumped, water. The follow-on to this Regional Workshop was the big event in Varese in March 1979. This was a bit of a circus, but the food and wine was out of this world. In Varese there were several British colleagues, fellow members of UK-ISES, like Peter Dunn, David Hall, John Page, Peter Musgrove and Jerry Leach. After the drinking, the networking was the most useful. David Hall introduced me to Derek Lovejoy of the UN Department of Technical Development Cooperation for

Figure 1. Girls get clean PV-pumped water, Mopti, Mali, 1984.

Development (UNDTDC). This was most important as Derek was preparing new solar projects, but solar experts were hard to find. In Mongolia he urgently needed wind and solar experts and immediately following he asked Peter Musgrove and I if we could fill these two positions. We went through the procedures and were selected, but then Derek called to say there was a problem. The Mongolian government had responded: "There must be some mistake with the two experts, they are not Russian". East Germans might have been acceptable. Derek argued, and the end compromise was the Mongolians would accept one non-Russian. The coin was tossed; Peter did wind and a Russian did solar. I met the Russian (who could not speak any English) together with Namjil English at the ISES Solar Congress in Brighton in 1981. Namjil has been for years the leading authority on PV in Mongolia. Derek contracted me soon afterward as PV expert for Romania, and as advisor on solar air-conditioning for the United Arab Emirates (UAE). For this I recommend to consider PV, but this was a waste of time as Germany donated a hugely expensive experimental solar-thermal system from Dornier.

So I had achieved what I had dreamed of, a professional trip to Africa. This was the opportunity I really needed. I also persuaded the British Overseas Development Ministry (ODM, now the Department for International Development, DFID), represented by Toby Harrison, to finance a trip to the other countries working on solar in West Africa (Niger, Senegal and Upper Volta (now Burkina Faso)); and to the African Solar Meeting linked to the ISES Congress in Atlanta in May 1979, which is where I met people who became lifelong friends, like Suresh Hurry and Dick Dosik. Dick later took charge of the World Bank Solar Pumping Project, which was the catalyst for the formation of IT Power.

6. Global Solar Pumping Programme

The Solar Pumping project was the brainchild of Steve Alison, a water resources engineer at the World Bank. His vision was of millions of small PV pumps on the millions of small farms on the Indian sub-continent. He tried to create a project to achieve this. But by the time the project was designed, changed, designed again, target countries chosen on political grounds, he had left the World Bank. Eventually, in 1978, this emerged as a UNDP/World Bank project managed by the World Bank, to do field testing of solar pumps in 4 pre-selected countries; India, Mali, Philippines and Sudan, and the Bank hired an Egyptian Project Supervisor, Essam Mitwally. The budget was $2 million, which was a lot of money in those days. The Bank published an advertisement calling for consultants to mange the project and for manufacturers to supply the solar pumps.

At this time, one of the ITDG Trustees was Sir Alan Muir-Wood, who was also a senior partner in the consulting engineers, Sir William Halcrow and Partners. Halcrow knew nothing about solar but they did know water engineering and had experience throughout the world. Alan knew of the "Solar Panel", so he suggested that Halcrow and ITDG make a joint bid for the World Bank contract. There was no-one actually in ITDG with any solar knowledge, just the loose group of volunteers like me. I was one of the few people competing for the contract to have actually seen and touched working and non-working solar pumps in "real-life" situations. Our team comprised full-time Halcrow water engineers led by David Wright and Michael Starr, Peter Fraenkel of ITDG (who thought it was crazy to spend the money on this solar project instead of developing simpler, less expensive mechanical windpumps (to do the same job), and part-time "consultants" from the "Solar Panel" including myself, Anthony Derrick of Cardiff University, Peter Dunn, Fred Treble and others. We won the contract. The structure of the contract was very favourable to Halcrow, as they would send a "Resident Engineer" full-time to each country to oversee the field trials, and with their project management this would consume most of the budget. But there was a bit left for the solar team to steer the work, select the equipment, train the "Resident Engineers" etc.

To the newer generation it may sound curious, but when we started the solar pumping project, the World Bank and much of the solar community expected that most of our work would be with solar thermodynamic systems, "not those expensive PV cells used in space"! The first part of the project in 1979 was to assess the State-of-the-Art, including existing installations and manufacturers. This was mainly done by Anthony Derrick and myself. Our first visits were to the solar-thermodynamic companies like SOFRETES in France and Dornier in Germany. The solar-thermodynamic pumps I visited in Africa and Egypt were not working. The only functional one I discovered was in Mexico, at the University of Monteray, and with a skilled team to keep it going. In the Philippines, the SOFRETES pump did not work and was quickly dismantled. In the US we met companies who had done a hard sell on the World Bank, but the contraptions they showed us were clearly useless. We also visited some large solar pumping systems using parabolic troughs in Arizona, and met researchers in Universities and Sandia Laboratories. We also visited a lot of PV pump suppliers. The leader was clearly Pompes Guinard in France, but there were also other very interesting PV products on offer.

We concluded that none of the solar-thermodynamic systems were ready for demonstration and field trial, as required by the project (although I did say casually that if the suppliers could provide free-of-charge and with a full-time team of PhD scientists/engineers to operate, we might consider).

The next step was to visit the four countries selected. My special interest was Mali (Why choose the most difficult place?!), described more fully later. Sudan was difficult, but the four pumps we installed were all at the same location, the Energy Research Centre in Khartoum. Philippines was straightforward as we were working with a well-developed research institute, the Non-Conventional Energy Center. India was easy, as we did nothing there. On the first visit the Energy Adviser "Guru" Guraraja told us that India would accept its share of the project budget and develop their own solar pumps, but it would not allow foreign experts to come and test foreign equipment in the country. So India exited. But India was not totally left out. Anthony Derrick illegally exported in his luggage some indigenous PV modules for us to test.

We managed to get one solar-thermodynamic system into Sudan — it made a few strokes, provided you packed ice around the condenser — not very practical. The Sudan test site is shown in Fig. 2. Figure 3 is an example one of the pumps we installed in Babougou, Mali in 1980.

Figure 2. Solar pump test site, Energy Research Centre, Khartoom, Sudan, 1980.

Figure 3. PV pump at Babougou, Mali, supplied by Photowatt, 1981.

Figure 4. Derrick, Fraenkel & McNelis at PV Pump trial on a farm at the Philippines, 1981.

Anthony Derrick, Peter Fraenkel and I had started work on the project in 1979. Figure. 4 shows us at one of the PV Pump test sites in the Phiippines in 1981. As soon as this project was underway we decided to form our own company, and we had the support of ITDG. The project I believe produced very good scientific results, despite all the logistical difficulties. We pioneered life-cycle costing for the

Figure 5. PV pump at the Desert Development Centre, Sudan City, Egypt. Supplied by ARCO Solar, 1983.

economic analysis. Another country joined, Egypt, thanks to our Project Supervisor, Essam Mitwally, who even persuaded ARCO Solar to donate the system, which we installed at the Desert Development Centre at Sadat City (see Fig. 5). The World Bank was persuaded to undertake a second phase of the work, and by then we had formed IT Power.

7. IT Power

The Solar Pumping Project was the catalyst for IT Power, which we incorporated in 1981. We had GBP 100 of capital and a contract from the World Bank (Solar Pumping Phase 2), so we managed to persuade our local Barclays Bank (which had no idea what the World Bank was) to give us an overdraft for working capital. The second phase should really have come before the first phase. In this we set up a test facility and tested pumps and components under controlled conditions, modeled performance, made optimized designs. This should have been done before the field trials (but we did not design the project, we were just the humble consultants). From this work, Grundfoss was the clear winner. We also visited other countries to assess interest and market potential; Bangladesh, Brazil, Kenya, Mexico, Pakistan, Sri Lanka and Thailand.

We opened an office in a shed on the premises of Architect Julian Keable, a friend and member of the ITDG Solar Panel. We hired more people, so by 1983 we

Figure 6. IT Power team and Headquarters, Mortimer Hill, UK, 1983.

had a team of seven, shown with one trainee from Mali, in Fig. 6 (note the solar roof on our shed). In our first year, 1981 we won a lot of projects such as from the World Health Organisation (WHO) to write the specification for, and test, PV vaccine refrigerators, and from the European Space Agency (ESA) to advise on how to power satellite ground station in Africa. It needs a whole book or more to tell the full story of IT Power, so enough for now. Some projects and experiences are described later.

8. Mali

In my first trip to Mali, I discovered important facts like nothing works, local beer is cheaper than bottled water, government agencies are dysfunctional. When we started the Global Solar Pumping Programme, I was the only member of the team who had even come close to a solar pump (thanks to my trips to West Africa in 1978), I was the only one who had been to Mali and Mali had more solar pumps installed than any of the other participating countries. So I took responsibility for Mali and immediately started to improve my French. My trip to try and initiate the project there was a total disaster (and not just because of my poor French). The hosting agency, the Laboratoire de L'Energie Solaire (LESO) was like a graveyard for equipment lying rusting and abandoned, even some crates of expensive instrumentation, marked "this side up" but upside down and overgrown with biomass. When I asked, I was told Italian experts had visited and subsequently shipped the pyranometers and other instruments. But the Malians were waiting for the Italians to return and unpack the crates. I also discovered that there was no point cleaning up the junk-yard, because the Americans were going to build and equip a new Laboratory. I visited the US Embassy to learn about this $5 million project to see if I could get some help. I found that USAID had appointed the Solar Energy Research Institute (SERI, now called the National Renewable Energy Laboratory, NREL) to undertake a wide ranging project to deploy solar, wind, biomass technologies in the country, as well as build a new LESO. But there was a snag; they needed an engineer to manage the project, and he or she should have relevant practical experience and speak French. They found there was no-one suitable (or willing) at SERI or in the USA, so they had hired a French speaking Canadian, but with no experience in Africa. He had arrived in Bamako, but could not cope with the conditions and left after two weeks.

The plan for the SPP project was to install and test four systems in each country. My counterparts at LESO wanted to install these at four very distant locations, all different directions from Bamako, requiring two or more days travel over non-existent roads. Maintaining my best political correctness, I tried to point out that this was crazy as we wanted to test and monitor performance of these. There were plenty of villages and small farms close to the capital. Of course it turned out that LESO wanted the project to supply four-wheel drives (while they waited for the USAID project to deliver). An even better reason to be far away was the established standard for international projects is to pay a daily cash allowance to the local staff working on projects outside Bamako.

Eventually, USAID did find someone to lead the project, a British Engineer, Terry Hart. Terry had worked for British Aerospace on Concorde, learned French in Toulouse, and was teaching engineering in Algeria. He wrote to SERI asking for publications on solar to help with his teaching programme, and someone smart at SERI realised here was a Franscophone engineer already working on solar in Africa. Hey presto they asked Terry if he would like to move to Mali, invited him to Colorado to discuss, and he quickly relocated to Bamako. He achieved wonders, considering the conditions of Mali, and the primitive state of the technologies. The designers of the project (and many other USAID projects in Africa) were certain that the technology was not a problem, and a key component was to send in teams of anthropologists and socio-bullshitologists to measure the impact of the technologies (even if nothing was working!). After the USAID project Terry joined IT Power.

Mali quickly became the target for more and more PV pump donations. I do not think any were delivered by parachute, but "parachuted-in" is the best description of projects where the donor sends the equipment (maybe with) engineers to install it, gets a photo-call with the Ambassador and President (or whatever diplomats/officials), then buggers off quickly. There is no provision of maintenance and the system fails quickly and is abandoned. I saw so many failed

Figure 7.   Tioribougou, Mali, water collected and transported to another village.

and abandoned PV pumps in Mali that I suggested we look for new funding to set up an operation to go and retrieve the PV modules and anything else useful, and then destroy the remaining evidence, with dynamite. Figure. 7 is a good illustration of the value of water, being taken away from the lucky village to the one next door.

9.   Dominican Republic

In 1981, The Inter-American Development Bank (IADB, or BID) called for proposals for a large technical assistance project to build expertise at the Instituto de Tecnologia Industrial, INDOTEC, and establish solar energy technologies in the country. So Halcrow, well experienced with BID and with a representative in the DR, together with IT Power as solar experts, put in a bid. As with the World Bank Solar Pumping Project, the largest part of the available budget would be consumed by having a Halcrow Resident Engineer based full-time in the country, with the actual solar experts supporting very part-time. We were selected and invited to come to Santo Domingo to negotiate the contract. A Halcrow partner and I went to do this. I had no experience at all in negotiating big contracts, but my Halcrow colleague did. For the first days all the discussion (argument) was about charge-rates, what was included, what was fixed and what was re-imbursable. This was totally boring for me, but my colleague, assisted by the formidable local representative lady, did a great job to get the final price agreed (the Halcrow colleague's name was Ivor Price). I had contributed nothing (but enjoyed the food, music, sun, sand and sea), but then the Dominican Client said "Oh there's one small point, we don't want the Halcrow Resident Engineer, he has no solar experience (although he has been one of the Solar Pumping Resident Engineers), we want McNelis here as the full-time guy for the project duration". I was not available, having interests broader than what was on offer (but see later), so the negotiations had to restart. Eventually we persuaded our client to accept the Halcrow engineer (we also kept secret that as well as his limited solar experience he could not speak Spanish; he started a crash-course as soon as we reported the successful outcome of the contract negotiation). My time inputs was increased, and a condition of the contract was that I would learn Spanish. I never did, and every trip they asked me: "Has aprendido espaniol? "and I answered: "No tuve la puta chance de hacerlo".

Having won the contract, we found that the technical difficulties began. The expected content of the project had been prepared years before by an Israeli consultant. So we were to build a solar pond demonstration, solar-thermodynamic air-conditioning and water pumping, with parabolic dishes, solar stills — no mention of PV. So I pointed out that the topography and climate of the island of Hispaniola did not suggest much potential for solar ponds, and maybe solar technologies had advanced, and what about PV? Initially they would not even discuss changes, and PV was definitely not to be included! So our first project was to build a simple solar still. Meanwhile I and Anthony Derrick eventually convinced them to include PV (and I just ignored all the noise (nonsense) about the solar pond demonstration. We explained how the World Bank Solar Pumping Project had struggled, and failed, to find a solar-thermodynamic water pump which actually

Figure 8. Wrede solar-thermodynamic pump and PV pump on test at INDOTEC, Santo Domingo, Dominican Republic, 1984.

worked. Then Anthony had a stroke of genius. He had been advising a Finnish inventor, Thomas Wrede, to develop a linear trough solar-thermodynamic pump. Finland was not a very suitable place to test the invention, so Anthony convinced Wrede to supply a pump and INDOTEC to be the first to test this, with a PV pump of similar rated capacity right alongside, see, Fig. 8 End result = the PV pumped, the Wrede did not.

An interesting side effect in the UK of this project is Halcrow were somewhat unhappy that they were deemed to have no in-house solar expertise and had to rely on IT Power. So they advertised for a junior engineer with some solar experience. To their great surprise and pleasure, a far more experienced individual was one of the applicants, Bill Gillett, the right hand man to Brian Brinkworth at Cardiff University. Halcrow grabbed Bill, and I commented: "If I had known he was interested to leave academia, I would have made him a generous offer".

Bill joined the team on the DR project, and thus Halcrow made a bigger intellectual contribution. I enjoyed the rest of the project, and, as well as the PV-skepticism, I had to contend with meringue, kinky sex and Christian fundamentalism, all on the same day.

10. China

My first trip to China was a side-trip from Hong Kong to Guangzhou and rural areas in 1983. The stark contrast between the tiny British Colony and the enormous People's Republic was overwhelming. I went to Beijing as solar expert for the EC in 1985 for a workshop. This was eye-opening, but the most memorable thing today is the person I cannot remember. Very few of the Chinese could speak English, so it is hard to recall who was who. The official group photograph was put on my office wall along with many others and that was that. In 1999 IT Power won a contract from UNDESA to help establish the China Renewable Energy Industry Association (CREIA). The (forgotten) gentleman next to me in the photograph, Zhu Junsheng was appointed as CREIA President, but more significantly, today he is my father-in-law.

In 1988, I was asked by Pierre Lequeux of the EC Development Directorate, and who I had been in Mali with in 1978, to be the solar expert for the EC-China Renewable Energy Development Project in Zhejiang Province. I made a lot of trips to Hangzhou, and after great difficulty getting permission, I visited China's biggest PV manufacturer in Ningbo. But I saw no production, all was top secret. It did not occur to me that China today would be the world's largest PV producer. For this project I was scheduled to be in Hangzhou in June 1989, but I also had a World Bank PV Project in Morocco and needed to go there. I was excited to go to China later because of the good news of the peaceful student occupation of Tiananmen Square. I was travelling in remote parts of Morocco and saw no news, but on the flight home from Casablanca I saw a newspaper — the tanks had moved in. When I got back to my office there was a telex (an ancient form of

Figure 9.   PV Shop in Lhasa, Tibet, 1990.

communication) from the EC to say no project travel to China should be under-taken. There was also a telex from the Zheijang Energy Research Institute (ZERI) to say they were expecting my visit and when would I arrive? I responded that I could not come by order of the EC because of the disturbances. I got a quick reply: "What disturbances? Everybody is hardly working". In the late 80s and 90s, I did a lot of travel all over China, for UNDTDC, UNDP, UNIDO and others. I enjoyed the travel, by steam train, I developed a taste for some but not all of the food, and many of the women. A particular PV success in China, and without government or foreign aid is small PV in Tibet, see Fig. 9, one of no less than 11 PV shops in Lhasa. My PV story in China would require a whole book.

11. Robert Hill

I met Bob as a fellow member of UK-ISES at one of the early national conferences. We quickly became good drinking partners, probably more thanks to our work-ing-class origins than solar energy, but the true academic and the bullshit consult-ant had one thing in common, a passion for using PV to change the lives of poor people in developing countries. We also had other passions-in-common, like red wine and loose women, but that is for another book (see Fig. 10). Looking back, we probably together appeared like a double-act, respectable Bob and bad-language me, but together fighting the British establishment, and others, to support PV. In 1988, Bob persuaded the UK Department of Industry to finance a study for a PV Centre in the UK. The government called a big meeting of all interested parties to

Figure 10. Bob Hill and Bernard McNelis in the company of Marisa, Cuernavaca, Mexico (for IEC/TC 82 PV Standards Meeting), 1997.

Figure 11. Developing Countries Workshop Panel at the 10th EPVSEC, Lisbon 1991. From left to right: Rolf Posorski (GTZ), Derek Lovejoy (UNDTDC), Toby Harrison (UK-ODA), Bob Hill, Bernard McNelis, Anthony Bromley (UNIDO), Mike Crosetti (World Bank), Michel Zaffran (WHO).

receive their response in 1989, which was, no big surprise "No". Straight after, Bob, Philip Wolf of Intersolar, Rod Scott of BP Solar and I walked across the street to the nearest pub, and over several pints of (warm) British beer, we founded the British Photovoltaic Associates (PV-UK), with myself as the Chairman-elect, to

Figure 12. Bunker Roy of the Barefoot College, India gives Robert Hill Award lecture at the 25th EPVSEC, Hamburg 2009.

fight the (damn) British establishment. PV in the UK needs a whole chapter, even a book. For the present, Bob Hill combined the qualities of respectability and the highest academic standards, with his PV in Developing Countries passion. For the 10th EPVSEC in Lisbon, Bob, assisted by me, organised a workshop on PV for Developing Countries, see Fig. 11. This attracted a big audience and was a great success.

In memory of Bob and his passion, and instigated by Wolfgang Palz, the Robert Hill Award for the Promotion of PV for Development was created. The first award was made to Anil Cabraal at the 20th EPVSEC in Barcelona in 2005. The 4th and most recent award was made in Hamburg (24th EPVSEC) to Bunker Roy of the Barefoot College in India. He received the award not only for his work bringing PV to villages in India, but also to Africa, interestingly starting with Mali, as well as other countries in Asia. Before the ceremony, Bunker asked if he could talk frankly about the crude realities of life for poor people in the developing world to the assemblage of distinguished scientists and engineers. I answered: "Yes please". From the stage as Bunker delivered his speech (see Fig. 12), I saw emotions and tears in the audience. The audience moved to a standing ovation! Not a very frequent occurrence, and never for something as far removed from PV science and engineering.

12. EPIA

The European PV Industry Association (EPIA) was launched at the 6th EPVSEC in London in 1985. The first Secretary General was John Bonda. He had lived and worked in Africa and his vision was of the electrification of Africa and developing countries with PV. But he also recognised that to make PV a success in Africa it also needed to be a success in Europe. He also had experience of the European aid money being poured into corrupt pockets and worthless projects; some of this diverted to PV would be a good cause. The founding and initial members of EPIA were PV manufacturers, none of which were really viable, let alone profitable. John asked IT Power to join EPIA and to help with promotion and market development. But we could not. One of the clauses of the EPIA Articles of Association called on members to work together to promote European products. A lot of our business was for UN Agencies involving selection of equipment suppliers; we could not be seen or perceived as being biased towards European companies. So EPIA created the Associate Member status, not requiring us to "sign the pledge" and IT Power became EPIA's first Associate Member.

Despite the miserable financial state of EPIA and almost all its members, the Secretary General managed to pull together exquisite dinners at all the associations' events. Invitations issued never met demand and some of the PV community almost begged for tickets. A most memorable occasion was the EPIA dinner organized by John at the 9th EPVSEC, chaired by Gerry Wrixon in Freiburg in 1989. This was in fact a well-lubricated medieval banquet with a whole cow consumed at a local castle in the Black Forest. Moreover, the local host, Director of the Fraunhoffer Solar Institute , Adolph Goetzberger, EPIA's first Chairman, Giovanni Simoni and Wolfgang Palz were elected the "Three Knights of Solar", and suitable attired, they climbed on the banquet table and posed for the EPIA members. See Fig 13.

Figure 13. The Freiburg Knights of Solar (left to right: Adolph Goetzberger, Wolfgang Palz and Giovanni Simoni), 1989.

Figure 14. John Bonda and Bernard McNelis at EPIA Workshop, Istanbul, Turkey 1992.

I went on most of the EPIA promotional missions with John Bonda, often including his wife Janka, and we created very little business but had great fun (see Fig. 14). When John sadly passed away in 1999 the (small, unprofitable) PV industry had no succession plan. Straight from John's funeral in Brussels most of the EPIA Board members, and I, walked to the nearest pub. Everyone was silent even after the first (or more) bottle(s) of wine. I said: "So what do we do now?" Everyone looked at me, and Chairman Tapio Alvesalo said "You!" So I became Interim-Secretary General of EPIA (part-time, as was John), charged with tidying up, holding-the-fort and starting the process to hire a full-time, permanent, Brussels-based, Secretary General. I started a fortnightly trip to Brussels. EPIA did not have a real office, but paid for space and secretarial support from an agency located in a residential block of flats. It was low-cost, cosy but primitive. There were no proper files, just huge piles of documents on the shelves, and the floor, which John knew intimately, but no one else did. EPIA had contracts with the EC, but finding the documents, reports and cost statements was a nightmare. I did this for one year.

Since then the whole PV world has changed. From having members many of which could hardly pay the annual membership fee, and most of the business being small off-grid systems, now the industry is booming (or has boomed), attracted € billions of investments and made € billion profits — but all from grid-connected systems in industrialised countries like Germany, thanks to mandatory feed-in tariffs. This has been good news for the PV industry. But the developing countries have been forgotten. A few EPIA members, led by the EPIA President at the time, Ernesto Macias, launched the Alliance for Rural

Electrification (ARE) in 2007, to try and get the developing countries back on the agenda. But even today the grid-connected PV experience of recent years has "corrupted" the thinking; with a lot of effort going into for example "feed-in tariffs for off-grid in developing countries" — but there is no grid to feed into!!, to even paying Management Consultants ATKerney to make in-depth studies and hence report that there is big PV potential in "Sun Belt" countries, including China and India, because they need more electricity and receive more sunshine than we do in Europe. Startling!

13. World Bank, Washington, Corruption

It's common knowledge that there is corruption in most, if not all, developing countries. I have been asked to pay money to government officials to win contracts across the world; Ethiopia, China, India, Indonesia, Ghana, Philippines, to name but a few countries. In Indonesia I had a government guy following me from bar to bar trying to get me to pay him off just for offering me entry to the bribery framework. Sometimes it was necessary to count your fingers after shaking hands with some characters. I know many countries where there could have been much more PV installed, but most of the money disappeared. But the stories of corruption in the developing countries are probably boring and repetitive, so in the limited space I will just record some close-to-home examples.

I was going to include a section on the World Bank (not because of corruption!), but there is a whole chapter devoted to this. From my first day pre-IT Power with the Global Solar Pumping Project I had been one of the tireless promoters of PV at the Bank. In the 80s Dick Dosik, who led the small renewable energy team, hired several engineers and economists, such as Ernie Terrado and Chas Feinstein, who were real supporters of PV.

We had more World Bank work and so IT Power opened an office in Washington, I had a girl-friend there and so was keen to visit, and we got (small) contracts from USDOE, USAID and others. I developed a good friendship with Bud Annan, in charge of solar at DOE, and he also had a passion for PV in developing countries. DOE contracts came through US National Laboratories like Sandia and NREL, or DOE's support company Meridien. At Meridien I worked for DOE with Anil Cabraal, and for example we made an assessment of what had been achieved with PV pumps in Mali. I convinced Bud that the World Bank would be key to serious and big deployment of PV in developing countries, and one (extremely useful) thing Bud did was to get DOE to pay for Anil to be seconded to the World Bank. Anil never returned to Meridien and has been the champion for PV at the World Bank ever since.

In the early 90s the world started to look seriously at PV. PV projects became a real possibility thanks to the creation of the Global Environment Facility (GEF) in 1993, as an outcome of the Earth Summit in Rio de Janeiro in 1992. The GEF had $2 billion which was available to the World Bank (and other

UN Agencies) to prepare and fund environmental projects such as renewable energy which would otherwise not be economically justified and hence could not be financed through the World Bank normal procedures. The World Bank started to prepare a variety of PV projects in countries across the developing world, with the GEF money "buying down", i.e. subsidising costs, so the projects became "bankable".

Another old friend and "PV promoter" Dana Younger had joined the International Finance Corporation (IFC), part of the World Bank Group. The IFC started to look seriously at a decent-sized project, initially called the "Green Carrot" (based on the "Golden Carrot" programme of the US Environmental Protection Agency, which had successfully incentivised the refrigerator manufacturers to develop products with hugely increased efficiency and replacement of chlorofluorocarbons (CFC's), the ozone destroying refrigerant.) The first idea was a competition for PV companies to win a "prize", but this evolved into the Photovoltaic Market Transformation Initiative (PVMTI). IFC considered all GEF-eligible countries, visited many, and selected India, Kenya and Morocco for the project.

The Green Carrot/PVMTI during its years of preparation attracted a lot of attention. So in 1996 when IFC were ready (at last!) to start to select consultants to manage and undertake the project, there was a lot of interest. I had commented on various papers relating to the concept for IFC, and naturally IT Power was most interested to win the management contract. We teamed with Impax Capital. There was a budget of $30 m from the GEF with $25 m to be invested in the three countries with the hope this would leverage another $75 m, and $5 m for the management. The procedure used by IFC was to invite expressions of interest, make a short-list for the tender, then the winner from this would be awarded a $250 k contract for the "Appraisal Phase", and if this was completed successfully the same consulting firm would be given the $5 m contract to do the project, without further competitive tendering. In 1997, IT Power/Impax were selected, I went with my Impax colleague to Washington to negotiate the contract, and we got started. We then discovered (discreetly) that one of our competitors had offered to do the $250 k Appraisal "for free", in expectation of the $5 m follow-on contract. It turned out the "generous" competitor was the British Energy Technology Support Unit (ETSU). Hmm...very interesting.

ETSU was part of the Atomic Energy Research Establishment (AERE) at Harwell. AERE was a government agency and ETSU was set up in 1973 as a special unit to advise the government on the "Energy Crisis", and subsequently to manage the government's various renewable energy programmes (and to maintain jobs for recycled nuclear researchers). So I was well into ETSU, as any government contracts for IT Power were awarded and managed by ETSU. But under the Margaret Thatcher government the utilities, including the nuclear industry, was to be privatised, including the R&D components, e.g. AERE — Harwell. But none of the investors would touch the nuclear component. So the Government had to become more creative. So, in the case of AERE — Harwell it created AEA Technology plc (AEAT) and awarded very generous consultancy contracts to make AEAT saleable. For ETSU this in practical terms included using government official connections to put ETSU in a good position to win business. For example, paid by the government (Department of Trade & Industry, DTI, nowadays the Department for Energy and Climate Change, DECC). ETSU staff were "seconded' to institutions like the European Commission (EC) in Brussels and the World Bank in Washington. Ostensibly these experts were sent to assist with programming, but in truth were to provide market intelligence and position ETSU to win contracts (in this example from the EC and the World Bank). The government also increased the level of "support" from ETSU to the Department for International Development (DFID).

Now to explain why this long digression, the Mechanism for ETSU to work for IFC on PVMTI "for free", was in essence, the ETSU official seconded to the World Bank had "arranged" for ETSU to make a proposal to the EC to do this. The proposal was evaluated at the EC by another ETSU secondee. But the EC only provided 50% funding. But, no worries, DFID provided the other 50%, thanks to another ETSU secondee at DFID.

We were, very politely, asked by IFC to creatively find something ETSU could contribute to our work. I was shocked but kept reasonably quiet because the official at ETSU who wanted the "work" on PVMTI was also the official overseeing contracts I was doing for the British Government! ETSU were desperate to contribute something, as their EC contract was entitled: "….support to PVMTI…". But there was nothing for them to do, as IFC had appointed IT Power/Impax as consultants through the correct procedures. But there was another project emerging at the World Bank, and this time right inside the World Bank, under the umbrella of the "Solar Initiative", where the ETSU secondee was positioned. Indeed DTI said he had been sent to Washington to strengthen the "Solar Initiative". This was the "Solar Development Corporation" project, with the aim pretty much the same as PVMTI, but with countries not pre-specified and $50 million as the target investment. Our famous ETSU secondee, by now describing himself as a "Banker" had more influence here. So he helped re-negotiate the EC contract to be a more general "support to World Bank" as opposed to the specific PVMTI project. But more important to me, this was another contract IT Power wanted to win. There was the same tendering/short-list procedure, and we got onto the short-list. Unlike IFC/PVMTI there would be an initial contract for the Appraisal (to prepare the SDC Business Plan), then there would be another tender for the management of SDC (and winner of the Appraisal phase would be ineligible). IT Power had the highest scored proposal, but our ETSU "colleague" was adamant that he wanted a Business Plan with the Coopers and Lybrand (Nowadays Price Waterhouse Cooper, PWC) imprimatur on the cover. So there was a forced marriage between IT Power and C&L, and moreover, history repeated itself, we did 90% of the work, but C&L took 90% of the fees! I do not know what ETSU ever did for their EC contract. I tried to get a copy of their report in Brussels, but I was told to request this "from the local OPET" (Organisation for the Promotion of Energy Technology). The "local OPET" was none other than ETSU. I did not ask.

There are many other examples of ETSU using its position as a presumed British Government institution, and not just IT Power were upset. Bob Hill, assisted by me, and including IT Power and other European collaborators made a proposal to the EC for a project to report on "Successful PV Projects in Developing Countries". There was a real need for this, as there were more reported failures than successes. We together approached ETSU to request UK Government co-financing with the EC. ETSU asked for a lot of supporting information, e.g. previous reports, a list of case-studies we could use. We complied, and one case-study I offered was the work we were doing in South Africa (providing unpublished data). Eventually ETSU responded that our proposal did not meet the criteria of the UK Government (DTI). But later we discovered that ETSU had been awarded a contract by DFID (where the Energy Advisor was "seconded" from ETSU) to study and report on "Critical Success Factors for Renewable Energy Projects in Developing Countries": To Bob and I, this was plagiarism of our proposal and work. Bob, the ever gentleman, even used bad language, as I did, to describe the corrupt UK establishment. The ETSU report was never released, but I did manage to get a bootleg copy from DFID — total bullshit. ETSU also got one of their "secondees" to the EC in Brussels further "seconded" by the EC to South Africa. He used his EC credentials (business card with Euro-logo) and IT Power's background work to solicit ideas for projects the EC could fund, and the interesting ones somehow metamorphosed into ETSU proposals.

But nowadays, Google ETSU and you won't find, except for a Japanese restaurant Chain and East Tennessee State University. ETSU changed their name to "Future Energy Solutions" (FES), I think to cover up their reputation. I wrote to New Scientist to compare the renaming of the town Windscale to Selafield, in order to cover up the bad publicity of the Windscale nuclear accident (the UK's biggest) = easy, just rename the place, nothing bad ever happened in Selafield (it did not exist at the time of the Windscale accident!).

14. Other Countries, People, Institutions...

If there were space, I would like to record other experiences, good and bad, trying to promote PV in developing countries. Some other countries could include India (the place you can love and hate at the same time), Indonesia (wonderful but destroyed by corruption), Laos & Sri Lanka (emerging on PV successes), Mongolia (exciting), Yemen (impossible, dangerous), and most African countries (with real needs but acute poverty), South Africa (someways fantastic, other ways a disgrace), Zimbabwe (had PV module assembly in 1982, now a disaster). I could report on EC projects in Africa and the Pacific where clear lessons are not learned from previous projects. Other agencies I have worked with include UNDP, UNIDO, UNESCO, UNEP, FAO where the extremes of quality of people (from super dedicated and knowledgeable, to complete idiots) and their inherent bureaucracy make achieving anything challenging. I have also worked a lot with the International Energy Agency (IEA). Perhaps another book or books are required.

15. Where do we Go from Here?

The challenge to electrify the developing world with PV is as important today as it ever was. The reader will how know have noted that promoting PV in Developing Countries has not been easy, and has achieved little, compared to the potential. The number of people in the world without access to electric light, clean water, primary health care etc. is not decreasing, despite advances in science and technology, and the €85 billion/year aid budget. With global climate change high on the world agenda, renewable energy accepted as part of the solution, and now a (almost) mature PV industry, we should have the ingredients to mobilize. Some people have become very rich from the PV business. The Clean Development Mechanism (CDM) of the Kyoto Protocol has also made millionaires from trading Carbon Emission Reductions (CERs), but not much for "clean development", or provided any incentive for PV electrification. What is needed is not more projects, demonstrations, studies, carbon trading. We need to get politicians activated and commitments made. A challenge for IRENA is to persuade member countries to support the "Power for the World" concept directly. Hence, I want to spend what is left of my career not doing more of what I have done, so now I need training as a political activist.

Bernard McNelis

Prof. Bernard McNelis is Managing Director, and was Co-Founder of IT Power 29 years ago, and is also Professor at Southampton University's Sustainable Energy Research Group. He has extensive experience with renewable energy technologies, particularly photovoltaics, in industry and consulting practice, and has been engaged as expert by most of the United Nations agencies, World Bank, EU and UK Government Departments, and has completed assignments in more than 50 developing countries. . Projects successfully completed have included comparative assessments of renewable energies, including biomass, wind and hydro-power. He has managed projects in solar, wind and micro-hydro for rural electrification, as well as energy efficiency and carbon emissions reduction. Prior to ITPower, he qualified in chemistry and then conducted research in advanced electro-chemistry for new types of batteries and fuel-cells, and worked for the world's first terrestrial solar-photovoltaic company, Solar Power Corporation, USA,

He has served as a Director and Vice President of the International Solar Energy Society (ISES) and received the ISES Achievement through Action Award in 2001. He was a founder of the British Photovoltaic Association (PV-UK) in 1992 serving as Chairman for several terms, and was the 2nd Secretary General of the European Photovoltaic Industry Association (EPIA). He was Chairman of the 17th European Photovoltaic Conference in Munich in 2001 and Vice Chairman of the 3rd World Conference on Photovoltaic Energy Conversion in Osaka in 2003. He received the Robert Hill Award for the promotion of Photovoltaics for Development, bestowed by the world PV community in Dresden in 2006. He is the British Representative on the International Electrotechnical Commission (IEC) Committee on Photovoltaics.

In 1998 he initiated, the International Energy Agency (IEA) PV for Developing Countries Project (IEA-PVPS Task 9), now renamed Renewable Energy Services for Developing Countries. This IEA-PVPS Task has produced Recommended Practice Guides which have been widely used, and has conducted Workshops in several developing countries.

He was co-initiator of the UK's first building-integrated, grid-connected PV installation. Within the private sector, he has advised major multi-nationals (including BP and Shell) on global market entry strategies in solar, wind and hydro, and has been engaged as expert by investment houses and banks in relation to investments in and finance for PV manufacturing.

Prof. McNelis is principally based in the UK, but he is also a Director of ITPower-Australia and ITPower-India, and was for some years US-based as President of ITPower, Inc. In addition to his directorships of ITPower, he is Board Member of the Institute for Sustainable Power (USA), The Global Approval Programme for Photovoltaics,(Geneva), and the Alliance for Rural Electrification, (Brussels). He currently focuses on advisory services, and divides his time mainly between China and the UK.

Chapter 29

A World In Blue

Bernd Melchior

1. From Butterflies to a World in "Blue" — How did this Happen?

As a student the most exciting question was how to get some additional money? During the holidays I worked in several companies and earned some money, but it never could be enough. At this time my father, a dentist, had a hobby to collect butterflies from all around the world. So I thought about what I can do to preserve these fragile collector's items better. I had the idea to encapsulate these butterflies in acrylics and arrange them in a nice picture frame. I started designing and producing these pictures, and hired two salesmen to sell them, whilst I went to the university to study.

Figure 1.

I organized the production on a long table, and I prepared the butterflies, pictures (Fig. 1) at one weekend, they were sold during the week by my salesmen, and next weekend I got the money. With this system, I earned at least 1800.00 DM net per week (which was much more than many people earned at that time). My dream was a Porsche and I bought a 365 Roadster in 1962 (Fig. 2), which was a famous

Figure 2.

car running 200 km/h, when a normal car would go only 120 km/h. During my University time I drove more than 280 000 km all over Europe.

My first professional activity after studying chemistry was to polymerize acrylic monomers to be used for different products. Now my main activity was to embed all kinds of "things" into acryl blocks; from butterflies to fossils — from flowers to minerals — and more, in all sizes, up to blocks of several tons (Fig. 3).

Figure 3.

2. New Treatment for Porous Materials to Conserve Monuments Like the Dome of Cologne

For this purpose, I developed a new polymerizing treatment, where the objects are soaked in liquid acrylics, and the materials are inseparably combined under pressure in an autoclave. At this time, in 1972, I had my company in an old historical building in Remscheid, where there were many plaster structures incorporated in the ceiling, and it was quite normal that some fell down because of the usual decay. During the preparation of 100 molds for my business, one piece of this plaster fell into a mold with the liquid acrylic, and cracked the mold. During the hardening process in the autoclave, the acrylic liquid flowed out of the mold and penetrated into the weak plaster. After the polymerization, the plaster became so hard that it was impossible to destroy it. At this moment I saw on television how the chief of the Cologne Cathedral, Dr. Wolf, demonstrated the decay of the figurines of the Dome, by easily pushing his fingers into the outer layer of a weathered stone, and he was calling on the chemical industry to save the cathedral. After the event with the plaster in my workspace, I knew that I had to penetrate the liquid acrylic into the stone and harden it in the pores, so no water and pollution can get into the porous materials. This method avoids corrosion and destruction, making the materials mechanical stable and weatherproof. This was the birth of my patent for the acrylic treatment of porous materials by acrylic penetration, which is now being recognized by all monument protection offices, and is a successful business today.

I presented my idea to Dr. Wolf and he organized a major program sponsored by the German BMFT, testing 16 different preservation methods on the 7 different sand stones used on the cathedral. My treatment was the only one passing all the tests, especially the salt-test (Fig. 4), without any failure.

Figure 4.

3. Process Steps for a Treatment with the Autoclave (Fig. 5)

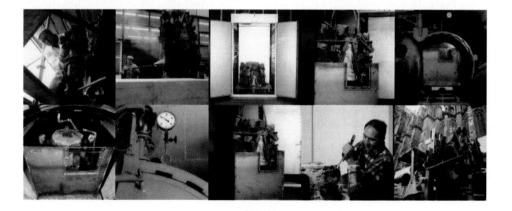

Figure 5.

The first stone from the Dome of Cologne weighed about 1,5 t, and we pressed about 20 kg of acrylic liquid into the pores. What price could I ask for the conservation? I calculated the price to be around 15 000.00 DM for my work, and asked the stone masons, what the price for a new stone in this design would be. They confirmed a price of 54 000.00 DM, but then again that would not have been the historical stone. So, I proposed 46 000.00 DM for the prepared old historical stone and they paid it.

Figure 6.

I founded with my partner Wolf Ibach a new company for the preservation of porous materials in Bamberg (Fig. 6), with the necessary machinery and equipment to process materials on an industrial scale. We received many orders, like the restoration of the castle "Seehof" with its many sculptures and cascades.

4. Translucent Insulation Material

Another acrylic development was a transparent insulation material (Fig. 7) saving energy. The basic idea of a transparent insulation came from Mr. Philipsen from the company August Hohnholz in Hamburg and Mr. Philipsen asked me if I could imagine creating a foamed transparent acrylic sheet? Optimistically I confirmed and started the development. I found a way to polymerize a foaming agent under pressure into the acrylic, so the acrylic sheet looks like a normal glass plate. In a second step I placed the acrylic sheet into an oven heating up the plate close to the melting point, and the agent started expanding, creating small or big bubbles as desired.

Figure 7.

The first client was Prof. Götzberger, who tested the transparent foamed plate on his house, and found out that in winter time, especially when there was sunshine, the wall was heated up, and the heat was captured, so the house was gradually heated. With his Fraunhofer Institute ISE, Prof. Götzberger initiated and constructed several buildings using this new future oriented technology.

In winter and with diffuse irradiation, the transparent insulating material allows more radiant energy from the sun to enter the room than heat energy to leave it. As a result, a "solar panel effect" is obtained and the surface of the siding is used as a contributor of energy; an optimum k equivalent value is the result. This is easy to explain: the transparent insulation foam is only lightly penetrable for the wavelength range of heating rays, but the transparent foam is penetrable by the short wave radiation of the sun, so convection is extensively eliminated.

In 1981 we received the innovation price of the German economy as the first winner against big companies. Additionally I got the award for a four star invention from Prof. Ludwig Bölkow in 1990 (Fig. 8).

Figure 8.

Some sample houses with transparent insulation initiated and constructed by Fraunhofer ISE Freiburg (Fig. 9). My idea was creating a transparent insulation roof single for heating the room below. In combination with electrochromatic glasses a future oriented product.

Figure 9.

5. My Start into Photovoltaic

As a renowned specialist for embedding all materials into acrylic sheets, I received the first contract in 1978-79 from the EC for the encapsulation of solar cells N0. 449-78-1 EDS to embed blue solar cells from AEG Wedel into acrylic sheets. This was my start and the beginning of a life thinking in "blue".

First I couldn´t imagine what to do with these blue acrylic panels, but then I used them to power a small Playmobil train. By exposing the panel to the sun, the train ran faster and faster; turning the panel on the back, the train stopped! A technology to power Playmobil?

At that time, my brother invited me to join a holiday trip to Fuerteventura on the Canary Islands, where he built the bungalow resort Altamarene in Jandia in the southern part of Fuerteventura. This bungalow complex was powered by a diesel generator running 24 hours a day with a constant noise in the background. My basic idea was now born: could we take this new photovoltaic technology to power a complete bungalow, instead of just Playmobils? Can we take the sun as the supplier of energy, and make homes or even resorts completely independent from fossil fuels? Especially in this area with 360 days of sunshine — can this be realistic?

I prepared several acrylic panels and placed them on the flat roof of an Altamarena Bungalow to test the efficiency. But, unfortunately, the dirt and dust was sticking on the surface of the acrylic panels, and they began yellowing under the constant hard radiation (Fig. 10). Was the dream of an autonomous electricity supply dead from the get go?

Figure 10.

At this time, EVA foils were introduced to the market for the encapsulation of solar-cells under a glass-sheet, and I prepared the first PV-modules in this new technology with some additional effects. I embedded a fiberglass net between the substrate and the solar-cells to have the same expansion coefficient for both the cells and the encapsulation material, and a triple foil Tedlar — Aluminum — Tedlar as a substrate, avoiding that any gas or water molecule could penetrate into the embedding layers. These modules have now been successfully working for 25 years outside, and still maintain an efficiency of almost 100%; this is unique in the Solar Industry. These modules still can be seen on a test field at the TÜV Rheinland main building in Cologne.

The next question was on how to fix the panels on the flat roof of the Altamarena resort? We couldn´t drill fixing holes in the roof, and during stormy weather conditions the panels would fly away. So I designed concrete blocks to adapt the solar-panels, which were placed on the flat roof (Fig. 11). This turned out be a brilliant solution, as we could face the panels in a southern

Figure 11.

direction and place the blocks according to the roof dimensions. The first **blue** roof was created.

To install the new photovoltaic block stone system on the roof of Altamarena I had prominent assistance from Dr. Jürgen Schmid, Fraunhofer ISE Freiburg, and Dr. Klaus Ullrich, BMFT (Fig. 12).

Figure 12.

6.   Changing DC Current into AC Current

The problem now was how to change the DC electricity produced by the Solar Modules into the normal 220 Volt AC current. There were not many companies that had developed this technology in 1982. After my own research, I installed the adequate charge controller, a battery station and an inverter for 220 Volts (Fig. 13). It was a great event to disconnect the diesel-generator in the switch board, and to connect the solar power plant on the roof for the sole energy supply of the

Figure 13.

bungalow. This was a world event: The first bungalow worldwide completely connected to the sun with a **blue** roof.

The Governor of The Canary Islands, Mr. Saavreda, and Senator Cabrera visited the installation (Fig. 14) at a time when there was Sirocco, a hot wind from the Sahara with up to 50°C, and in Morro Jable there was no electricity, because a truck broke down the grid. Nevertheless, I could present the governor solar cooled beer and music, both of which were not available in Morro at this time. Mr. Saavreda couldn´t image that the sun was the motor for the electrification, and he curiously climbed on to the roof, following the cables from the switch board. He stated that this is the future technology for the world to generate electricity; decentralized at any place of the world. Nobody can switch off the sun!

I had built a small cigarette-pack sized box, with a solar cell on one side and the other was a small music box. When you opened the box, you could hear tunes, or better "solar music". I gave such a solar music box to the governor, and like a child he left the bungalow moving in the rhythm of the music. I prepared about 20 of these solar music boxes with different melodies and put these boxes into the flower pots during the night. On the next morning, when the sun rose, all these flower pots started to play different melodies (Fig. 14).

Figure 14.

7. Diffuse Light Concentrator

The photovoltaic industry has one future oriented question: how to concentrate diffuse light as no conventional system is able to concentrate diffuse light on a point or a line. I was fascinated from the idea and created acrylic plates with a special fluorescent molecule, absorbing radiation, and rejecting the light in the acrylic sheet. Now the light was in the medium acrylic sheet, and reflected by the inner layer, so that all radiation was directed to the edges of the acrylic sheet (Fig. 15). We achieved a concentration factor of 7-10×. We got the luminescent color from Bayer Leverkusen, and called the plates LISA-CONCENTRATOR. On the edges I affixed GaAs (Gallium Arsenide) — Solar cells and the system was working fine. The basic idea was to use these systems in greenhouses. We prepared large acrylic sheets with LISA, and installed them on a greenhouse. It was running well until it started raining. It was fascinating to watch that all rain drops were illuminated, but no energy was flowing onto the edges and the system failed. The Fraunhofer Institut ISE in Freiburg also performed extensive work with this technology, but in the end in real life it turned out not to be successful.

Figure 15.

8. Tracking and Concentration Systems

In 1982, I measured on the roof of Altamarena that the energy production was increased by 20–25 % when the modules were automatically directed to the sun, following the sun's inclination, and I started to construct the first tracking systems with four sensors around a square pole. It was running well initially, but mechanical problems stopped the application (Fig. 16).

Figure 16.

One other question was how to concentrate direct radiation onto a center to save space on the roof? All concentrating systems are working with Fresnel lenses or mirror systems. In praxis many problems stopped these applications. I was invited to a company in Los Angeles aiming to generate electricity from the sun in the Nevada desert to split hydrogen from seawater to power cars. The big mirror installation worked nicely, but then a sand storm filled up the mirrors, and after the necessary cleaning the efficiency was reduced to 60%, as small scratches on the surface produced diffuse light.

The next system to concentrate sunlight was a Fresnel system with GaAs cells, which we produced ourselves. The acrylic Fresnel-lens was put on top of an acryl case 35 cm down on the ground. In the center we deposited the GaAs cell. But the air in the box expanded during daytime and contracted over night. This "breathing" effect absorbed finest dust in the Fresnel structure, and after half a year of operation, the efficiency dropped down to 50%.

In 1984, I then developed a new concentrating system with a massive acrylic block with a back side mirror and a GaAs cell placed on the surface covered with a glass-plate (Fig. 16-left). This innovative concentrating system can now be put in a tracking system on any surface, and no sand storm or dust can damage the system.

9. The ADS Concept — Autonomous, Decentralized, Sustainable

After the first success with the photovoltaic energy supply, I said that a bungalow with only the electrification by the sun is "half a bungalow", because we also have to add water supply and sewage treatment in a biological way to make it complete, meaning totally Autonomous, Decentralized and Sustainable. I explained this idea to the senator of Fuerteventura; he was excited and we tried to find the right place for an ADS bungalow, which I had designed in 1983.

Figure 17. Centro Piloto **"CASA DEL MORRO"** para la investigacion y application del sistema fotovoltaico.

When exploring the areas on the coast line of Jandia, we came to the bungalow Acebuche, which was being built by a Russian lady as a pension-home, and she had no money to finish the complex. Three Germans bought the bungalow but didn't get the building permission. The senator convinced the community of my ADS concept, and I got the building permission immediately. This happened only because of my basic new idea to combine various innovative technologies to build a complete infrastructure with energy — water supply and sewage treatment. I call the new milestone project "Casa Del Morro or Acebuche.

Figure 18. This ADS project Acebuche in Jandia was a milestone project, which was started in 1984. All renewable energy technologies were integrated in a completely autonomous, decentralized and sustainable bungalow.

10. The Blue Mountain

The new mounting system using corrosion-free concrete, with a channel for hot air ascending like a chimney behind the modules provides a cooling effect, thus increasing the efficiency (Fig. 19-right). This provides an additional plus of kWh´s for a more economical power plant installation and is applicable on all flat and hilly areas. In combination with the self cleaning and dust rejecting Bluenergy FluorSilicone-Encapsulation, for the first time in history, these solar power plants can be realized directly on any seashore (Fig. 19).

Figure 19.

11. ADS Robinson Club on Fuerteventura

In Jandia, TUI had built the first Robinson Club, so I went to the managing directors explaining the ADS concept. First they were very suspicious and their comments were: "This is impossible; what will happen if the technologies fail?" But at last they were convinced, because the figure "Robinson" was the lonely, isolated man on an island, supplying him selves, and based on this philosophy, they accepted my ADS concept, and we started to develop the first holiday resort in my ADS concept (Fig. 20).

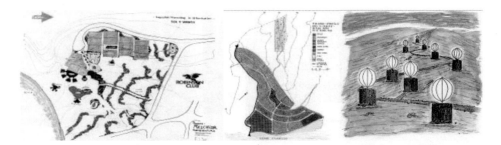

Figure 20.

With the assistance of the senator of Fuerteventura I found an excellent site of one million square meters to build the second Robinson Club as a completely autonomous development. With the combination of roof integrated photovoltaic, the blue-mountain and special wind turbines we would have generated more than enough energy to power the whole complex. On the mountain I intended to install Darius wind generators with golden blades with the effect that the sunshine was reflected and it would look like a disco illumination. This fantastic project failed, because the financing industry was lacking the vision and sense to invest in future oriented technologies; they rather find all kinds of excuses not to get involved. On Fuerteventura, I started many more additional ADS activities on different beach areas to realize my dream, but failed because of the non-performance of financing institutions.

12. First Bungalow in the World Realized in ADS: "Casa Solar", Almunecar, Spain

Figure 21.

In Almunecar, Spain, on an exposed plot, I realized the first bungalow worldwide built according to my complete ADS concept, starting in Spring 1986 (Fig. 21). I developed a new photovoltaic shingle in an overlapping design with the function of a roof. The electricity was stored in a battery station, and we could take out energy as we needed depending on the consumption. We changed the AC current into 380 Volt DC current, so all technical machinery could be supplied with the right voltage (Fig. 22).

Behind the bungalow we found a well in 45 m depth with fresh water from the Sierra Nevada hills. We pumped the water approx. 30 m to the tank in front of

Figure 22.

the Casa Solar. It was a moment of success when the water started flowing into the tank, after having installed all pumping systems and the water pipes to the house (Fig. 23). What will happen, when we connect the pump to the solar grid of the house and how long can we pump water? We heard the water flowing through the pipes into the tank, first the water was dirty but after 10 minutes we got crystal clear fresh potable water with the highest quality, and to our surprise the water flow didn´t stop after two days of steady pumping; the water flow was enough to supply a whole settlement.

Figure 23.

It took a lot of effort to find a biological waste water recycling system using low energy. But at last, we were successful, and all waste water from the bungalow, from the kitchen, the toilettes, the shower cabins, and so on, flowed into a preinstalled biological recycling tank, and we pumped oxygen into the waste water creating bacteria to clean the waste water. We stored the clean water in a

Figure 24.

large tank behind the house. We connected this tank with a small pumping system to feed an underground irrigation system for the garden behind the house (Fig. 24). During the hot summer time, our garden was green and the grass in the neighborhood was yellow and dry. We created an oasis in the desert with the connection to the sun.

We achieved the realization of an autonomous bungalow, connected only to the sun, with the three main infrastructure functions: energy and water supply, waste water cleaning, and irrigation with the recycled water.

Figure 25.

The inauguration of Casa Solar was on the 28th of September 1987, and Dr. Ludwig Bölkow participated in this milestone event. Dr. Bölkow was my mentor and good friend, and we both shared our vision for a "blue" future. I had equipped a standard golf caddy with solar panels to be powered by the sun (Fig. 25). In this caddy we drove from Casa Solar to the beach of Almunecar to celebrate the opening of Casa Solar with friends. During the drive, we got stuck in the sand. When we had left it to walk to the beach, Dr. Bölkow looked back, and recognized that the caddy was on fire. We ran back and extinguished the flames with the champagne, which we had with us for our celebration. A stone had blocked the accelerator, and the motor was on full current, so the caddy went up in flames.

13. Solar Powered Container — 3000 km Trip to 7th EU Photovoltaic Solar Conference and Exhibition in Seville, Spain, October 1986

In October 1986 we organized a 3000 km trip from Wermelskirchen Germany to Seville, Spain, to present my latest development, the solar powered container. This trip was sponsored by Dr Bölkow, Wacker Chemie and Ford in Cologne. I prepared one regular shipping container with all necessary medical equipment needed for general medicine, powered completely by my Solar Modules. The other container was equipped with a universal x-ray station, and included

Figure 26.

air-conditioning and a disinfection system for germ-free air. The electricity from the solar modules was stored in a battery station, so the containers also worked throughout the night and on days without sun.

The first stop on the trip was in Freiburg to attend an exhibition organized by Rolf Disch, with whom I constructed the first solar powered glass pavilion. We picked up the first solar powered bicycles and started the trip to Seville (Fig. 26).

The next stops were in Almunecar in front of the Casa Solar, in front of the University of Barcelona, and in Malaga at the company Isofoton (Fig. 27). The mayor of Barcelona took a trip through Barcelona with the new solar powered bicycles, followed by many journalists. Unfortunately, the mayor was not a good bike rider, and the wheels of the bike got caught in between the tram rails, and he got stuck. With a hard pull we managed to pull the bicycle out, but now the axles were crooked. Nevertheless the mayor continued to "limp" with the bicycle through the streets.

Figure 27.

After five days we arrived in Seville at the European Photovoltaic Solar Conference and Exhibition, and placed our solar powered x-ray container and the medicine container aside the conference hall (Fig. 28). This was a milestone event: the world's first fully functional containers completely independent, autonomous, decentralized and sustainable — a much needed product for the world. All necessary equipment is pre-installed in the containers, and with the solar energy, you can start working immediately, one of the main advantages of the containers, which are absolutely independent from conventional energy supply. This is especially important in areas where there was a catastrophe, like an earthquake.

Figure 28.

14. A Solar Powered Orthopedic Workshop Container for a Hospital in Tanzania

In 1985, my partner, Peter Schein, and I received the order to construct a solar powered 20′ container with a fully equipped wood workshop to build prostheses for a leprosy hospital in Shirati, located at the foot of the Kilimanjaro in Tanzania, Africa (Fig. 29). The energy requirement was calculated to be 3 kWh / day, which we could achieve with the 1 kWp photovoltaic installation. After six weeks of transportation, the container arrived at the Tatkot Hospital for testing. The technicians considered it as the eight's world wonder, as they could produce prostheses with only energy from the sun. This was a milestone for an autonomous decentralized and sustainable application in the third world. The solar-electric container had been designed as a mobile service unit, but it could also be also used as a stationary workshop without the chassis. The unit could also be delivered as a 40-foot container with extended energy supply.

Figure 29.

We successfully tested the solar orthopedic container for three months in front of the Tatkot hospital. The Minister of Health of Tanzania, Prof. Sharungi, visited this mobile orthopedic workshop, and was so excited that he wanted to order 10, 40′ containers for other hospitals without any orthopedic applications (Fig. 30). After we successfully completed the testing of the container, we started to a dangerous 800 km trip through the Serengeti to bring it to its final destination. A Mercedes Unimog pulled the container in four days to the hospital in Shirati, located on the Victoria sea, which had no facility to build prostheses for orthopedic care (Fig. 31).

Figure 30.

It was a dangerous trip, which included attacks by lions, but after many hassles, we arrived at our destination, the leprosy hospital run by a Baptist organization.

Figure 31.

We placed the container between two houses, set up the solar, and immediately a worker could start building the first prostheses out of a piece of wood. It was for these people like a miracle. The container worked for six months, but then the people from hospital send a message that the container didn´t work. What had happened? After a little bit of investigation, I found out that they had forgotten to add distilled water to the batteries, and they dried out. We replaced the batteries and added an automatic system to replenish the distilled water. The container worked for over six years without any failure, until rebels from Uganda destroyed it with a bomb.

15. Integration of Photovoltaic into Roofs "Sunflate"

I constructed the first house in the world covered with photovoltaic roof tiles in the shape of normal, traditional roof tiles in 1985 — The Junction Box to The Sun (Fig. 32).

The electricity production from sunlight was achieved with 432 polycrystalline PV-tiles "Sunflat" and 432 mono-crystalline PV tiles "Sunflate" from Telefunken Electronic, Heilbronn, with a capacity of about 8 kWp, producing approx. 7800 kWh per year. The project was sponsored by the EU Commission under contract No. ESC-R-096-D.

Figure 32.

The PV Sunflate tile is a combination of a polymer tile with the regular size of the concrete tile "Tegalit" from "Braas", and a special adapted PV module. It allows an optimum utilization of the roof surface to generate electricity from sunlight. An increase of the electrical power can be achieved step by step through the exchange of replacement of conventional concrete tiles with PV Sunflate tiles.

The installation of conventional solar modules damages the roof, resulting in leaks, danger of wind and corrosion damages, and also disturbs the architecture of the roof.

Figure 33.

The new photovoltaic on the house "Remscheid" demonstrated, how an innovative energy supply can be harmoniously combined with a representative architectural design. Solar energy generation and conventional roof architecture can now be harmonically combined.

The modules can be produced with other types of solar cells, which will influence the module´s performance as well as its appearance (colors black or blue). The connecting wires on the front side match the color of the cells. On the back side of each module there is a bypass diode; in case one module fails, the others still produce power.

I founded the company Sonnergie Inc. on July 15th 1991, and the organized the incorporation celebration in the house "Remscheid" with prominent participants like Mr. Niels Peter Dr. Ludwig Bölkow, Erzherzog Josef Arpad von Habsburg-Lothringen, Sieger from Lisbon and Dr. Hermann Scheer.

Figure 34.

16. SUNCLAY + SUNERGY — A Two-Component Photovoltaic System for the Harmonic, Aesthetic and Flexible Integration into the Architecture of Roof

As standard Photovoltaic Modules are not always a pleasant sight when placed on a roof, I developed PV modules, which are clipped into roof-tiles for a harmonic, aesthetic and flexible integration of photovoltaic into the architecture. I went to seven different manufacturers of roof tiles and presented my solution. Finally, I found the company Gebr. Laumans in Brüggen. Laumans was excited to realize a clay-tile with integrated photovoltaic, which I had developed in 1994; the combination of a normal clay tile with a special photovoltaic module to be clipped into the tile once the roof was installed.

Figure 35.

The "SUNCLAY" is a normal, slightly modified clay-tile with indentations on the upper side to accept the PV-modules, providing the support (Fig. 35), and the usually expensive support construction for conventional modules is not necessary. The pre-wired "SUNERGY" PV-modules, which are manufactured in my innovative, patented, mono-fluorpolymer encapsulation technology, are supplied in stacked packets for easy installation. Starting at the top of the roof, the modules are laid out according to the layout-plan, and then clipped into the tiles. The top and the bottom of the module-chain are quickly connected to the cable strings provided via special connectors, which I designed. The system is easily expandable in case there is further supply of power needed for the house. SunClay + SunEnergy — roof elements of the future.

Figure 36.

My new technology was presented to the public during Gebr. Laumans' 100-year anniversary celebration on October 25, 1995, and many celebrities out of the political arena were invited. Minister Töpfer gave an inspiring speech, Dr. Herrmann Scheer congratulates Mr. Laumans, the owner of the tile manufacturer Gebr. Laumans and Mr. Laumans thanked Dr. Ludwig Bölkow for his participation in this future oriented celebration in Bracht-Brüggen (Fig. 36).

The first reference roof with the new photovoltaic tile system was built in Bracht-Brüggen, followed by the "Märkische Haus" in Potsdam. The PV-Modules can be mounted quickly and cost effective, are nearly maintenance-free, and can be expanded without problems at any time.

Figure 37.

First installation in Zadar, Croatia, on a traditional roof with the new blue mini modules, allowing for excellent ventilation behind the modules for cooling (Fig. 38).

Figure 38.

17. Next Generation Photovoltaic

With the experiences on Altamarena, Fuerteventura, in 1983 in mind, I developed the next generation photovoltaic with many new features for more safe and main-tenance-free applications.

In the evening and during the night, the salty humidity was condensing on the surface of the glass of the modules. The next day, the sun burned, and small white structures were building on the glass, etching it (Fig. 39). This rough surface attracted dirt and dust, and as this happened day after day, a thick layer built on the surface, thus nearly completely blocking the light penetration into the solar cell. To be able to maintain the efficiency, the roof had to be cleaned almost every day, which was very time consuming and inefficient. The necessary consequence

Figure 39.

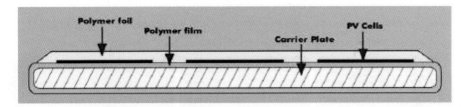

Figure 40.

would have been to install solar at least 2 km away from the seaside. But the dirt also reduces the efficiency of photovoltaic in our area, so I needed a solution, and invented a new way to encapsulate photovoltaic.

I created a new encapsulation method, and instead of glass used fluorpolymer, which is a material with no chemical reaction and many additional features (Fig. 40), enhancing the efficiency of the modules: no dulling of the surface caused by salt — **dirt-repelling, self cleaning** — highest transparency of 96% — highest resistance to atmospheric corrosion — best heat rejection — gain in efficiency — no yellowing — no framing required — no biological contamination with fungi and spores — recyclable — great possibilities of alignment in architecture — self extinguishing — pleasant, diffuse, soft — no water absorption — highest electrical insulation — thermoplastic — doesn't curl. In addition to all these excellent features, fluorpolymer allows the imprinting of a dendrite structure with three main features: absobing light from all angles, no reflection, and a lotus flower effect for additional dirt rejection, with higher yield and cost-reduction.

I tested this new encapsulation method with success on Fuerteventura, and improved the features of the modules with the new combination of films; there

Conventional Modules   Figure 41.   Bluenergy

Bluenergy Modules: lamination around the edges — no framing necessary — edge protection — no dirt on the frame possible (Fig. 41).

was nearly no water absorption and the modules were electrically resistant up to 8000 Volts.

18. New Generation of Solar Clay Tiles

Following the principles that the tile is the support construction for the photovoltaic panels, I developed new solar tiles with an optimized optical appearance by using back contact cells, which do not have the bus bars on the surface, providing more surface area and a better aesthetic appearance. It is also now possible to deliver the modules in any colour with very low losses in efficiency (Fig. 42).

Figure 42.

There are many new markets for this optimized system, and especially the Department for Monument Protection and the Ministry for the Environment accepted and promoted my modules, which are according to their regulations in regards to the architectural integration (Fig. 43). If you calculate only 6 sqm per person with adequate available photovoltaic roof area, the potential in Europe alone is about 300 000 MWp decentralized and sustainable — the equivalent of 300 nuclear power plants.

Figure 43.

With the high feed-in tariffs in Europe, I created a new marketing concept: the roof pays for itself or will be paid by the sun and once amortized, the income will provide for a secured additional pension plan.

Stainless steel as the substrate for the new photovoltaic elements I created a new pv-modules. The house can be covered with normal shingles in the traditional way and then the new photovoltaic elements can fit in. This system has the advantage that the big al-shingles have a weight of only 600 g. Exporting building integrated

PV system for the Caribbean or the Seychelles is attractive as those countries have no strong roof construction for the heavy weight concrete or clay tiles (Fig. 44).

Figure 44.

19. The Combination of Solar and Wind BSWT

The new fluorpolymer encapsulation made it possible to use solar modules in applications which cannot be built with conventional Solar Modules. I had the idea to combine photovoltaic and wind, and I created the Solar Wind Turbine, a vertical axis double-helix turbine design (Fig. 45), where both sides of the wings are covered with solar cells, generating electricity using both solar and wind energy for commercial, home and government use. The turbine is a first-of-its-kind hybrid, capturing wind energy through the rotation of its vertical vanes and collecting solar energy from the photovoltaic (PV) encapsulated directly onto the vanes.

Figure 45.

The turbine has several advantages over conventional wind propeller turbines and traditional solar panels. For the same output rating, the BSW turbine surpasses solar PV and wind systems because it:

- Generates power in the day when solar is plentiful, and also at night when winds are most prevalent,
- Creates no noise; traditional small-wind propeller turbines emit high-pitched sound when generating power,

- Works safely at low (4 mph) and high wind speeds (up to 100 mph and more); wind propeller turbines operate in a limited wind speed range (12–45 mph),
- Requires a small footprint: the 5 kW model has an 8-foot diameter footprint (60 square feet); a traditional 5 kW solar panel installation will have a footprint of roughly 350 square feet,
- Is aesthetically beautiful as a sculptural element with a trademark shape and color (a piece of art where the art works!),
- Has a dendrite structure on the surface of the modules, absorbing light from all angles optimizing the power generation from the sun.

Figure 46.

I built prototypes, which were placed on our office building, and tested them for over three years (Fig. 46). During a strong storm in 2008, which uprooted many trees and created damages in the billions, the Solar-Wind turbine was fully functional, because the strong winds actually stabilize the mast; it produced energy, when conventional wind-propellers had to be shut down. This proved that my system is much more efficient than the conventional wind turbines; it produces energy when conventional systems was not able to.

In 2006, a new Bluenergy company was created in the U.S.: Bluenergy Solarwind, Inc., based in Santa Fe, New Mexico, to commercialize the Solarwind™ product line, starting with the 5 kW model for commercial installations and then the 2 kW model for residential installations.

The SolarWind Turbine is ideal for many applications in the near future, like a SolarWind powered cell tower, the patented Bluenergy Seawater desalination ship, and the desalination and medical containers (Fig. 47).

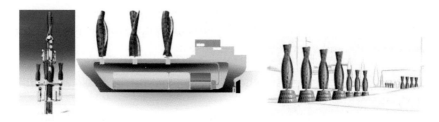

Figure 47.

I hope to realize many of my visions in near future with new creative financial systems and motivated partners. The acceptance of solar energies is increasing and necessary for the surviving of mankind. The time is now. I will do my best to assist with my vision of a world in "BLUE".

Bernd Melchior

Bernd Melchior, married to Susann, born Hädrich — 2 daughters — (Profession). Chemist — developer of new technologies and concepts — founder of several companies — inventor, created about 142 patents and patents applied — founding member of 5 organizations — creator of the ADS-biogeophotarc concept — recipient of 12 awards — realized about 17 R & D contracts — founder and leader of the Bluenergy structure worldwide contact: melchior.de@bluenergy-ag.net;

www.bluenergy-ag.net — www.bluenergyusa.com — www.usinasolarusx.com.br

Chapter 30

The History of Renewable Energies in the Canary Islands, Especially in Tenerife

Ricardo Melchior

President of Cabildo de Tenerife

Manuel Cendagorta

Managing Director of ITER

Ricardo Melchior carried out his studies of industrial engineering in the University of Navarra, Spain, and later moved to Germany for a period of three years motivated by a scholarship awarded by the German government to study Financial Engineering in the Aachen University (RWTH), where he also started his PhD. in industrial engineering. During this period, Mr. Melchior came into contact with researchers and scientists who studied the use of renewable energy as an alternative to fossil fuels as a result of the fourth Arab-Israeli war of 1973, also known as the Yom Kippur War, which caused an energy crisis that quadrupled oil prices.

The same conflict steered Manuel Cendagorta´s concerns towards the field of renewable energy while studying in Barcelona. During this period, Mr. Cendagorta had his first technical experience with this discipline, after finishing his studies in telecommunications engineering in the Polytechnic University of Barcelona in 1980, which culminated with a final thesis project that was actually an electric vehicle design powered by solar energy.

Once he finished his university education, Mr. Cendagorta moved to Germany for a period of three years, during which his awareness of alternative energy intensifies due to the mature and important environmental movement in that country. During these years, he built a 1 kW wind turbine model and conducted studies in aerodynamics and mechanics of wind turbine blades in the city of Bremen. At this point the only wind installations in Spain were designed for mechanical pumping, and a sixth of them (about 550) were located in the Canary Islands.

Power for the World by W. Palz
Copyright © 2011 by Pan Stanford Publishing Pte Ltd
www.panstanford.com
978-981-4303-37-8

Figure 1. Melchior at the centre, Cendagorta on the right.

Coinciding with the early 80s, Spain began its wind expansion through the activities of the Centre for Energy Studies [which later became the Governmental Institute for the Energy Savings and Diversification (IDAE)], within the now dissolved Ministry of Industry and Energy. Another agency of the ministry, the Centre for the Development of Industrial Technology (CDTI), established in 1982 a special credit for the development of wind machines for PIMES, crucial for the development of the Spanish wind industry within which the turbine ECOTEC-NIA 12/20 of 25 kW was developed (the first conducted by this cooperative, now integrated into the Alstom group), an 8-kW wind turbine of the GEDEON company, (Natural Origin Energy Study Group), the wind turbine CEFIR 12 of 12 kW of the company TECNER Engineering PLC and a 35 kW wind turbine of the STS Company (Solar Thermal Systems S. Coop Ltd.).

In 1982, the Canarian Institute of Astrophysics hired Mr. Cendagorta to work in Tenerife in the design of a 10 kW wind turbine. This project failed in its fulfilment, although the preliminary studies of that wind turbine design were the stepping stones for the construction of a 5 kW wind turbine, which Mr. Cendagorta finally finished autonomously within the DAURA Company, which was later sold to the utility of the Canary Islands (Unelco). It was also during this period when Mr. Cendagorta started to project for the Krupps Company, and developed the first major desalination project powered by wind.

These two projects motivate Ricardo Melchior, Head of the New Energies Department of Unelco, to hire Mr. Cendagorta in 1985 to join the department. Initially his work was focussed in the design of a 1 MW wind turbine for water desalination in Fuerteventura. However, the two most important projects that developed during his participation in the company were the autonomous supply of the island of El Hierro using wind energy and hydraulic storage, project that was started handed over several times until it started running 23 years later; and the 20/20 Project, which sets the foundations for the installation of 20 MW of wind power for the island of Tenerife and another 20 MW for the island of Gran Canaria. The wind, solar and geothermal potential resource maps of the Canary Islands were also developed during this period. Manuel Cendagorta and Ricardo Melchior attended a key meeting for the national wind research: the 'First National Conference on Renewable Energy: Solar and Wind Energy, organized by the Governmental Institute for the Energy Savings and Diversification (IDAE) in Tenerife, where the research lines of the Renewable Energy Plan (PER) were defined.

In 1986, the first wind park is installed in Tenerife (in the locality of Granadilla), constituted by four GESA-MADE wind turbines of 25 kW, four ECOTECNIA wind turbines of 30 kW and two ACSA wind turbines of 55 kW. This project was financed by the Governmental Institute for the Energy Savings and Diversification (IDAE), the Canary Islands Utility (UNELCO), and the Government of the Canary Islands.

Halfway through 1987, Mr Melchior joined the island government as Vice President of Cabildo de Tenerife, the island's administrative authority. He had combined this position with Island Councillor of Economic Development and Agriculture. It was then he was able to promote the creation of Instituto Tecnológico y de Energías Renovables (Institute of Technology and Renewable Energy — ITER), aimed at starting a new research field in the islands in order to reduce the dependence on exterior energy supply and to allow a cleaner and sustainable development. With that in mind, Mr. Cendagorta was hired by Cabildo on 1990 for assessment on telecommunications and renewable energy.

Figure 2. Visit to ITER of EU Energy Commissioner, the late Loyola di Palacio.

Figure 3.   PV plant Solten 20 MW in Tenerife produced in 09 pro Watt 1.859 kWh a world record.

Because the installation of a power station is planned in the same settlement where the existing wind park was installed, a few months before Mr. Cendagorta joined Cabildo, it was decided to dismantle the wind park and erect a new one two kilometres away from the old one. This removal was used to upgrade the installation to an Experimental Wind Platform, where the old turbines were replaced with new ones, each of them from different manufacturers and technologies, in order to find out which one was best suited for that specific area. The Wind Park was inaugurated on the 23rd of October 1990. The Platform was upgraded with new wind turbines in several occasions during the following years.

The creation of ITER, funded and initially financed by Cabildo de Tenerife, was consolidated on April 1991. The institute is a non-profit organisation that has been in existence for two decades, promoting technological research and development in renewable energy implementation and integration of energy systems to provide efficient management of natural resources, so as to reduce greenhouse-gas and polluting emissions, while at the same time carrying out dissemination of sustainable energy strategies and policies. As ITER was expected to be self-financed, it is given the management of the Experimental Wind Platform, so the income derived from selling the energy generated would allow the institute to participate in research and development projects. Simultaneously, Manuel Cendagorta was appointed as managing director of ITER.

After 12 years being vice president, Mr. Ricardo Melchior was elected as President of Cabildo de Tenerife, and he's been holding that public position since then. During four years, starting in 2004, he was also appointed senator in the Spanish Government.

The institute went into operation with only three people as staff and had two offices borrowed in the headquarters of Cabildo de Tenerife. Its growth in these 20 years has been astonishing. Presently, the institute has a multi-disciplinary team of more than 232 professionals, comprising of three R&D areas such as renewable energies, engineering and environmental sciences, which are complemented by a Dissemination Department. It has its own headquarters in Granadilla, owning a terrain of over 400 000 m², where a large showcase of RE installations and facilities were installed.

With Ricardo Melchior as President of ITER and Manuel Cendagorta as its Managing Director, the institute has become a reference on RE installations. ITER is a member of several major European renewable energy associations and networks, and has been appointed by UNESCO as a "Centre of Excellence" for dissemination and training in Renewable Energy.

Since its creation, the institute has two main lines of action: the generation of electricity with renewable energy (particularly photovoltaics and wind energy) and achieving research and development projects in the areas of renewable energy, environment and engineering.

ITER has installed and is managing three wind parks (13 MW) which will be increased to six with a total power of approximately 63 MW. It has also promoted several PV plants, with almost 40 MW installed.

The institute launched its foray into Europe by being one of the founding members of the European Renewable Energy Research Centres Agency (EUREC Agency) in the early 90s. From then on, it actively participated in several RE projects co-funded by the European Union.

Figure 4.   General view of ITER on Tenerife.

As a result of Mr. Melchior's interest on adapted technologies and the constant work being carried out by ITER in promoting sustainable development policies in rural areas, the European Commission has appointed ITER as technical advisor for development of the Euro-Solar Program, a project led by D. G. Europe AID, whose main objective is to furnish isolated rural communities in various parts of Latin America with 600 renewable energy installations. ITER is also currently carrying out projects to supply electricity to rural settlements in North African countries, and aims to achieve a 50% electrification rate in the country of Senegal by the year 2012. At the same time it is developing new methods for the prediction of natural disasters though seismic-volcanic surveillance in areas where high risks are known to exist, in countries such as El Salvador, Rwanda, Italy and Japan, along with various other projects within a framework aimed at fighting poverty through the promotion activities that can generate sustainable employment and wealth.

The development of ITER is based on the transfer of knowledge or "know-how" through the projects and research work it carries out, and within this context the project "25 Bioclimatic Dwellings for the Island of Tenerife" was born. The project began with an international competition, endorsed by the International Union of Architects, in which 397 teams of architects participated with the goal of creating a "laboratory" of dwellings using bioclimatic architectural criteria, energetically self-sufficient, and capable of optimizing their adaptation to the environment. The 25 projects selected were built on ITER terrains. The results obtained will be used in the composition of designs for warmer climates, which will help to reproduce the techniques developed and provide sustainable construction initiatives in the future, using proven methods that are easy to apply, and are suitable for use in other areas with similar climates.

During the last 25 years, Mr. Melchior has been working on activities related to the research, development and demonstration in the renewable energy scope on a European level. He's been actively participating in international seminars, congresses and conferences. He's also a member of the Scientific Council of the Institute for Solar Energy Supply Technology (ISET), and founding member of the European Solar Council, Communities of Europe for Renewable Energies (CERE).

Among the several acknowledgements of his outstanding career, he was given the degree of Doctor Honoris Causa in Sciences (National University of Ireland) in a solemn ceremony that took place in Cork's University College. He has also been awarded with the Chevalier dans l'Ordre National du Mérite by the Presidency of the French Government in 2003.

Since 2004, he has attended the "Clinton Global Initiative" Forum as a guest of the former USA President Mr. Bill Clinton. He committed himself to work for the renewable energy development and its introduction and growth in the African continent.

Chapter 31

Why was Switzerland Front-Runner for PV in the 90s but Lost the Leadership After 2000?

Thomas Nordmann

TNC Consulting AG CH-8706 Erlenbach, Switzerland

On September 23rd 1990, the popular vote in Switzerland approved a national initiative of a ten-year moratorium for new nuclear power plants with a Yes-Fraction of 54.5% of the Swiss voters. This political signal was interpreted by the authorities to start a national programme with the name "Energie 2000". This programme included political targets especially for the new renewable energies and energy efficiency.

Within the framework of this national programme, Energie 2000 PV had a prominent place with the background of a number of successful pilot and demonstration projects. A target was set of a total achievable PV installation capacity of 50 MW by the end of the year 2000. This was a compromise between the representatives of the utilities (22 MW) and the Swiss national solar industry organisation, SOFAS, which was in favour of 120 MW total PV capacity.

To develop this market and to fulfill the goal, Thomas Nordmann was commissioned by the national office of energy as the leader of the photovoltaic market development programme 1992.[1] In the same year, Switzerland was hosting the 11th European PV Conference in Montreau. All the programme committee members got the slide set "A PV Tour through Switzerland", showing 24 attractive PV installations all over Switzerland including 15 pilot and demonstration installations all grid tied.[2] By 1991, Switzerland had an installed PV capacity of 2 MW total or a capacity of 0.34 W/capita equivalent 5.4 cm × 5.4 cm. The target of 50 MW is equivalent to 8.5 W/capita or an area of 27 cm × 27 cm. The approach of giving the installed capacity per capita was a successful and popular way to demonstrate the achievability of the still low PV targets of the time.[3]

Power for the World by W. Palz
Copyright © 2011 by Pan Stanford Publishing Pte Ltd
www.panstanford.com
978-981-4303-37-8

Because of the lack of financing through the parliament and the attitude of most of the Swiss utilities, the target was only partly achieved.[4] The utilities have seen Energie 2000 as a voluntary programme with voluntary targets. From the promised annual budget of 170 Mio. CHF by the government the budget was cut first to 50 Mio. CHF per year and later to 27 Mio. CHF per year.

Under this financial regime, PV, at that time, the most expensive form of new renewable electricity, was suffering. Nevertheless Switzerland was developing at that time a number of innovative ideas and concepts including the first 1:1 application of a feed-in-tariff. In 1991, the city of Burgdorf in the canton of Bern had adapted a law, which promised a financial compensation for each kWh PV fed into the grid of 1 CHF and this for a period of 12 years. This compensation was given for the total production, not only for the surplus. Burgdorf had already reached at the end of 1991 an installed capacity of 16.7 W/capita in comparison to Switzerland's 1.62 W/capita and Germany's 0.66 W/capita. Sacramento (SMUD), as a US pioneer, was at 5.45 W/capita, US as a nation was at 0.38 W/capita.[5]

As president of the Swiss solar industry association, Thomas Nordmann was trying to convince in 1993 that the Swiss utility organisation, VSE, adopted as a voluntary measure, the concept of the national feed-in-tariff to meet the target of the 50 MW. The prominent newspaper, Neue Zürcher Zeitung, published on 13th September 1993 an article, which gives all the technical and economical details of a successful feed-in-tariff concept, which was pioneered in Burgdorf nine years earlier before the German Bundestag adapted a national feed-in-tariff.[6]

Some prominent Swiss PV pilot and demonstration projects 1989–1993 included in Reference 4.

World's first PV-Soundbarrier — 100 kWp, Domat/Ems (near Chur, Switzerland), 1989.

Alpine BIPV-Dairy — 13.3 kWp, Alp Findels Pfäfers, 1991.

PV-Sunshelter — 18 kWp, Kirchberg, 1992.

PV by Swiss Parliament — 24 kWp, Bern, 1992.

PV-Powered Church — 19 kWp, Steckborn (Canton of Turgau, TG), 1993.

After the establishment of an attractive feed-in-tariff in Germany in 2000, Switzerland could not follow the successful expansion of PV in the German market but many ideas, especially of buildings integrated with photovoltaic pioneered in Switzerland, have been adapted in Germany and other markets in Europe.

References

[1]   Paper about Switzerland's 50 MWp photovoltaic-program presented at the 11th European PV Conference Montreux, Switzerland 1992.
[2]   Annual report of Swiss PV promotion program 1994.
[3]   PV Target of the Swiss Energy 2000 program 50MW -> 8.5 W/Capita.
[4]   A PV-Tour through Switzerland — a 24 slide set of pilot and demonstration PV projects in Switzerland, published by TNC Consulting AG 1992 in Switzerland.
[5]   Comparison W/Capita between different countries 1997/98.
[6]   Article in the Neue Zürcher Zeitung (NZZ) by Thomas Nordmann about a national rate-based PV support concept, September 15th 1993, seven years later introduced in Germany as EEG law.

Thomas Nordmann

Thomas Nordmann has been working for solar energy for more than 31 years now:

- In 1974 he was working in the development, research and conversion of thermals and later photovoltaic applications in the government research (Paul Scherrer Institute) and for the Swiss energy industry.

- In 1985 Nordmann founded his own enterprise named «TNC Consulting AG»

- He developed and built the world's first 100 kW PV noise barrier installation along a Swiss motorway (1989) and has since planned, engineered and installed a total PV-capacity of approx. 5 MWp in Switzerland, Germany, the Netherlands, and Luxemburg.

- In 1992, the Swiss Federal Office of Energy commissioned him as the program coordinator of the Swiss Federal Promotion Program for photovoltaic.

- He was president of the Swiss Solar Industry Association (SOFAS) from 1992–1999 and Vice President of the Swiss national organisation «Swissolar» till March 2002.

- For his contribution to the development of solar energy he received three times the Swiss solar price (1994, 1998 and 1999) and the European solar price (1997). At the 3rd PV world conference Osaka in May 2003, he received the paper award for the best presentation/paper in section 7 national strategies and policy of PV. He was an invited keynote speaker at all PV world conference.

- Since 2005, Thomas Nordmann is member of the supervisory board by Sunways AG, Konstanz/Germany.

- Thomas Nordmann has published more than 110 Papers and publications in his career since 1974. He was the first non-german Chairman of the PV Symposium Staffelstein 2008.

Chapter 32

A World Network for Solar R&D: ISES

Monica Oliphant

President International Solar Energy Society (ISES), International HQ,
Villa Tannheinn, Wiesentalstr. 50, 79115 Freiburg, Germany

In 1954, US solar visionary Professor Farrington Daniels attended a "Symposium on Wind and Solar Energy "in New Delhi, India. On a field trip he saw bullock-powered irrigation pumps that his wife, Olive, was to paint in oils. He often stated later that this was the picture that changed his life and led him to look at the role of solar energy in the service of humanity.

Not long after this experience, he, with others, formed the Association for Applied Solar Energy, AFASE, incorporated in Arizona, USA. In 1964 this became the Solar Energy Society and in 1971, the International Solar Energy Society (ISES) with the headquarters (HQ) in Melbourne, Australia. In 1995, the HQ moved to its current location in Freiburg, Germany.

Since its AFASE days and through the initial encouragement of Farrington Daniels, ISES has kept, as its major function, the importance of informing, and alerting the public to the need of demonstrating, and commercialising renewable sources of energy. Activities have been based on encouraging and fostering research in renewables, organizing congresses and publishing articles on the results of research in the society's scientific journal, the highly rated "Solar Energy Journal" and popular magazines as well as external publications.

ISES is a membership society and our last survey showed that we have,

- Members in over 100 countries and in 50 sections round the world,
- Almost 70% of our members are scientists, engineers, architects and technical people.

In addition, ISES is a non-profit NGO accredited with the United Nations since 1973.

Figure 1. Farrington Daniels.

Figure 2. ISES HQ in Freiburg Germany.

Our vision is *"A rapid transition to a renewable energy world"*. And our mission is, *"To provide scientifically credible and up-to-date renewable energy and energy efficiency information and networking opportunities to the global communities of scientists, educators, practitioners, industries, policy makers and to the general public."*

Being the sole international renewable energy organisation for many years can probably enable us to boast that for the first 30 or more years of our existence a high percentage of new solar energy research and development came from ISES members. In fact, since this is a book on the history of PV, it is pertinent to note that two other contributors — Professor Karl Böer and Professor Adolf Goetzberger have both played very important roles within ISES.

Professor Böer was instrumental in establishing the American Solar Energy Society, ASES, when the ISES HQ moved to Australia. He also instigated the Karl W. Böer Solar Energy Medal of Merit Award for significant contributions to the promotion of solar energy through scientific, development or economic enterprise — an award supported also by the University of Delaware and ISES. Recent Böer Award recipients from the field of PV include, Prof Adolf Goetzberger (Germany), Prof Martin Green (Australia), Prof Yoshihiro Hamakawa (Japan), Dr Lawrence Kazmerski (USA) and the 2009 recipient, Dr Hermann Scheer (Germany) whose involvement in establishing the German Renewable Energy Act has led to widespread growth of PV.

Professor Goetzberger has been involved with ISES and the German ISES Section (Deutsche Gesellschaft für Sonnenenergie) for over 30 years. Not only has he received the Karl W. Böer Solar Energy Medal of Merit Award for his outstanding work in PV, but he has held many positions within the ISES Board including being President from 1992 to 1993 and currently he is an Honorary Director.

Though ISES has always promoted all forms of renewable energy and our congresses still have papers representative of a broad range of renewable energy sources, we are now predominantly a Solar Energy Society. This has become more so in recent times as the range of renewable energy technologies have broadened and became more commercialized. As a result, each technology has formed its own association or society. This can only be seen as a positive development in that it shows that renewable energy has come of age and is moving into the mainstream of energy sources. However, recognising the need for the various technology groups to work co-operatively together, in 2004 ISES became a foundation member of the International Renewable Energy Alliance or REN-Alliance, together with the International Hydropower Association and the World Wind Energy Association. Since then, the REN Alliance has expanded and includes the International Geothermal Association, and the World Bioenergy Association. The REN Alliance was established to advance policy and information on renewable energy by providing a combined voice for renewable energy technology and practice.

As we move into the future, Solar Energy Research networking is the core strength of ISES. "The Solar Energy Journal" is a premier publication in the field with an ever-increasing impact factor and a new journal "Progress in Solar Energy", soon to be released, will provide in-depth technical review articles by leading experts in their field. ISES, through its unique network, is also instrumental in fostering the application of renewable energy and has produced White Papers for the Developed and Developing World (translated into many languages) and Pocket Reference Books on specific topics such as Solar, Wind and Passive House Design.

Recognising the need to encourage and increase the participation of Youth in Solar Energy, a Renewable Energy Research Exchange Programme is to be implemented in the near future and also a Young ISES initiative was launched in 2006 to create a network among students and young professionals in renewable energy.

Figure 3. Commercial Solar Food Processing.

ISES Congresses have always encouraged students to participate through reduced fees, and refereed papers and posters.

ISES facilitates and manages projects in developed and developing countries. In 2009, we completed a project on Solar Food Processing culminating with the organisation of the International Solar Food Processing Conference in Indore, India. This conference was the first of its kind to focus on solar food processing for commercial applications and income development, the focal point of the overall project. Also, as part of the European funded project PREA (Promoting Renewable Energy in Africa), ISES organised a series of workshops on sustainable architecture in South Africa, Tanzania and Uganda in 2006 and 2007. Projects such as these are examples of how ISES encourages and fosters information and know-how exchange and promotes good practices. In 2010, we will manage a UNEP project "Quality Criteria for Solar Resource Assessments".

In addition, ISES strives to build links between renewable energy research and industry, and significant effort is placed developing these links. ISES recognizes the importance of remaining relevant to industry and through its corporate membership the Society aims to create benefits and services that are mutually beneficial to both sectors

After years of planning, the International Renewable Energy Agency (IRENA) was officially established in Bonn on 26 January 2009. Currently, 137 countries are members. ISES, together with the REN-Alliance has recently signed a MOU with IRENA to work together in order to speed up the utilization of renewable energy sources worldwide and the areas of cooperation and research will take advantage of our large network and research capabilities into the future.

Solar Energy is the greatest renewable energy resource available to humanity and mainstreaming its use and enlightening governments to ratify policies to enable large scale integration into the energy mix is a worthy a challenge for ISES and its membership. Many scientists and technologists are working to show that a strategic mix of renewable energy sources can and will provide reliable and clean power. As this book has shown the intellectual capacity of those working in the solar field is great, ISES aims to maintain its scientific and research affiliations with these in order to aid the transition from yesterday's sunshine to today's.

Reference

This article has drawn on information taken from "The Fifty-Year History of the International Solar Energy Society and its National Sections" edited by Karl W Böer. (See ISES web site Bookshop at www.ises.org.)

Monica Oliphant

Chapter 33

Early Work on Photovoltatic Devices at the Bell Telephone Laboratories

Morton B. Prince

Formerly with Bell Telephone Laboratories and later the Director of the PV Programme of the US Administration in Washington

The first observation of the photovoltaic effect was made by Alexander-Edmund Becquerel in 1839. The effect was observed later in many different forms with different materials. However the conversion of solar energy into electrical energy with these structures and devices never got above 1% until the 1950s. John Shive of the Bell Telephone Laboratories (BTL) in 1952 reported on a grown p-n junction photocell of silicon that used the very small area of the region where the junction came to the surface of the crystal. It was not thought of as a power device but as a sensor. In 1953, Daryl Chapin (an electrical engineer) was looking for a power source to replace batteries that operated remote communication systems. Once such source was a thermopile heated by burning a gas such as propane. At the same time Gerald Pearson (a physicist) and Calvin Fuller (a chemist) were investigation large area p-n junction structures for power rectifiers. Fuller studied the diffusion of lithium at moderate temperatures (~500°C) into silicon to form a large area junction and Pearson would contact the two sides of the diffused wafer and observed that the resulting device was very sensitive to light. However this device changed its electrical characteristics in time due to the rapid diffusion of lithium into the silicon even at room temperature. Fuller then initiated the study of boron diffusion into n-type silicon, which required a much higher temperature (~1000°C) and the resulting p-n junction was stable at room temperature. When Pearson made the measurements in sunlight and observed the photovoltaic effect, he called me into his laboratory to witness this observation (in December 1953). Pearson recognized the importance of this development and three of them (Chapin, Fuller and Pearson) worked together to improve the characteristics of the

Power for the World by W. Palz
Copyright © 2011 by Pan Stanford Publishing Pte Ltd
www.panstanford.com
978-981-4303-37-8

device for solar energy conversion and published their important letter in the *Journal of Applied Physics* in May 1954, where they reported a conversion efficiency of approximately 6%. In February 1954, I was asked to take that boron-diffused device, optimize it and make it reproducible so that it could be used for some demonstrations. Two months later, I and my technologists had assembled sufficient cells to demonstrate these Bell Solar Batteries (as they were called) at Murray Hill New Jersey (the location of the BTL) and a couple of days later at the National Academy of Sciences in Washington, DC. in a public announcement. During the next few months I performed some calculations and analysis of experimental data that I had accumulated and submitted a paper that was published in the *Journal of Applied Physics* in May 1955, pointing out that the series resistance of the device was the most important parameter to control as well as showing that the energy band-gap of the semiconductor material determines the maximum efficiency that might be expected form such materials. Silicon was near the maximum of this curve. Carl Frosch and his team improved upon Fuller's diffusion technique, which permitted us to diffuse several large slices of silicon simultaneously. By mid-1955, we were able to make 8 to 9% efficient cells, which we reported at the Conference on the Use of Solar Energy held in Tucson and Phoenix, Arizona in November 1955. Later that year, Ed Stansbury of the BTL was able to make an 11% efficient small cell. The first application of these silicon solar cells was their use in a telephone application in Americus, Georgia, where a trial was held for six months in late 1955 and early 1956. Even though the results of the trail were successful from a technology point-of-view, no further use was made in the Bell system at that time due to cost considerations.

Morton B. Prince

Chapter 34

Leaders of the Early Days of the Chinese Solar Industry*

Qin Haiyan

DG of China, General Certification Center, Beijing, China

I was born in 1970, the year China woke from dormancy and began to move and prosper. In that year, PV cells were used for the first time in Dongfanghong 2, China's telecommunications satellite. Since then, various PV cells have been gradually used in marine navigation, railway signals and other terrestrial facilities. Like scattered flames, they have worked together brightening solar energy's prospects in China in the 21st century.

With the effort of many insightful individuals, by 2002 the solar energy industry in China had acquired a considerable scale. In that year, I resigned from a public institution to set up the China General Certification Center (CGC), to create third-party certification for renewable energy products in China. Through this route, I became involved in the solar energy industry. From my personal experience meeting visionaries of the industry through the developments of the past years, I can truly understand the joys and woes of the pioneers in the early development of the Chinese solar energy industry. Some of their histories I share here.

1. Turning A Dream into a Reality: The Story of Huang Ming

In 2004, my colleagues and I successfully completed our certification in solar product testing and authentication techniques. With the referral of Wolfgang Palz, the most important friend in my life, I joined the Global Approval Program for Photovoltaics (PVGAP) as an executive member, and thus was in a better position

---

* Translation from Chinese by Jodie Roussell, USA.

Power for the World by W. Palz
Copyright © 2011 by Pan Stanford Publishing Pte Ltd
www.panstanford.com
978-981-4303-37-8

to further study international PV certification practices and to gain wider recognition for our work. Subsequently my colleagues and I implemented the *Solar Water Heating System Building Program in China* sponsored by National Development and Reform Commission and UN Development Programme and the *PV Product Certification System Program in China* initiated by National Development and Reform Commission, Global Environment Facility and World Bank. In the course of this work, I have been able to witness the rapid growth of solar energy industry in China and to work closely with many significant individuals; Huang Ming is one of them.

Huang Ming was among the earliest college students after college entrance examinations were resumed. He majored in petroleum studies and was told by his Professor Chen Ruheng that the oil resource was quite limited and might be exhausted within 50 years. The time for China to act as an oil-deficient country was much shorter. Huang Ming was shocked. If this were true, then he would lose his job or the worth of his education when the oil was exhausted 50 years later. From then consciously or unconsciously he began to look for information related to new alternative sources of energy.

By 1982, Huang Ming was studying drilling pit in the Drilling Research Institute under the Ministry of Geology and Mineral Resources. More rapid oil development meant less energy resource for this planet. Huang Ming gradually felt a sense of guilt over the course of his study. He was inspired by the American author Beckman's book, *Solar Energy Thermal Processes*, which lit up his solar energy dream.

Following instructions given in the *Solar Energy Thermal Processes*, in 1988, Huang Ming made his first solar water heater and gave it as a wedding present to

Figure 1. China Wind Power Conference, Beijing 2009; left Qin Haiyan, Conference Chairman with the Director of the company Goldwind.

a relative. Surprised, many people did not believe that this device would work — but they were shocked when they were burned by the hot water it produced!

In the following seven years, Huang Ming continued to make solar water heaters for relatives and friends for free. After giving away about 20 heaters, a factory manager asked him to install bath-water heaters for their workshop. His heaters that had been originally made as experiments began a period of commercialization.

In 1995, he resigned from his job, established Himin Co. Ltd. and embarked on the difficult journey of starting a business. Ten years later, he had turned Himin into the largest solar product manufacturer in China, which produces solar water heaters and PV systems, with a valuation of RMB 5.1 billion (750 million USD).

On May 5, 2006, the United Nations held the 14th Symposium on Sustainable Development in New York. As a Chinese entrepreneur, Huang Ming was invited to share with the world his business model and wisdom. His 17-minute speech won the curiosity and applause of over 300 leaders in sustainable development from across the world.

2. The Richest Man in China: The Story of Shi Zhengrong

In August 2006, the *China PV Product Certification System Program* was put into use. In the course of pilot certification, my colleagues and I visited Suntech Power Holdings in Wuxi, and had the opportunity to get to know Shi Zhengrong, a PV leader in China.

In 1988 when he was 25 years old, Shi went to the University of New South Wales to study under Professor Martin Green. Green is and was an international expert in solar cell development, the winner of Alternative Nobel Prize in 2002 and was often called "the father of solar energy". In Green's program, Shi obtained a doctoral degree with just two and half years and subsequently became a researcher in the Solar Energy Research Center.

In 2000, Shi gave up the high quality of life and world class research facilities in Australia, to bring back 14 solar energy patents and start a solar business in Wuxi. In January 2001, with the support from Wuxi municipal government, Suntech Power was formally established. With 5% cash and a 20% technology contribution, Shi became the majority shareholder and acting general manager. Suntech began its journey from near Lake Taihu, to the world.

By 2002, Suntech's first 10 MW product line was put into operation, with a capacity equal to the total of the nation's solar cell production over the past four years. Suntech went public in December of 2005 on the New York Stock Exchange, making a new record as the largest capital raise on a foreign exchange of any non-state owned Chinese enterprise (SOE) to date.

Shi is consistently innovative and challenges himself. He said that he liked the subversive quality of innovation, and only through consistent, unbroken innovation

Figure 2. Great Wall Renewable Energy Forum, Beijing 2006; from the right Qin Haiyan, Mike Eckhard (ACORE Washington), Shi Dinghuan, China State Council, W. Palz.

efforts could the solar energy industry move forward. Through his passion and innovation, he became a PV leader in China.

To become a success, a person has to ensure unimaginable hardship and trial. At the hardest times, he suffered financial difficulties, had to face violent creditors threatening to confiscate equipment, and partners and friends departing, one after another. But he chose to persist.

From the perspective of corporate and financial management, Shi possesses strong perceptive abilities, a unique vision and the capacity for swift action. From the very beginning, Suntech looked to mature international markets. Listing abroad on the NYSE in 2005 not only raised capital for corporate growth, but made him the richest man in China.

From his international experience, Shi knew very well the importance of international certification. In the first year of production, Shi focused on international certification over market expansion. He subsequently obtained almost all internationally recognized certifications for solar technology, clearing the way for the international market.

Shi personally is not driven by the pursuit of wealth. He lives a simple life and advocates frugality in the company's operations. In his own name, he set up the Sunshine Charity Foundation and the Shi Family Charity Foundation to help the needy. In March 2007, Suntech donated PV power systems to the Bird's Nest, the main stadium for the Beijing Olympic Games to appropriately represent the vision of a "green, hi-tech and humanistic" philosophy.

3. Internationalization and a Traditional Chinese Soul: The Story of Miao Liansheng

On March 17, 2007, CGC, the National Development & Reform Commission and the Renewable Energy Program Office jointly held the *Conference for the Release of PV Product Certification* in Beijing. Yingli Green Energy Holding Company from Baoding was one of the firms passing this certification. Its President is Miao Liansheng.

Miao retired from military service in 1980 and subsequently worked in the cosmetics industry for over ten years. In 1993, he began to import solar neon light components from Japan to assemble lighting systems in China, and thought about the developing PV industry. *"Heaven has given us the Sun, and we are obliged to turn sunshine into energy for the benefit of mankind."* Inspired by such a simple thought, Miao embarked upon this difficult path without hesitation.

In 1998, the Chinese government planned to build a 3 MW multi-crystalline system; Miao Liansheng saw the first beam of light. Despite his partners' withdrawal, Miao Liansheng still tried to obtain the approval for this program and became the first to overcome great difficulties and achieve success in China solar energy industry. This demonstration program passed national testing and approval, but received little recognition. Miao had demonstrated his modules widely, but struggled with little initial market interest. Despite this difficult start, Miao never lost heart. He spent much time working to remedy the public's

Figure 3. Qin Haiyan with Jodie Roussell.

illiteracy about solar energy, educating citizens about the benefits of PV to increase public acceptance. He simultaneously realized that international market was a force pulling the new energy industry in China forward. As a result of this international pull, he changed strategy.

Miao started the expansion into foreign markets in 2004, first in southern Germany, then Spain and Italy. When the European rush to buy solar modules began, Miao decided to make use of the popularity of football to establish Yingli's brand in European markets. Yingli provided PV modules to the Kaiserslautern Stadium for the 2006 World Cup, and sponsored leagues teams in Spain greatly increasing Yingli's brand recognition.

On June 8, 2007, Yingli listed on the New York Stock Exchange, raising 319 million USD. The next day, Miao became the first businessman in the 170-year history of the opening bell of the New York Stock Exchange that did not wear a necktie! Miao said, though his firm chose an international path, he never forgot he was Chinese. Just like having meals with chopsticks instead of knife and fork, on this special day, he preferred to let Western society remember Yingli as a Chinese business, by not wearing an overtly Western garment like a necktie.

Along with the successful overseas listing, Yingli's Phase 2 and 3 expansion projects were also successfully completed, and by 2009 annual production capacity grew to 600 MW.

4. Development Led by Technology: The Story of Gao Jifan

My colleagues and I understand very well that to participate in the PV cause in China, beyond honest market service, CGC has to rely on technology to promote PV growth in China. In the past several years, CGC together with many PV companies, has studied the Chinese market conditions and international technical standards, and prepared briefs such as *Specifications for the Certification of Lead Acid Battery of PV System*, and *Technical Conditions and Testing Methods for 400 V Less Grid-connected Special PV Inverter*, and filled in many information gaps in China. In the course of preparing these specifications we have met many outstanding PV entrepreneurs, Gao Jifan, the president of Trina Solar Limited, is such a person.

One of Gao's early ventures was establishing a research institute in Shenzhen focused on industrial cleaning products, which didn't turn a profit until 1992. Then in 1998 he learned from a classmate that the United States was promoting solar energy cells, and by 2012 it was projected that over 1 million households would use such systems. Gao felt making products every family could benefit from would be better than his current business venture, producing exterior wall construction materials. As a result, Gao registered Trina Solar Ltd in Changzhou. By August 2000, Trina Solar had successfully produced the first PV powered house in China. Subsequently in 2002, Trina Solar became the first private business included into the *China Brightness Program*, and installed 39 PV power stations in Tibet.

Figure 4. Great Wall Renewable Energy Forum, Beijing 2006; from the left Qin Haiyan, Dominique Campana (ADEME, Paris), W. Palz, Gallus Cadonau (Switzerland), a colleague from ADEME.

Gao said his PV business made considerable progress in 2004. That year the German government announced the feed-in tariff for solar systems. Gao realized that the time for PV had come. He closed his other businesses and devoted himself full time to PV. By 2005, Trina Solar had completed its module assembly line expanded its mono-crystalline production and integrated the two.

The following year in 2006, Merrill Lynch and other international strategic investors began to support Trina's efforts, and the firm established an international management team. Trina Solar went public on December 19, 2006, on the New York Stock Exchange, becoming the second private solar business after Suntech Power to be listed on a foreign stock exchange. Trina's scope at that point covered a larger portion of the entire PV industrial value chain, from silicon wafers, batteries, modules and other mid- and down-stream links.

In 2008, Trina Solar sold 201 MW modules, 164.8% higher than 2007, and net profit amounted to 6.1 million USD. In the second quarter of 2009, despite the market downturn, Trina Solar still realized a net profit of 18.9 million USD.

To become a successful businessman, a scholar has to possess both special knowledge and business wisdom. Gao is focused on being an expert-type entrepreneur rather than innovation-focused expert.

With the guidance of insightful and bold entrepreneurs like Huang Ming, Shi Zhengrong, Miao Liansheng and Gao Jifan, the Chinese solar industry has grown tremendously in the last 15 years. Their business ventures, through hardship and success, have inspired a generation of young entrepreneurs, who will continue to grow the opportunities for solar energy both domestically in the burgeoning Chinese market, and abroad.

Qin Haiyan

Qin Haiyan, Secretary General of Chinese Wind Energy Association (CWEA) & General Director of China General Certification Center (CGC). I got to renewable energy sector by chance in 1999, and from then on, I fell in love with it and devoted my life to this industry. I always believe renewable energy will take the place of fossil fuel. With this belief, I have been working in the area of wind, solar and other renewable energies for a decade, mainly for policy study and technology R&D. Through working at CGC and CWEA, I have been involved in many research projects, either as a team leader or a team member, and initiated a variety of renewable conferences with a goal to stand for the development of the industry. All these work are well recognized by the industry.

Chapter 35

Illiterate Rural Grandmothers Solar-Electrifying Their Own Villages

Bunker Roy

Barefoot College, Rajasthan, India

1. Ground-breaking Innovation in the Field of Technology

The Barefoot College was established over 38 years ago in the deserts of Rajasthan in India. The objective was to use a powerful combination of the "technology of the people" with traditional knowledge and skills to improve the quality of life of the very poor, the impoverished, the economically and socially marginalised and the physically challenged living on less than $ 1/day. The goal was to empower the poor with technology skills and knowledge to enable them to stand on their own two feet.

It's the only college in India built by the poor and only for the poor. The application of every day science and the old existing technology of the people — not identified, not recognized and not even applied — is what is reflected and internalized in the practical beliefs of the college. Along with the scientific temper of the 21st century, Mahatma Gandhi's practical beliefs and his spirit of sacrifice still live in the lifestyle and work style of the college. It has been shown that if the learning and re-learning environment is relaxed, non-structured and informal, the poor rural men and women are capable of technological wonders.

There is a scientific and spiritual dimension in the college because working relationship depends totally on equal respect for basic out-of-formal schools and out-of-the-box scientific skills, mutual trust, tolerance, patience, compassion, equality and generosity. The idea is to apply the scientific knowledge and skills the impoverished rural poor already possess for their own development thus making them independent, promoting quality of life, letting them live with self respect and dignity. The goal is to use these powerful tools in their hands to free themselves of exploitation, injustice, discrimination and fear.

Power for the World by W. Palz
Copyright © 2011 by Pan Stanford Publishing Pte Ltd
www.panstanford.com
978-981-4303-37-8

2. Sustainable Development: Now and in the Future

The college believes the very poor have every right to have access to, control, manage and own the most sophisticated of technologies to improve their own lives. Just because they cannot read and write there is no reason why the very poor and illiterate women cannot be barefoot technologists, water engineers and architects constructing rain water harvesting tanks, solar engineers solar electrifying their own remote villages, fabricate low cost radios, run community radios in villages reaching 50 000 people, conduct water quality tests and feed data into computers. They have shown the impossible is possible and confirmed what Mark Twain said, "Never let school interfere with your education."

3. Innovation and Its Practical Application

The Barefoot College believes one of the key innovators to demystify technology is the illiterate rural woman all over the world. Given the opportunity, she is strong and confident enough to change their immediate world. Over the last 15 years, thousands of illiterate and semi-literate rural women have been trained in non-traditional occupations changing the lifestyle and mindset of village India and indeed the world.

The Gandhian message through the Barefoot College has spread to 13 states of India and by end 2009 will have spread to more than 21 of the least developed countries in Africa and around the world.

Figure 1.

4. Demystifying of 21st Century Technology in 19th Century Conditions — Management, Control and Ownership in the Hands of the Rural Poor Around the World

1. The Barefoot College is the ONLY fully Solar Electrified College based in a village in India. Forty kWs of solar panels and 5 battery banks of 136 deep cycle batteries were installed by barefoot solar engineers from 1989 onwards. The solar components (invertors, charge controllers, battery boxes, stands) were all fabricated in the college itself. Installed by a Hindu priest who barely passed secondary school.

2. Two-hundred-and-eighty-nine barefoot solar engineers (128 semi-literate rural women) have installed 23 solar power plants of 2.5 kWs each. The women have fabricated 17 parabolic solar cookers: 97 solar water heaters have been fabricated and installed in the Himalayas: trained rural communities to establish 2 Micro-Hydel power stations of 100 kWs each: established 23 rural electronic workshops. The total population being reached is over 100 000 people all over the country.

3. The first ever solar powered Reverse Osmosis (RO) (desalination) plant in the country powered by 3 kWp solar power plant producing 600 litres per hour potable (sweet) water from salt water measuring 9000 TDS was installed by the college around the Sambhar Lake, the largest inland salt lake of India in 2006. There are over 100 villages with a population of over 250 000 facing acute shortage of drinking water. Ten more such solar powered RO plants are being installed with funds from the Coca Cola Foundation and the Government. (http://www.youtube.com/watch?v=Q7oVmz_yqVI.)

Figure 2.

What is innovative is that the Barefoot College has given priority to the ideas and thoughts as the rural poor see as immediate and relevant and respect their wishes. It is also the main reason why it is the only college that gives no paper degrees or diplomas after training. No one comes to work for the money, power, position or security. Very few people leave because in spite of their incredible skills they are "unemployable" in the eyes of the outside world.

5. Present and Future Impact of Innovation: Number of People Affected

> " Science is simple common sense at its best."
>
> Thomas Huxley

The Barefoot College through the basic barefoot services they provide in India and globally reach nearly three million men, women and children annually who live on less than 50 cents a day.

5.1 Renewable Energy

Two-hundred-and-eighty-nine illiterate (213 men and 76 women) barefoot solar engineers have solar electrified 599 villages generating a total of 550 kWp

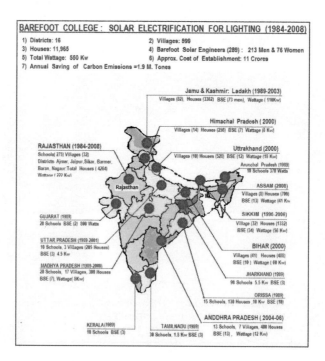

Figure 3.

electricity per day reaching 11 900 families in 14 states of India: 274 night schools, distributed 4736 solar lanterns. Thus saving thousands of litres of kerosene from being used for lighting, several thousand tons of wood from being cut and drastically reduced the use of torch batteries and candles. As a result, the college has prevented 1.86 million tons of carbon emissions from polluting the atmosphere. Illiterate women are fabricating parabolic solar cookers water heaters.

In the Himalayas 3500 metres up in Ladakh (Kashmir) with temperatures reaching −30°C where solar light had reached remote villages for the first time a woman was asked what benefit she received from solar lighting. She thought for a minute and then said, "It is the first time I can see my husband's face in winter."

5.2 *Continent of Africa*

In 21 of the poorest countries round the globe — Ethiopia, The Gambia, Sierra Leone, Mali, Mauritania, Cameroon, Benin, Malawi, Uganda, Tanzania — illiterate women (many of them grandmothers) have been trained to solar electrify their own remote inaccessible villages, construct rain water harvesting tanks in the schools where their children go and the traditional male-dominated society look on these formidable women with shock and awe.

For the first time in the history of aid being provided to Africa by the Government of India, the Ministry of External Affairs under a scheme called

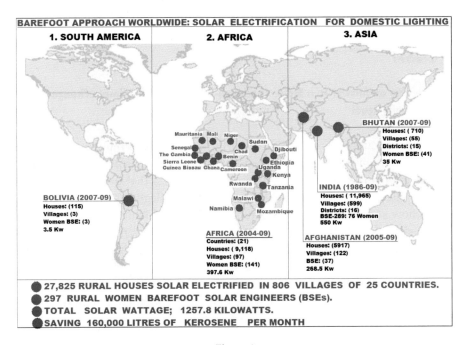

Figure 4.

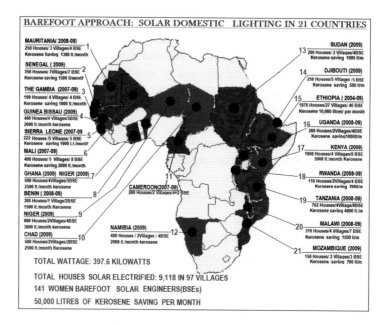

Figure 5.

India Technical Economic Cooperation (ITEC), has approved the Barefoot College as an official training institute to train illiterate rural grandmothers as solar engineers in six months. The extent, impact and reach all over Africa has been significant. By 2010, the barefoot approach of solar electrifying villages will have covered the Least Developed Countries in the whole continent of Africa.

5.3 *Global*

5.3.1 Providing an Answer to a Major Challenge-tackling Global Climate Change from the Community Level

A "barefoot revolution" is sweeping rural India and indeed the world. Very ordinary people written off by society because they are labeled as poor primitive, backward and impoverished are doing extraordinary things that defy description and cannot be rationally explained. What the Barefoot College has effectively demonstrated is how sustainable the combination of traditional knowledge (barefoot) and demystified modern skills can be when the tools are in the control and in the hands of the virtually written-of-poor. The "barefoot approach" has been recognized globally. It can be replicated anywhere in the world by the poor and for the poor.

"First they ignore you, then they laugh at you, then they fight you and then you win".

Mahatma Gandhi

Chapter 36

The Kick-off PV Programme in Germany: The One Thousand PV Roofs Programme

Dr. Walter Sandtner

German Federal Ministry of Research and Technology (BMFT)
now in Vienna, Austria

A number of factors have contributed to the remarkable development of renewable energies, especially of PV, in Germany: the oil crisis of 1974, the widespread anti-nuclear attitude of the German population, the climate issue, the dream of entirely environmental-friendly energy forms, to some extent also the lack of economic considerations etc.

The development took place in 3 phases:

- From 1974 to 1989: the research phase,
- From 1989 to 2000: the demonstration phase,
- Since 2000: the commercial phase.

The 1000 Roofs Programme which came into effect in 1989/1990 marked the beginning of the demonstration phase. At that time, the German public and the politicians requested with growing impatience that after 15 years of intensive research, there should finally be something *visible*, a "real programme".

I will briefly report on my personal experience in initiating the 1000 Roofs PV Programme, which at the time, was the biggest PV programme in the world.

It was exactly in this moment of public unrest, in 1989, when the then German Federal Minister for Research and Technology, Mr. Riesenhuber, asked me to take over the responsibility for Renewable Energies. He mentioned the widespread request for renewable energy programmes by some members of the German Parliament.

So I pondered over the issue, considered a number of options and finally I conceived the basic structure of the "1000 Roofs Photovoltaics Programme". I chose "1000", because this figure is easy to remember. In fact, in the end, nearly 2200 PV installations were funded, systematically spread over the whole territory of Germany, in close cooperation with the 16 German Länder and some research institutes.

After having conceived the programme, I invited some German specialists in order to discuss with them the concept and to find out, what had to be done to make the programme work. At the time there existed only six PV roofs in Germany, but it was my belief, if there are six roofs, there could also be 1000 roofs or more.

The specialists came together, an intensive discussion took place, but some days after the meeting they sent me a three-page letter with explanations not on how the programme could work, but on the contrary they affirmed that the programme could not be realized, because many technical things were still missing, from inverters to DIN norms.

So I reconvened them and told them that the decision had been made to work out such a programme. I asked them not to tell me that and why the programme will not work, but rather what had to be done to put it in practice.

First slightly reluctantly, but then with a steadily growing enthusiasm, everybody made constructive contributions. Already after some months, the general public and the specialists became really excited till to a degree which I have rarely seen before and afterwards. We sent — together with the Länder — more than 60 000 information sheets to individual citizens. Technical craftsmen and mothers with crying babies in their arms called to get more information on this "mysterious" solar energy and what was planned to get electricity out of it. There was hardly any renewable energy seminar, conference etc. where this programme was not debated in a detailed manner.

The basic features of this programme were the following:

1. In order to spread the photovoltaics roofs all over Germany, I cooperated with the 16 German Länder in such a way that I accorded quotas to them, that means, the major Länder received a share of up to 150 roofs, the minor Länder up to 100.
2. The PV installations were heavily subsidized, that is with 70% of the costs of an installation. Fifty percent of this subsidy was borne by the Federal Ministry, 20% by the Länder. One Land thought that a 50% subsidy is sufficient and did not pay the additional 20%. I accepted this as a means to enlarge our experience. The result was that this was the only Land, which could not find the necessary quantity of interested citizens to fulfil their share of the quota. In all the other Länder the interest was higher than the distributed quota.
3. There was a number of technical criteria to guarantee the success of the programme. It was foreseen that all photovoltaics installations should be grid

connected, they had to be installed on the roofs with a certain distance to the roofs, no shadow during the day-time should impede the electricity production etc. It was only later that roof-integrated and façade photovoltaics installations were developed.

4. From the outset, aesthetic considerations were also of great importance. Once we organised a competition and gave prizes to the most beautiful solar roofs. All the technical necessities, which were missed by the specialists at my first concept meeting, were developed very quickly. Already after a couple of months there were 40 inverter types on the market, DIN norms were formulated etc. There were many new scientific studies and improvements. The price of the photovoltaics installations during the period of the programme fell by a third. One of the reasons for this reduction was that we accepted in the second half of the duration of the programme not only German made photovoltaic installations, but also American, Japanese, British and Spanish installations.

Overall this programme effectively demonstrated that photovoltaics works well also in northern latitudes like Germany. One frequent criticism could effectively be met: it was often claimed that photovoltaics would use up a lot of natural landscape. The programme proved that no additional landscape was needed. One can install photovoltaics on roofs, alongside motorways, railways etc. For instance on the roof of the Munich trade fair centre, which was the former Munich airport, a 1 MW PV installation was installed.

The programme contributed also considerably to our international cooperation. In Spain, near Toledo, a joint Spanish-German 1 MW photovoltaics installation was erected. Generally, Spain was one of our most important partners in the renewable energy field. But we cooperated also with other European countries, with the US, with Russia and with a number of developing countries like Brazil, Argentina, India, China, Indonesia and others.

As already said this programme had a substantial impact on the PV development in Germany. Some years later the Federal Parliament decided to create a "100 000 Roofs Photovoltaics Programme" and finally the Renewable Energies Law from July 2004 foresaw, for photovoltaics especially, high subsidies between 45 and 57 cent/kwh. So, excellent prospects were created in order to expand the photovoltaics development in Germany and beyond. The example set by Germany was especially followed by Japan, Spain and some other countries.

In conclusion I should like to say that the creation of the "1000 Roofs Photovoltaics Programme" was a very exciting experience.

The development initiated by this programme was a very rapid one and has led to results, which were difficult to foresee in the beginning.

The programme was a milestone in a development which has since become worldwide and has found its most recent expression in the creation of IRENA in June 2009. Many pioneers and promoters have contributed to this development and I should not like to terminate this brief souvenir without mentioning MdB Dr. Scheer and Dr. Wolfgang Palz who both played an important role in this field. They both took, from the beginning, an intensive interest in this programme and used its results for further and far-reaching measures. In this manner, the "1000 Roofs PV Programme" was certainly only a small programme, but it had many consequences which continue till today.

Walter Sandtner

Chapter 37

History of Technologies, Development for Solar Silicon Cost Reduction

Frederick Schmid

Crystal Systems, Inc., 27 Congress Street, Salem, MA 01970, USA

1. Introduction

It has been extremely rewarding to be involved in the development of photovoltaic (PV) energy technology, a technology that continues to prove its tremendous potential for future growth as a clean and sustainable energy source. Over the years, it has become evident that fossil fuel resources are not sustainable; rather, they leave a carbon footprint that contributes to the growing concern of global warming. It can be said that PV is the most environmentally sound form of energy generation currently available, hence the tremendous amount of satisfaction that comes from working with it.

The goal of our R&D was to reduce the silicon material cost since it is the primary barrier between PV technology and grid parity. Success in the advancement of PV technology and the reduction of costs is highly dependent upon the open communication of talented people selflessly working together for the greater good. It has been extremely encouraging to see increasing numbers of people around the world, particularly the immense number of university students, becoming involved in PV research.

I have been fortunate enough to work with many talented people for over 38 years at CSI, as well as at many other organizations. Without the help of these people, I would not be able to find my way through the unchartered technical maze of silicon processing technology.

This paper will discuss the history of how solar silicon evolved from a new technology developed for an entirely different purpose: to produce ceramic ingots

Power for the World by W. Palz
Copyright © 2011 by Pan Stanford Publishing Pte Ltd
www.panstanford.com
978-981-4303-37-8

via solidification of the ingots from the melt. This technical innovation was taken from concept stage to commercialization. The R&D developments to reduce Si material costs have often led us down a technological road less traveled. Many knowledgeable people in the field believed that it was too far removed from the understood technology base.

A new crystal growth technology to produce large sapphire crystals was invented in 1969, leading to the formation of Crystal Systems, Inc. (CSI) in 1971. CSI's focus was to be on the commercialization of technology for sapphire production and the development of growth technology for other crystals such as Si for PV applications. This new crystal growth technology was the result of an R&D program at an army research laboratory near Boston. The program was designed around the production of sound ceramic parts via solidification rather than sintering or hot pressing ceramic powder into ceramic parts. At this point in time, solidifying ceramics to produce ceramic parts was not practiced commercially due to the fact that most ceramics melt at much higher temperatures than steel, making containment difficult. In addition, the lower thermal conductivity of ceramics makes heat extraction difficult for directional solidification.

In starting this research, aluminum oxide was chosen as the ceramic material because of its stability at the melt temperature. It was a well-understood material available in high purity form with large commercial markets. There were no simple directional solidification approaches for producing fully dense ceramic ingots or parts from the melt; directional solidification to produce metal parts from the melt was just being commercialized for metals. All processes for producing sound ceramic bodies from the melt involved complex crystal growth processes such as Czochralski (CZ) for sapphire growth.

The development of a high temperature heat exchanger for heat extraction was necessary to directionally solidify ceramics from the melt since water-cooled copper heat exchange used for metals was not feasible at these high temperatures. Helium gas was chosen as a cooling medium because it was an inert gas with high conductivity and high specific heat. Based on this, I developed a helium gas cooled refractory metal heat exchanger that involved injecting helium gas into a closed end tube (Fig. 1). The heat exchanger was incorporated into a graphite resistance furnace. The heat exchanger was used to control heat extraction by regulating the flow of helium gas through it. This proved to be very effective in directionally solidifying the aluminum oxide melt into a fully dense ingot with three large grains. It appeared that single crystal alumina, i.e. sapphire, could be solidified from the melt if a seed crystal was placed in the bottom of the crucible.

The regulated helium flow through the heat exchanger prevented melting of the seed crystal, but allowed the liquid to melt into it to nucleate a single crystal. Single crystal growth could be achieved by increasing the helium flow to extract heat and by reducing the furnace temperature, thereby directionally solidifying the aluminum oxide melt to produce a single crystal of Al_2O_3, or sapphire. The goal was to directionally solidify large full density multicrystalline ingots, but it was

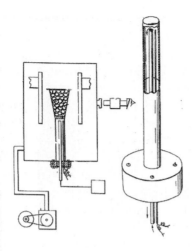

Figure 1. Heat Exchanger Method (HEM) Furnace. Schematic of graphite resistance furnace with refractory metal helium cooled heat exchanger at right inserted through bottom of furnace to extract heat by increasing the flow of helium gas.

quickly demonstrated in 1969 that sapphire crystals could be produced with this technology. The first crystals were 2 cm diameter × 4 cm tall, 80 gm. The challenge was to grow large diameter crystals using a flat bottom crucible that was much larger than the centrally located heat exchanger, as shown in Fig. 2.

The ability to grow a single crystal across a flat bottom crucible was unique. This new technology was called the Heat Exchanger Method (HEM) because crystal growth was achieved with a heat exchanger. This was the first directional solidification technique that required no movement of the crystal. It quickly became apparent that HEM had significant advantages over existing crystal growth technologies. Unlike all other crystal growth techniques, the heat input could be controlled independently of the heat extraction, and this was possible without movement of the crystal. By shutting off the helium flow through the heat exchanger, an isothermal zone was created at a temperature slightly below the melt point for annealing. This removed stress in the crystal after solidification.

There were many other advantages that came with this new technique, such as hemispherical growth with an expanding solid-liquid interface as opposed to linear growth. Growth from the bottom of the crucible to the top results in a stabilizing temperature gradient, which minimizes convection. In addition, the solid-liquid interface is protected from vibration and temperature perturbations by the surrounding liquid.

I was convinced that HEM was a breakthrough crystal growth technology mainly because of its simplicity and ability to independently control the heat input and extraction without movement. Although I had no business experience at that time, I had a dream of forming a company based around this unique technology. In the summer of 1971, I discussed this concept with my sailboat

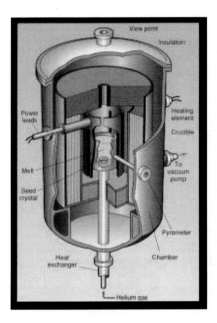

Figure 2. Heat Exchanger Method (HEM).

racing friends who had money to invest and who believed in my dream. A sailing competitor and his friend believed I could commercialize HEM for growth of sapphire. CSI was subsequently established in August 1971 to commercialize HEM for the production of sapphire crystals. I leased commercial manufacturing space in a refurbished textile mill on the waterfront in Salem, MA, north of Boston. The space was conveniently located, only 15 minutes from my home, which allowed me to spend much of my time there. To this day, CSI remains an independent company and is still located in its original home in Salem, MA.

My goal was to commercialize HEM for production of sapphire for an evolving electronic market called Silicon on Sapphire (SOS), and to develop HEM for growth of other important crystals. In three years, HEM was commercialized for production of 20 cm diameter sapphire crystals and I had a small breakeven business. The company was able to produce 20 cm diameter, 20 kg, high quality sapphire in a three-year period. However, slicing sapphire into wafers for the evolving SOS (Silicon On Sapphire) market was a problem. Existing slicing technology, which used diamond-covered blades to slice one wafer at a time, was producing a large amount of waste. I thought a better method would be to use diamond plated wire to slice multiple wafers at a time. This new technology was called Fixed Abrasive Slicing Technology (FAST) because it used diamond fixed on wire. I soon came to realize that there were significant technical obstacles involved with the development of this new process, and that such an undertaking would require outside funding. Once HEM was commercialized for sapphire in 1974, I wanted to develop HEM for growth of other crystals as well as to develop

FAST to effectively slice hard materials. In the early 1970s PV was being considered as an alternative renewable energy source. To make it a true alternative, the cost of Si had to be reduced. I believed that HEM and FAST could be developed to reduce the cost for producing Si wafers for solar cells.

2. Development of HEM and FAST for Reducing the Cost of Silicon Wafers

There was a major energy supply problem in the early 1970s due to the Iran oil embargo, making the US and the world vulnerable to energy supply by unfriendly oil producing nations. At that time the US oil reserves were running low, and it quickly became apparent that the long-term supply of oil was not sustainable. Due to the shortage, there were long lines of cars trying to get gas wherever possible. These factors made people aware of the need for developing renewable energy technology. Agencies around the country, such as the National Science Foundation (NSF), began to look at various renewable energy technologies. Solar energy was high on the list since the National Aeronautics and Space Administration (NASA) used it to reliably power satellites in space with solar electric power via photovoltaic (PV) energy using crystalline solar Si cells.

In 1976, Jimmy Carter was elected president, and one of his major initiatives was for the US to achieve energy independence using renewable energy sources. A new agency, the Energy Research Development Agency (ERDA), a predecessor of the Department of Energy, was formed to develop renewable energy technologies, terrestrial solar energy being one of the mainstays. The Jet Propulsion Laboratory (JPL) in Pasadena, CA, administered the R&D program since it had experience in the development and commercialization of PV for space. The renewable solar energy program involved cost reduction in four R&D areas: Production of Silicon Melt stock, Production of Silicon Sheets for Solar Cells, Production of Solar Cells, and Production and Testing of Solar Modules. The goal of the program was to reduce the production cost of terrestrial solar modules to $0.50/watt based on a 20-year lifetime of the solar modules.

My goals when forming the company were to commercialize HEM for sapphire growth and to develop HEM for growth of important crystalline materials. HEM crystal growth was a low-cost directional solidification process for successfully growing large high-quality sapphire crystals. It appeared that the HEM process was well suited for growth of large Si crystals at significantly reduced cost. In addition, multiwafer FAST slicing of Si into wafers using diamond on wire could significantly reduce slicing cost and waste. On the basis of these concepts, CSI submitted a proposal to JPL in 1975 to grow large Si crystals with HEM and to slice Si bars into wafers with FAST. CSI was selected along with many large semiconductor companies including

RCA, IBM, GE, Honeywell, Texas Instruments, Motorola, Westinghouse, and Mobil Solar, to develop its technology for production of low-cost Si sheets for solar cells.

It was very interesting to see what technologies the large companies proposed to reduce the cost of Si sheet production. CSI was the only small company in the program, and it was a great opportunity to stand on the same stage as recognized leaders in the semiconductor business. We were the only company proposing to grow ingots and slice them into wafers. Czochralski (CZ) was not chosen for the program as it was considered a high-cost nature technique for producing Si single crystals. Many of the programs involved producing Si sheets directly from the melt or through sintering to avoid the cost and waste of slicing.

Our proposal was to melt Si meltstock in a silica crucible with an HEM furnace, as shown in Fig. 2. A seed crystal was placed in the bottom center of the crucible above a helium-cooled heat exchanger that extracted enough heat to prevent melting of the seed while allowing the Si liquid to melt into and nucleate from it. Directional solidification was achieved by lowering the furnace temperature while simultaneously increasing the helium flow through the heat exchanger to extract the heat of fusion for solidification. We proposed to grow the crystal in a vacuum to avoid the use of argon. A thermo chemical analysis was undertaken to determine the stability of silica in contact with liquid silicon, and silica in contact with graphite, as a function of pressure and temperature. The analysis showed that at the melting point and at vacuum below 20 torr, silica reacts with Si and graphite. Reactions occur between Si liquid and the silica crucible to generate SiO, and therefore CZ Si is grown in an argon atmosphere to suppress this reaction. However, silicon could be grown in a vacuum with HEM because the stabilizing temperature gradients prevent turbulence of the melt that breaks the boundary layer between the melt on the crucible.

A fundamental problem that had to be solved was the shattering of the Si ingot due to attachment to the crucible. The mismatch in thermal expansion of silicon and silica caused the crucible to shatter during cool down. The solid Si is intimately in contact with the crucible and forms a tenacious SiO_x bond to the inside surface of the silica crucible. During cooldown, the solid Si contracts 10 times faster than the silica and is in tension while the crucible is in compression. Silica in compression is stronger than silicon in tension, and at about 750°C the tension forces exceed the elastic strength of the silicon and it literally explodes and shatters the ingot into small fragments about the size of the starting material. This was an unsolved fundamental problem going back to 1950s when the semiconductor industry was developed. At that time, one approach to prevent the crystal from shattering was to solidify the one inch diameter Si crystals in silica crucibles with the thickness of a light bulb. During cooldown, the thin walled silica crucible shatters rather than the ingot. As the size of the crystal increases, this approach is unreliable. The ability to solidify silicon in a silica crucible without shattering had not been solved. The CZ technique, which eliminates contact of the Si crystal with the crucible, was developed in the late 1950s to produce

Si single crystals. However, the idea of using a thin crucible sparked an idea to form a thin glaze layer on the interior of a slip cast crucible with an oxy-hydrogen torch. This idea worked as the glazed interior layer shattered during cooldown leaving a crack-free crystal. This technique was used to make a 30 cm square with nearly single crystal structure weighing about 35 kg.

The ingots that were produced were characterized to determine diffusion, length, lifetime, and above all high solar cell efficiency. The material ranked high on all accounts. Solar cell efficiency is proof of the pudding, and the material exceeded expectations even in multicrystalline areas.

3. FAST Development

The FAST technology was being developed concurrently with HEM Si growth. The focus was to slice Si with diamond fixed on wire. Diamond plated on wire had been attempted and abandoned by the Si semiconductor business since it would not effectively cut Si. For diamond to slice effectively, there must be a high enough force per diamond particle to penetrate into the work so it can plow out the Si kerf. Slicing with fixed diamond requires high force, unlike slicing with loose abrasive where high force is not required. With multiwire slurry (MWS), particles tumble under the wire on the high edges to indent and chip through the workpiece. This is a slow process that uses SiC particles in slurry. Diamond is normally fixed onto the blade or wheel, where high enough force can be achieved to drive the diamond into the work to plow it out. The force necessary to slice Si effectively could not be achieved with wire because it is not rigid like a blade or wheel. It was found that if the wire was well supported with guide rollers, Si could be effectively sliced with diamond wire if the contact between the diamond wire and the work was minimized to achieve high force at the contact point. We were able to minimize the contact length by rocking the workpiece to create a tangential contact point. With this arrangement, the force was high enough for the diamond on wire to cut Si effectively. Other plating technology developments were made to increase force on the diamond by reducing the kerf width. Diamond was plated only on the cutting edge of the wire. Tear-drop shaped wires were developed to be thinner than round wire with diamond on the cutting edge, thereby increasing the force between the diamond and the work and reducing the kerf loss. With these approaches, larger diamond could also be used without increasing the kerf width.

With these developments, high enough force was achieved to effectively cut with diamond on wire. However, the cutting ratios were low because the surface speed with the reciprocation approach used for FAST was low.

To make FAST commercially viable, the cutting rate had to be increased. To achieve high cutting rates, both high force and high speed are required. The force was high enough but the speed was too low for high cutting rates. In 1995, about 20 years after FAST was developed, high-speed rotation of the work

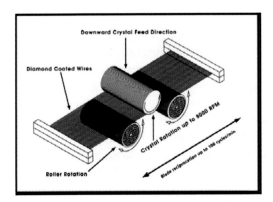

Figure 3.   Fixed Abrasive Slicing Technology (FAST) Schematic.

piece shown in Fig. 3 was conceived to create force at the tangential contact point.

The slicing rate using high-speed rotation was three times faster than multiwire slurry silicon. In addition, using high-speed rotation resulted in less sub surface damage. However, MWS was fully established for slicing PV wafers. The FAST machines were too small, and larger machines needed to be developed to rotate 125 mm square Si bars for slicing into wafers. Despite this, the R&D FAST machines were successful at slicing up to four-inch diameter sapphire wafers for Light Emitting Diodes (LEDs).

By 1979, the fundamental technical problems related to directional solidification of Si crystals in silica crucibles and slicing Si bars into wafers with FAST were solved. The Si material produced by HEM was very uniform throughout the ingot, and solar cells made from this material exhibit high efficiency. There was a down selection process, and funding for most silicon sheet programs was terminated. Programs that had shown the best potential for reducing costs were chosen for further funding. CSI, Westinghouse and Mobil Solar were selected for continued funding. CZ, which came to the program later, was also selected for continued funding.

In 1980, there was a change of administration. Jimmy Carter lost the presidency to Ronald Reagan who was not an advocate of renewable energy. President Reagan's idea of renewable energy was to drill more holes for more energy. As a result, renewable energy programs were on shaky ground. "Drill baby drill" became the new energy policy. Our program was progressing well, and we secured funding from outside commercial sources to commercialize our technology and optimize it with our ongoing R&D program. One morning early in 1982 when I arrived at work there was a facsimile on my desk ordering CSI to cease and desist from performing any further work on our program. The rationale was that the government should only support long-range high-risk technology, and that ingot growth and slicing were not considered high-risk technology and therefore should be funded by the industry and not the government.

The solar energy program was significantly reduced by the termination of funding to successful lower risk programs. More importantly, the enthusiasm for PV and renewable energy was destroyed. Many companies with successful programs discontinued their PV work as the market potential disappeared.

This was a challenging time for CSI, since neither HEM nor FAST was commercialized. Significant PV R&D was occurring in Europe, particularly in Germany. Heliotronic, an R&D division of Wacker Chemie, developed multi-crystal casting technology that reduced ingot production cost. It was shown that multicrystal solar cells with large columnar grains were more efficient than expected, but only slightly less efficient than single crystal solar cells. We had to make some difficult decisions at that point: whether to continue single crystal or multicrystalline growth and whether to go forward with FAST despite the fact that it needed further development. We did not have the resources to continue FAST R&D, so we decided to put FAST on the back burner and convert HEM for multicrystalline growth. The cycle times for multicrystalline growth of columnar grains by directional solidification would be significantly reduced, because growth was nucleated across the entire bottom of the crucible rather than from a seed. Therefore, multicrystalline growth would be more economical than HEM single crystal growth. Conversion of HEM multicrystalline crystal growth involved replacing the helium cooled heat exchanger with a heat exchange block on which the crucible was seated. The graphite block was supported by a central rod for lowering the crucible out of the heat zone for growth.

Growth was controlled by lowering the crucible out of the heat zones. This approach was successful in nucleating large columnar grains across the entire bottom and growing the grains vertically to the top of the ingot at fast growth rates. In 1984, after converting the single crystal growth process to a multicrystal growth process and commercializing it, the multicrystalline Si bars were sold to Solavolt, a Motorola — Shell joint venture. Solavolt was developing their own multicrystalline technology using lasers but wanted to enter the multicrystal market with multicrystalline Si made with an ingot technology while their technology was being developed. Solavolt's technology did not succeed, and they decided to liquidate and auctioned off their inventory. In 1987, our HEM multicrystalline bars in their inventory were sold to BP Solar, who wanted to expand into producing their own Si wafers. BP found that the solar cells made from the HEM bars were very efficient. On the basis of this, BP entered a development program in 1988 with CSI to scale up our HEM growth and FAST slicing processes for commercial production of Si wafers.

The HEM growth program was scaled up from producing 33 × 33 cm ingots to producing 44 × 44 cm ingots, and FAST slicing was scaled up to slice one hundred millimeter square bars that were mounted end to end and were up to three hundred millimeters long. This program was successful, and BP set up a small wafer manufacturing plant nearby in Massachusetts. In 1992, there was a worldwide recessionary period and BP actually lost money. BP decided to get out of the material end of the business and to buy available wafers from Russia, which was no longer under communist rule. The HEM furnaces and FAST machines came back to CSI for further development. More work was carried out to improve the reliability of crucibles and we worked with Vesuvius to develop their crucible for HEM ingot growth. With silicon nitride coating, the crucible proved to be very reliable. HEM was successful for producing 44 cm square ingots. CSI recognized that it was too small a company to be in the Si material business, but that it could sell the highly efficient HEM furnaces it had developed. Furnaces for producing 44 cm square ingots were sold to Scanwafer and others. The technology was also being developed to produce 69 × 69 cm, 240 kg ingots. It was soon apparent that CSI was a materials company and not an equipment company since we were not prepared to service the furnaces in the field. A corporate decision was made to build on our extensive intellectual property base for reducing the cost for solar silicon and license the technical developments to PV companies. In 1999, the HEM technology to produce 69 cm square multicrystalline ingots was licensed to GT Equipment Company to manufacture, sell, and service the furnaces. This successfully put GT into the solar furnace field, and they sold HEM furnaces around the world.

The HEM furnace was a top-loading furnace, which involved movement of a crucible and charge weighing over 300 kg. It was recognized in 2002 that a bottom-loading furnace with no crucible movement would be more efficient for production of larger ingots. Lowering the crucible proved difficult with an increase in ingot size and weight. Therefore, a stationary crucible would be desirable. However, the problem with a stationary crucible is that the heat flow to the solidified material breaks down the gradient for directional solidification. This slows down the growth rate and limits the height that can be grown.

To overcome this problem a system was modeled in 2003 with a stationary crucible and moveable heat shields. Growth was controlled by raising the heat shields between the heating element and crucible to minimize the heat flow to the grown ingot and to prevent breakdown of the gradient for directional solidification.

The heat flow model in 2003 showed that this gradient control system with moveable heat shields would be very effective. The feature that was unique and important for directional solidification was that raising the heat shields to reduce heat flow to the grown ingot simultaneously opened the path for increased radiational heat extraction from the bottom of the ingot. This feature would allow for constant growth rate from the bottom to the top of the ingot, as well as for the growth of taller ingots. A patent was issued in 2008, and the system was implemented for growth of 69 × 69 cm ingots weighing 240 kg. The ingots produced have very large vertical gradients with a flat interface. A bigger furnace is being designed for larger sizes, 85 × 85 mm 450 kg ingots and 100 × 100 cm, 900 kg ingots.

4. Development of Technology for Reducing Silicon Meltstock

The biggest cost driver for crystalline Si solar cells is the cost of the Si meltstock, which is mostly semiconductor Si produced by the Siemens process. When the semiconductor market is strong, the availability becomes a problem and the cost skyrockets. Production of semiconductor silicon is expensive since the process must reduce impurities to one part per billion (one ppb). In addition, the process is energy intensive, leaving a big carbon footprint. The purity requirement for PV Si is much less stringent and is in the range of 1 to 5 ppm. The goal is to process MG Si that contains 5000 to 10 000 ppm impurities to produce SoG Si with 1 to 5 ppm at significantly reduced cost.

Significant efforts have been made by many others to refine MG Si. Many variants of the classic pyrometallurgical technique, such as slagging, have been developed. However, it has been difficult to remove B and P because of their chemical stability in Si, and therefore they cannot be significantly removed by directional solidification, like metal impurities. B and P impurities must be reduced to 0.5 ppm levels to produce high efficiency solar cells. The removal of B and P impurities from MG Si by traditional pyrometallurgical methods involves a number of steps in order to produce SoG Si. We have studied the results of other technologies and our own, and have concluded that it is possible to produce SoG Si using classic pyrometallurgical techniques, but in most cases it is not cost effective.

On the basis of our high temperature thermo chemistry analysis, it was shown that a non-traditional technology used by the steel industry could potentially refine MG Si to produce SoG Si. In 2008, it was decided to internally fund experiments involving the unique variation of the pyrometallurgical approach to produce SoG Si. This one-step approach involved using a furnace to refine and then directionally solidify the liquid silicon. A modified furnace was used to refine commercial grade MG Si in the liquid state to remove B and P. This process was followed by a rapid directional solidification. P was reduced from 26.2 to less than 0.13 ppm, B from 19 to 6.3 ppm. To our own surprise, the process removed not only B and P, but also most metal impurities including Al and Fe, which are the major contaminants in Mg Si. This work is ongoing in conjunction with other refining work to reduce the cost and environmental concerns for producing high purity silicon. The goal is to optimize the production of SoG Si melt stock with this one-step process. The cost could potentially be reduced from $75 to $15/kg if each impurity including B, P, Al and FE could be reduced to less than 0.5 ppm. We intend to develop this technology so PV companies could buy MgSi at $3 to $4/kg and refine it to produce SoG Si at a cost in the range of $15/kg.

In 2003, CSI developed a directional solidification process to purify the fine silicon powders and to directionally solidify to produce meltstock. Meltstock made from fine powder was processed into solar cells and the efficiency was equivalent to virgin electronic grade Si. The source of powder became limited, but it appeared that this process could be used to process Si swarf into meltstock, which could reduce the Si usage by up to 50%. A large fraction of the Si produced for photovoltaic PV applications ends up as swarf when slicing ingots into bars and wafers. Paradoxically, this fraction increases as wafers become thinner to reduce Si usage. Current technology for slicing Si, multi-wire slurry saws (MWS), uses a loose SiC abrasive, which mixes with the swarf, and cannot be removed cost effectively. In contrast, FAST uses a fixed diamond abrasive that is nickel plated onto the wire and does not become mixed with the swarf, therefore producing swarf that is not contaminated and that is potentially recyclable. The diamond also protects the wire from wear and prevents contamination of the swarf by metallic impurities.

Swarf generated by FAST was processed into meltstock with the CSI purification process. Resistivity and Glow Discharge Mass Spectroscopy (GDMS) analysis indicated the suitability of this material to serve as high-grade SoG Si feedstock. This raises the prospect that the large and growing fraction of Si that now ends up as a disposal problem in swarf can be recycled if Si is diamond-wire sliced.

By recycling the swarf, one of the largest waste streams generated by the PV industry will be eliminated. Additional benefit to the industry would be the reduction in consumables with diamond-wire slicing. The most obvious benefit to the industry is the long-term impact of possibly doubling the supply of Si with little added cost. Future work will focus on scaling up the process, producing large volumes of swarf by FAST, demonstrating this suitability directly, and using this material to produce solar tests cells and evaluate their properties.

Crucibles for processing silicon are another cost driver since they are used only once. Most crucibles are made out of silica and are not reusable. CSI's goal throughout the program has been to develop a reusable crucible technology that would be more cost effective than silica crucibles. Reusability has been demonstrated for several non-silica crucible materials. The work is ongoing to scale up the reusable crucible technology to improve material quality and reduce cost.

5. Summary

Progress has been made in all facets of PV during the 35 years Crystal Systems has been involved in PV silicon R&D. The price of silicon, which is the major impediment to PV energy generation, has not decreased as fast as all other processes for producing solar modules. However, the technical advances that have been made to reduce Si cost have opened up more avenues for solar silicon cost reduction. With further silicon cost reductions in store, PV can and should reach grid parity. This will be a tremendous accomplishment for future generations since oil and gas reserves are not sustainable and the carbon footprint of fossil fuels must be reduced in order to minimize global warming.

The development of technology to commercialize sapphire production and to reduce PV silicon production cost has made CSI a small start-up company well known worldwide. During this period, CSI has also commercialized and advanced its HEM technology to produce 38 cm diameter, 85 kg high quality sapphire crystal, and has adapted HEM for production of 20 cm diameter, 15 kg Ti sapphire crystals of high figure merit for high power lasers.

Since 1971, the company has been involved in developing and commercializing material production technology for small sapphire and PV silicon markets. These small markets are now experiencing explosive growth.

It has been a very rewarding experience to take R&D from concept through to commercialization and to advance the state of technology to reduce the cost of production of environmentally clean sustainable solar energy and efficient LED lighting.

Acknowledgements

I would like to acknowledge many of the people who I worked with to develop and commercialize new technology over the years. Because of the large number of people, I will only be able to acknowledge people I worked with directly at Crystal Systems.

Larry Ingbar, a sailing competitor and his friend Richard Kanter, believed in me and my dream to start a company based on new technology and put up money on the basis of a handshake.

Martin Zombeck, a friend who worked with me to put a winning proposal together for developing our technology for reducing the cost of silicon to make terrestrial solar energy viable.

Chandra Khattak, who worked with me for 30 years to adapt HEM for production of silicon ingots and to refine MG Si into SoG Si feedstock.

Maynard Smith, with whom I worked with for 25 years to effectively slice silicon and other hard materials with diamond fixed on wire and bring FAST from concept to commercial reality.

David Joyce, who has worked closely with me to model heat transfer for the Gradient Controlled Crystalline (GCC) furnace, and to analyze thermal chemical reactions for different operations for refining MG Si. Based on this analysis, we developed a more efficient way to refine and purify MG Si and to recycle silicon swarf.

Dan Betty and Frank Banacos, who manage Crystal Systems Commercial Operations so I could spend more time on technology development.

The many department of energy agencies listed below whose funding was important for the technology development. The Energy Research Development Agency (ERDA), Solar Energy Research Institute (SERI), National Renewable Energy Laboratory (NREL), and especially Small Business Innovation Research (SBIR), helped commercialize FAST and to develop purification and refining technology.

Frederick Schmid

Frederick Schmid founded Crystal Systems, Inc. in 1971 to commercialize a new crystal growth technology, the Heat Exchanger Method (HEM). He has commercialized HEM for growth of large, high-quality sapphire crystals and adapted HEM for growth of Ti:sapphire crystals. He has directed R&D since 1975 to reduce the cost of silicon for PV. He developed a new multicrystalline ingot growth technology for solar energy. To complement crystal growth, he developed a multiwire slicing technology using diamond plated on wire, the Fixed Abrasive Slicing Technology (FAST), to more effectively slice silicon and other hard materials. He has been heavily involved in development of technology to produce solar silicon meltstock from silicon swarf and metallurgical grade (MG) silicon.

Chapter 38

The Story of SunPower

Richard M. Swanson

SunPower Corporation, San Jose, CA, USA

1. Introduction

The direct conversion of sunlight to electricity using solar cells has had an interesting and colorful history. Today, the use of solar cells is finally emerging as an important tool in man's quest for new sources of energy, particularly now that the negative environmental impact of burning fossil fuel is becoming so apparent. It was not always so. I began working on solar cells at the time of the first oil crisis in 1973. You could say that I was part of the second generation of solar cell researchers. The first generation had invented the practical silicon solar cell and developed it as the standard source for powering earth orbiting satellites. This article presents my personal perspective on the development of PV from 1973 to the present. (For some reason the rather difficult-to-pronounce term "photovoltaics" has taken hold to describe this endeavor. Those in the field usually just call it PV, which rolls nicely off the tongue, at least in its English pronunciation, "pee vee". So that is the term I will use here.)

2. The Beginnings of Terrestrial PV

Soon after silicon solar cells became established for powering space satellites in the 1960s, some visionary researchers began looking at what would be needed to make it cost effective for use on earth. They found that something like a 200-fold cost reduction would be necessary in order for PV to compete with conventional fossil fuel sources. This was certainly a big challenge. The silicon solar cells used for space were indeed a very expensive way to generate electric power. It is just that they were basically the only way to do it in small space satellites. The conventional

Power for the World by W. Palz
Copyright © 2011 by Pan Stanford Publishing Pte Ltd
www.panstanford.com
978-981-4303-37-8

view was that standard silicon solar cells were incapable of the 200-fold reduction needed for earth use. This lead most researches into one of two approaches, thin films or concentrators.

The idea behind thin-films was to bypass the expensive silicon purification, crystal growth, and wafer sawing steps by depositing thin layers of semiconductors directly on a low cost substrate such as glass or steel. Certainly a compelling idea — in principle. The idea behind the concentrator approach was to concentrate sunlight on a solar cell using lenses or mirrors so that each expensive cell produced more power. Clearly a 500-times concentration, easily available with a parabolic mirror, could in principle make for the needed 200-fold cell cost. The key phrase here is "in principle." Both of these approaches have been pursued since the late 1960s, and one or the other may yet prove to be the ultimate solution for generating the massive power needed to replace fossil fuel. What actually happened, however, is that engineers discovered how to vastly decrease the cost of conventional silicon solar cells, and they have continued to dominate the market to this day. The cost of PV today is about one-one hundredth of what it was in 1970 in inflation adjusted dollars, so that it is now only roughly twice the cost of competing fossil generation. The remaining 2-fold cost reduction is expected to be achieved in the next five years. This surprising outcome forms an interesting case history in technology development.

Sometimes today, one gets the impression from reading the popular press that thin film solar cells are the new big thing, surely about to displace conventional silicon cells. Perhaps the title of a paper presented at the Photovoltaics Specialists Conference in 1967 will give some perspective, "New work on CdTe thin-film solar cells," by researchers from General Electric. By 1973 the world's largest PV installation was actually a thin-film CdS/Cu_2S array that powered the University of Delaware's Solar One demonstration house. One thread woven into this chapter is a look at why it has taken so long for thin-film PV, and indeed concentrators too, to become competitive with conventional silicon.

There were a handful of intrepid entrepreneurs who did not want to wait for researchers to develop thin-film cells into a manufacturable product, and who forged ahead making conventional silicon cells despite the prevailing view that they were hopelessly too expensive. Among these were Dr. Elliot Berman, who founded Solar Power Corporation in 1973 with support from Exxon, Dr. Joseph Lindmeyer and Dr. Peter Varadi who founded Solarex, also in 1973 (sold to Amoco in 1983), and Bill Yerkes who founded Solar Technologies International in 1975 (sold to ARCO in 1978). These companies developed simplified manufacturing processes that reduced the cost to below $20 per watt in 1975 ($66 per watt 2009 dollars) and found ready markets in remote applications such as mountain top telecommunications repeaters and vaccine refrigerators in the developing world.

3. My Early Years

In 1973, I was finishing my doctoral work in Electrical Engineering at Stanford University specializing in integrated circuits (silicon microchips) when we all found ourselves spending long hours waiting in line to buy gasoline as part of the first oil crisis. It was in one of these lines that I decided to switch fields and work in solar energy. I, like many others, realized that the earth would eventually run out of the fossil fuels upon which we are so dependent, and that their use was polluting the atmosphere. This was before concern about global warming became so prevalent. Solar cells had been scientifically proven, but the required 200-fold cost reduction seemed to me like a big and exciting engineering challenge, one that could last a career. In fact, it did. The field was clearly in its infancy and wide open compared to the maturing integrated circuit business.

In response to the oil crisis, the National Science Foundation convened a workshop of 130 experts in PV that was held in October of 1973. Its purpose was to assess the current state of PV generation and develop R&D strategies for lowering its cost to the point where widespread terrestrial use was feasible. This meeting became known as the Cherry Hill Workshop. It is generally considered the birthplace of the US terrestrial PV program. I was not yet a PV expert and so wasn't invited; however, several people from Stanford University were. These included Prof. Gerald Pearson, one of the inventors of the silicon solar cell while at Bell Labs, and my friend Dr. Alan Fahrenbruch. Alan, who was working with thin-film pioneer Prof. Richard Bube, came back from the conference very excited. That set the hook for me.

But how to enter an almost non-existent field? One day a representative from IBM came to Stanford to look for candidates for the IBM postdoctoral fellowship, a two year appointment. I interviewed for the position, and when the interviewer asked what I would do if awarded the fellowship I replied, "solar energy research." He indicated IBM had no interest in solar energy, so I held little hope of getting it. Imagine my surprise when I did. It was during these two years of academic study of solar energy at Stanford that I formed much of the conceptual foundation that would serve throughout my career. Foremost among these was the realization that for widespread use, solar systems would have to be more efficient than those at the time. This is because of the large area needed to generate power. These areas would need concrete foundations, steel supports, glass, and the like, all of which had to be paid for by the energy generated, notwithstanding the solar cells themselves of course.

I had little of the background necessary for researching thin film solar cells, and in any case did not believe that thin films would be capable of high enough efficiency, but my chip experience left me with a good understanding of how to make silicon devices. I had bought into the conventional wisdom, however, that silicon cells were hopelessly too expensive and, as an academic researcher, had little exposure to the new companies entering the field. So I studied the literature on solar energy conversion concepts in the hopes of finding one with higher efficiency potential than solar cells. These included conventional steam turbines plus newer conversion concepts such as thermoelectric and thermionic converters that were then a subject of research. A brainstorm came to me to increase efficiency by shifting the solar spectrum so that silicon cells would be more efficient than they are using the color spectrum emitted directly by the sun. This was to be done by heating a radiator to incandescence with concentrated sunlight, and then converting the incandescent radiation to electricity using a silicon cell. An analysis showed that in principle this two-step approach was capable of very high conversion efficiency, perhaps twice that of the standard procedure of illuminating the cell directly with sunlight — and it utilized my understanding of silicon. Imagine my chagrin when I found that this had already been invented and researched ten years earlier by researchers in France and at MIT. They called it thermophotovoltaics, or TPV. Considerable work on TPV had been funded by the Department of Defense, not with solar input but rather looking for silent alternatives to small gasoline powered generators for use in the field. Swallowing my pride, I worked up a research plan to investigate TPV conversion in solar energy.

Getting financial support for the research was another matter. Spurred by the energy crisis, the Carter Administration formed a new agency within the Atomic Energy Commission called the Energy Research and Development Administration (ERDA). (In 1977 ERDA was merged into the newly-formed US Department of Energy.) ERDA said that they would not fund my research because it was not clear if TPV would actually work. They suggested I talk with the National Science Foundation, which funded more speculative projects. The NSF said that they could not fund my work because all solar research was now to be done by ERDA — a nice bureaucratic catch-22 indeed. Fortunately, the US electric power industry saw the need for energy research and formed the Electric Power Research Institute (EPRI) in 1973. I approached EPRI in 1975 and they actually liked my ideas about the need for efficiency. Thus began a supportive relationship that lasted until 1993. At the end of the IBM post doctoral fellowship, Stanford decided it would be good to have some research in PV. I was offered a job as a new junior faculty member in 1975, and accepted. At that time, Stanford was about the only university with state-of-the-art laboratories capable of making silicon transistors and integrated circuits. It was the perfect venue, indeed perhaps the only venue, to continue developing TPV.

I settled in as a professor and began to meet the global community of people working in PV. This included Profs. Roger Van Overstraeten of the University of Leuven in Belgium, and Adolf Goetzberger, founder of the Institute for Solar Energy in Freiburg Germany. Both spent sabbaticals working in my lab at Stanford. I also developed a close relationship with Prof. Antonio Luque at the Polytechnical University of Madrid. The PV community was an amazing group of visionary and dedicated people that helped create today's multi-billion dollar solar industry. It has been a wonderful pleasure and honor to have worked with so many of these people over the years as we strove to make PV cost effective and practical.

Meanwhile, work in the lab revealed the TPV idea was basically dumb. Nice on paper, but hopelessly flawed due to numerous practical issues. As part of the development effort, however, we had invented the so-called point-contact solar cell. This cell had its metal contacts where current was extracted from the silicon reduced to small points. This was done to make the cell more reflective to the unused infrared portions of the spectrum coming from the TPV radiator. Surprisingly, however, the cell was working better than we ever imagined it could. By throwing away the hot radiator and illuminating the cell directly with concentrated sunlight we achieved a conversion efficiency of 20%, which was over 50% greater than that of conventional silicon solar cells of the time. In 1982, the decision was made to drop TPV and work on high-efficiency silicon concentrator cells. In other words, we threw away the radiator and let the concentrated sun directly hit the solar cell. We also decided to move the top metal electrode to the back of the cell, with positive and negative leads forming an alternating interdigitated grid across the back side. This type of cell had been researched in the 1960s by Prof. Richard Schwartz of Purdue and his group in conjunction with TPV converters. He called it an IBC cell, for interdigitated back contact cell. Such a cell is particularly suited to high concentration use. This is because high concentration also results in high electrical current. In the IBC cell it is possible to make the current carrying metal leads wide and thick to reduce electrical resistance, without obscuring the front of the cell where the light enters. We basically merged the point-contact and IBC cell concepts. That worked out quite well. Figure 1 shows the structure of the point-contact cell.

In 1981, there was an administration change in the US with the result that funding for solar energy was drastically reduced. Many of my research friends

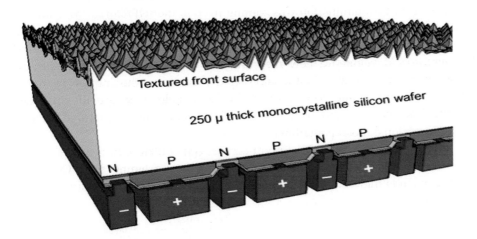

Figure 1. Cross-section view of the point contact cell. The metal conductors on the back touch the silicon through small holes in a backside passivation layer.

and colleagues were forced to leave the field for more mainstream endeavors such as microchip research. It was a sad time indeed. Fortunately, my funding came from EPRI, and was unaffected. The utilities still held out some hope for solar. Nevertheless, since there was very little employment opportunity in PV, I designed my research program so that the PhD students learned basic semiconductor skills that would suit them well in the microchip world. This approach had the added benefit that we improved the basic understanding of the physical process operating in solar cells as well as silicon chips. Such improved understanding provided a continuing foundation for continuously improving the cell efficiency.

Many very talented people joined my group at Stanford. These included 13 PhD students, each of whom increased our understating of solar cell operation. Most notably there was Dick Crane, the engineer who built our cell processing and characterization equipment, Dr. Ron Sinton, who became a key member and research associate after completing his PhD on solar cell physics, Dr. Richard King, who helped us apply our technology to non-concentrating cells,[1] and Dr. Pierre Verlinden, who spent a year in my group as a NATO Fellow on leave from the Catholic University of Belgium. When Pierre returned to Belgium he said "If you decide to start a company, let me know and I'll come back."

We methodically attacked each factor that limited cell performance. One such factor was metal contamination in the silicon. Some metals such as iron and copper can poison cell performance, even in very small concentrations, by creating so-called recombination centers. These catalyze the loss of the charge carriers generated by sunlight, carriers which we wished to remove from the cell and coax into the metal wires connected to it. Unfortunately, due to the very low concentrations involved, conventional methods of measuring metal concentration were very laborious, expensive, and time consuming. We could not find where the poisonous metal contamination was coming from in the cell manufacture, and so eliminating it was proving hard. We called this the "lifetime war" because the contamination reduced the lifetime of free charge carriers. After many false starts, we eventually developed a new measurement technique called "eddy current photoconductivity decay" that could rapidly measure the impact of metal contamination. With the help of this tool, we finally won the lifetime war. Actually, it is not completely won, for even today solar cell manufacturers have to fend off occasional metal contamination skirmishes.[2]

By 1985 we had increased the point contact cell efficiency to over 28% under concentration. We also broadened our research with funding from DOE and NREL to include high efficiency non-concentrating cells. These are usually called

---

[1] Richard King subsequently joined Spectrolab , where he led the team that developed record-setting 41% conversion efficiency triple-junction solar cells.

[2] Later, Ron Sinton formed Sinton Consulting to manufacture and sell lifetime measuring instruments. Today, Sinton testers are standard tools used in all silicon solar cell manufacturing lines.

one-sun or flat plate cells. Our main competition in high-efficiency one-sun cells was Prof. Martin Green's group at the University of New South Wales in Australia. His group tended to hold all the efficiency records for one-sun cells whereas we held them for concentrator cells. We did set one one-sun record of 22.7% in 1989, but it was subsequently eclipsed by UNSW, which holds the current record of 24.7%.[3]

We also greatly improved our detailed understanding of how solar cells work, and it was becoming clear that there was little room for continued efficiency increases. We were approaching theoretical limits. This caused something of a crisis with me. Should I head in some other research direction? Should I continue blindly ahead in silicon cell research even though it seemed that there was little fundamental research left to do? Or should I leave the university and continue with the next stage — commercializing the point contact cell? In the end, that is where my dream was, and that is what I decided to do.

4. Formation of SunPower

Dick Crane and I formed a company called Eos Electric Power in 1985.[4] We had the dream of making solar energy cost-effective on a utility scale by using high efficiency silicon cells in concentrating systems. In actuality, however, since we had no money we were basically a consulting company. EPRI also came to the conclusion that it was time to commercialize high concentration PV. We did some consulting for the EPRI high concentration program, but Eos mostly lay dormant in our filing cabinets. The times were very poor for raising venture capital to do PV, and we found none interested in us. There emerged a period of considerable foment as EPRI forged ahead with its commercialization thrust. EPRI introduced us to an experienced entrepreneur named Bob Lorenzini. Bob founded one of the first silicon wafer companies, Siltec, and had recently sold it to Mitsubishi. He was looking for the next big thing, and we hit it off. Bob joined as our president and immediately doors began to open. He loaned us money to hold us through until we could find venture capital. (This is what would be called angel funding today.) Bob's first official act as president was to change our name from Eos to SunPower. This was a very good move, as the name SunPower proved to have real brand appeal. Still, finding investors was difficult. During 1988 and 1989 we gave over 40 pitches to potential investors, all replying with a big "no." The view within the venture capital community was well summed up by the comment of one well known VC: "Someday PV will make someone a lot of money. I just don't know whether it will be in 20 years, 50 years, or 150 years. In any case, it is more than

---

[3] SunPower's eventual commercial success actually came from one-sun cells, but that was far in the future at this point.

[4] Eos is the goddess of the moon whose job it is to open the gates of heaven each morning and let the sun, Helios, out.

the 5-year horizon we have for our investments." In the end he was nearly right, it took 17 years.

As part of its high concentration program, EPRI released a request for proposals to develop commercial concentrator systems using the Stanford work as the starting point. We, of course, responded with a proposal, as did many others. EPRI was inclined to give us a contract, but only if we found VC funding to sufficiently capitalize SunPower. Simultaneously DOE announced a new program called the Concentrator Initiative. DOE had funded concentrator work since 1975, and had little commercial success to show for it.[5] But there was still reason for hope in their view, so the Concentrator Initiative was conceived as a last shot at seeing if a viable concentrator technology could be developed. SunPower submitted a response proposal to the DOE Concentrator Initiative as well. DOE was also inclined to give us an award, but with the added caveat that SunPower now had to get both VC *and* EPRI funding.

During this period we finally located two VC firms with an interest in what was then called "environmental investing." Today it would be called "clean tech," and many firms are active in this segment. These firms were Associated Venture Investors and Technology Funding. They agreed to invest $800 000 in SunPower with yet another caveat, we needed to get both the DOE and EPRI contracts that they knew we had applied for. They felt it was too risky to fund the whole project. This was a dicey situation. We had three parties each willing to invest, provided all the others did too. Eventually in August of 1990, all was worked out and we were off and running with over $5 million in commitments.[6] One of the conditions that Bob placed upon joining SunPower was that when we got funding I would resign from Stanford. He wanted my full commitment. So I resigned from Stanford, as did Dick Crane and Ron Sinton. Now we really had to succeed.

5. Concentrators

We began 1991 with full of great hope and expectation. Pierre Verlinden came back from Belgium to join us. We had substantial EPRI and DOE research contracts to develop a capability for making concentrator cells in pilot production, as well as develop a reflective parabolic dish concentrator in which to put them. We signed an agreement with Stanford University to use their laboratory facilities while we built our own pilot line. The Northern California utility, PG&E, started an innovative program called PVUSA, which stood for PV for Utility

---

[5] Conventional silicon cells continued to dominate the field, and thin films became the favored new technology. More on this later.

[6] In the end, EPRI decided to maximize the chance for success by funding "two horses in the race," as they put it, SunPower and Amonix. Amonix would eventually become a leader in concentrating PV systems.

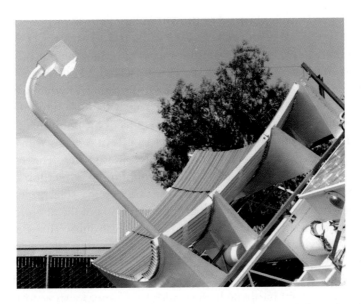

Figure 2. SunPower's dish concentrator test in 1992. Glass mirrors focus light on the cells, which are located in the receiver at the end of the boom.

Scale Applications. They were testing and experimenting with grid-connected PV systems, and SunPower was awarded a contract to put a 20 kW concentrator at the test site. By May of 1992 we had a one-third scale model of the dish completed and operating successfully at SunPower, which was now located just south of Stanford in Sunnyvale, California (Fig. 2). Our pilot line was also up and operating.

There were clouds forming on the horizon, however. Our research contracts required us to co-fund the project using SunPower's internal funds. The investors put another $1.3 million into SunPower, but our cash was rapidly depleting and we needed more. They said it was time to find additional investors. Once again we were on the road making investor pitches. After 23 presentations we still had no takers. We had the first of many staff reductions throughout SunPower's history. We were forced to cancel our PVUSA contract because we didn't have enough cash to proceed with the system. Ron Sinton left and formed Sinton Consulting. Things were very tense. I became President, and Bob Lorenzini assumed the role of Chairman. Then in February of 1993, DOE decided to cancel the entire Concentrator Initiative, and our contract with it of course. EPRI then followed suit, saying that we were now sub-critically funded. We went from over $100 000 per month in contract revenue, and all our revenue was contract revenue, to practically zero almost overnight. 1993 started out very dim indeed.

In the midst of this, it was becoming clear to me that the concentrator market was going to be a long time in coming, if it ever did. The price of natural gas declined precipitously following its deregulation by President Regan, and utility interest in PV was waning. Instead of the $10 million, five-year project I had sold to the VCs, it now looked more like a $150 million, 20-year project. No VC was interested in that. So we began a painful process of rethinking our business plan. The basic idea that emerged from our searching was to find other, non-solar activities that would be profitable enough to keep SunPower alive while we continued to work on our long term solar dream. We began to look into silicon power devices, the switches used on all sorts of power electronic equipment like power supplies and motor controls. I fervishly tried to invent devices we could make, and several patents were applied for and granted in this area. This resulted in a sort of bait-and-switch tactic that allowed us continue to attract top engineering talent — come to SunPower and you can work on making solar practical, but in the meantime you have to work on other things. It also got the attention of utility that was skeptical on solar, but interested in power electronics. That was Northern Indiana Public Service Company (NIPSCO), who warmed to the idea of investing in power devices that could improve utility operations. Based on our new plan, NIPSCO, AVI and TFI invested a combined $2.2 million in June of 1993. Then another important event happened. Honda Motor Company showed up.

6. Race Cars

Honda was interested in building a solar powered race car for the 1993 World Solar Challenge, a race for solar powered cars 3000 kilometers across Australia from Darwin to Adelaide that is held every three years. They had participated in the second such race and come in second, and very much wanted to win in the third race to be held in the fall of 1993. They had heard about SunPower's high-efficiency concentrator cells, and said, "We know you make the most efficient concentrator cells, and that they are quite small at 1 centimeter square. Can you make enough to cover our race car?" They were willing to pay a substantial sum if we could. We said "sure." Honda gave us a nice down payment to keep the lights on.

Of course, we had never made anything like this number of cells as yet. We needed to cover about 12 square meters with cells. Actually, we had had some experience making non-concentrating cells from the Stanford days, but that was only on a laboratory scale.[7] We began preparing for the new challenge, making 7000 cells for the Honda Dream. As we were attempting to plan the production, I had a scary realization. We had no clue what we were doing. None of us had ever worked in industry, or made anything in production before. We were all academics. That is when, through happenstance, I ran into TJ Rodgers, the CEO and founder of Cypress Semiconductor. I had known TJ from our days together as Stanford students, and relayed my concerns about our ability to make enough

---

[7] All one needs to write a nice academic paper is one working cell.

Figure 3. The Honda Dream finishes first in the 1993 World Solar Challenge.

cells for Honda. TJ put us in contact with Mark Allen, who had just left Cypress and was preparing to go to business school in the fall. Mark was a well known Silicon Valley chip manufacturing executive. He agreed to come onboard for the summer and help us get the production done. At that time there was a serious downturn in the chip business, and many experienced semiconductor production people were available. Mark knew lots of them, brought them on board, and within a month we were up and running with three shift production. The factory was abuzz with activity. We produced the cells, Honda put them on their car, called the Dream, and went on to win the race by over a day ahead of the second place car (Fig. 3). It was a huge shot in the arm for SunPower.

Many of the people who came to SunPower for the race production are still there. One, Bobby Ram, would play a key role in keeping SunPower going. Mark went on to a stellar career in semiconductors and venture capital. TJ Rodgers would later figure prominently in turning SunPower into an industry leader. Besides purchasing millions of dollars of cells, Honda eventually would invest $3 million in SunPower. This definitely helped keep us solvent at a time when our original investors were pretty much tapped out.

When we embarked on the Honda production our plan was to have a big party after the race, after which our production people would go get a job in a real semiconductor company. SunPower would revert back to an R&D company. We had so much fun making and selling an actual product as opposed to doing contract R&D work, however, that we all looked at each other and said, "Let's find something we can make and sell and keep the production going."

7. Optical Detectors

After the hubbub of the race production died down, we were simply a company with a small specialty factory designed for making small quantities of very expensive solar cells, and no products or customers. The specialty solar cells we sold to Honda cost about 100 times that of a conventional solar cell. Honda was quite willing to pay that in order to win the race, but the market for these was clearly very limited. We were back to zero revenue.

The closest product to a solar cell is a light, or optical, detector. We could make those. Bobby Ram scoured the globe for potential detector customers. It was hard work convincing companies to take a chance with an unknown solar wannabe company. We had two secret advantages, however. We had attracted some very bright engineers to SunPower, and we had fought and won the lifetime wars. We could process silicon wafers and introduce very little unintentional contamination. In solar cells this meant high efficiency. In detectors this meant low leakage current and superior low light level performance. Soon customers began trickling in. We started with the detectors used in grocery store check-out registers that read the bar codes. This expanded to a wide range of products. The largest volume products were the detectors used to read infrared beams that transmit data between hand held devices. Here we established a dominant market share position. These ended up in Palm Pilots, HP laptop computers, cell phones, and digital cameras. We made nearly a billion of them before closing down what became known as our opto business. The product we were most proud of were very sophisticated detectors used in the latest generation of GE X-ray CAT scanners. These enabled detailed images of internal organs.

The opto business basically funded SunPower through the period 1993 to 2000, and kept our solar dream alive.

8. The PV Business Takes Off

Meanwhile, SunPower was being left behind in the PV business. The concentrator market had not materialized, but the conventional solar module business had. Early on, most PV modules were used to power remote needs such as telecommunication towers on mountaintops, now a new market began to emerge. This was residential and commercial roof mounted systems that were connected to the power grid. This proved to be PV's next killer application. The market began growing quite rapidly after Japan instituted an incentive program for residential roofs. At that time, the leading PV companies were all owned by major oil companies. Rumor had it that they were losing money on PV, but then they had lots of money to spare. We periodically looked into trying to enter the terrestrial module business, but couldn't see how to compete. Not only were our high efficiency cells way too expensive, but we couldn't figure out how to make even conventional cells at the market prices. We could only watch in horror. Through this period,

SunPower was viewed in the PV industry as a tiny boutique manufacture of very efficient jewel like solar cells that were useful only for things like solar race cars and solar powered airplanes. In other words, nothing serious.

It is interesting to observe how the PV industry developed quite differently than originally conceived. Neither thin film modules nor concentrators gained significant market share. Instead, engineers, scientists, and manufacturing people found ways to vastly decrease the cost of conventional silicon cells. For example, the cost of refined silicon decreased from $300 per kilogram in 1975 to $25 in 2000. This was gratis of the vastly expanding semiconductor chip industry which had driven technology development. In effect, the chip industry completed the original DOE goals set in 1975 for PV silicon of $5 per kilogram ($20 in 2009 dollars). Thank you chip industry. Another seminal innovation was wire sawing. In 1975, each wafer was individually sawn from the silicon ingot — a very slow process indeed. This single factor probably did most to create the reputation that conventional PV as hopeless. Modern wire-saws, however, saw 4000 wafers at a time in a very cost-effective manner. So much for that problem. The more modules that were made, the more people figured out how to decrease cost. It was found that every time the cumulative number of modules manufactured doubled, the cost decreased by 20%. This has been going on since the beginnings in 1975 to the present. Because the industry has grown more than 10 000 fold since 1975, all this accumulated learning has reduced the cost of PV modules to around $2 per watt today. And this learning shows no sign of letting up. In effect, new technical approaches faced a moving target of ever decreasing prices. The larger the mainstream technology (standard silicon solar cells) became the more resources that could be applied to further decrease cost through R&D and increased scale.

Another aspect that turned out differently from the original vision was the end market. In 1975 it was thought that the big market for PV would be giant solar farms in the desert. As it turned out, it was the rooftop market that actually materialized.[8] Unfortunately, SunPower's concentrator concept was not suited to rooftops, it was conceived for giant solar farms in the desert. Meanwhile, SunPower was stuck making optical detectors.

9. Airplanes

SunPower's next PV breakthrough came from NASA. NASA was developing a solar powered airplane. Their vision was for a solar powered "eternal aircraft" that could stay up for extended periods doing things like monitoring atmospheric pollution. For the concept to work, they needed the highest possible efficiency, SunPower's specialty. NASA knew about us from the publicity surrounding the

---

[8] Today, large solar power plants are just beginning to be practical now that the rooftop market has helped decrease module prices so dramatically.

Figure 4. NASA's solar powered airplane over Hawaii.

Honda Dream, and we began making cells for their airplane, Helios. Fortunately, Helios was much bigger than a car, and needed lots of cells. In fact its wingspan was the same as a Boeing 747's. Over the period 1997 to 2000, SunPower made 60 000 high efficiency cells for NASA.[9] In 2001, Helios set an altitude record of 96 800 feet, the highest flight ever for any non-rocket propelled aircraft (Fig. 4). We may still have been a boutique cell manufacturer, but we were now a bigger boutique.

10. Project Mercury

Since NASA was interested in making lots of solar powered airplanes, they asked us what we could do to decrease the cost.[10] It seemed $6 million for each airplane's cells was too much. We had gained valuable experience in making so many high efficiency solar cells for Helios, and so the engineering team launched a project into how we might use that experience to decrease manufacturing cost. This project, which was headed up by one of our engineers, Dr. Bill Mulligan, was named

---

[9] The actual customer was AeroVironment, Inc., the company that built Helios. AeroVironment was founded by Paul MacCready, the first person to build a successful human powered airplane, for which he won the Kremer Prize in 1977.

[10] Actually, Helios eventually crashed into the Pacific Ocean on a test flight, and the whole solar powered airplane project was cancelled.

Project Mercury. It was a very successful project, exceeding our wildest expectations.

We had hoped to decrease the price from $200 per watt to $60 per watt. Ways were found to eliminate all the expensive photolithographic patterning steps that we inherited from the concentrator cell manufacturing process without unduly decreasing cell efficiency. We built a manufacturing cost simulation computer program to compute what the cost would be in a large factory, and to our surprise the computed cost came in below that we suspected to be the case even in the large, terrestrial PV companies. The reason for this was simple. Our high efficiency (50% more than a standard cell) resulted in more watts per expensive silicon wafer, more watts for the same amount of glass and other encapsulation materials, more watts for the same installation cost, and so forth. The added process complexity of our high efficiency cell did not appear to increase the cost enough to offset these advantages. On a dollar per watt basis, we felt we could be competitive. Of course our factory was just a computer spreadsheet fantasy, and we had no money to build a real factory to see if we were right.

11. A New Plan

We knew that the Helios project was just that, a project that would soon end. Then where would we be? Our existing investors had become tired of SunPower. Armed with our low cost, high efficiency cell idea we decided to once again look for new investors. We put together a business plan that encompassed all of our products — optical detectors, concentrator cells, and one-sun cells — and began talking to venture capitalists and investment bankers.

Concentrators were included in the plan because we had kept a small concentrator cell business going, mainly providing cells for the big parabolic dish concentrators that Solar Systems, Inc., in Australia was building. We also had worked on a large R&D grant from the National Institute of Standards. We doubted we would be able to raise enough money to build a large one-sun cell factory for some time, so these products would have to carry the weight for awhile.

Once again, the investment community was totally uninterested in solar. Throughout 1998 and 1999, we talked to bankers and VCs until we were blue in the face. The common refrain was: please bring us some optical fiber or internet concepts and then we can talk. This was, after all, the internet bubble time. After a year of fruitless search we were beginning to run out of money. The tech bubble burst and our opto business took a big hit. These were again difficult times for SunPower. We had more layoffs, and many of us became burned out or lost faith. It had now been over 12 years of struggle. Dick Crane left and went back to Stanford. Pierre Verlinden left. Then I had another talk with TJ Rodgers, CEO of Cypress Semiconductor.

12. Cypress

TJ immediately grasped the proposition — getting into the power generation business turning sunlight into electricity with silicon solar cells. Cypress was, after all, a maker of silicon chips and understood that material very well. He instinctively knew that PV was going to be a big business, and that silicon was the path to success. He liked the concentrator idea too. Cypress had been forged into a very lean and capable manufacturer by 25 years of competition in the brutal computer chip business. He quickly sized up the PV industry and realized that the manufacturing sophistication of the existing companies was quite limited. TJ felt that by combining our cell technology with Cypress' manufacturing knowhow, we could be a winner. His initial view was much more intuitive and conceptual than that of the banking community, which seemed to focus more on financial spreadsheets. The time for very careful number crunching came later.

I explained how we were about to run out of money, and that I was going to have to lay off most of the staff soon. TJ immediately offered to loan us $750 000 of his own money to tie us over until he convinced Cypress to invest. TJ now had to convince the Cypress board of directors of the wisdom of investing in PV. Their initial reaction was something like, "you have to be nuts," and turned him down twice at two separate board meetings. But he is not one to be easily thwarted, and put together a massive presentation for the next board meeting. He gave a truly inspired and impassioned two-hour speech about the importance of solar, and the role that existing chip companies can play in making it happen. He presented it as the logical extension of what silicon manufacturers should do next. At the end there was hardly a dry eye in the house. The board said, "TJ, if you want your solar company, go ahead." In June of 2002, Cypress invested $9 million, and got half the company in exchange. They also had rights to purchase the remaining shares at a pre-negotiated price, should we meet certain milestones.

Prior to investing, Cypress commissioned PowerLight to install 350 kW of standard PV modules on the roof of two of their buildings in San Jose, California. They were interested in seeing just what this PV thing could do. At the time, it was the largest privately-owned PV system in the world. The other large systems tended to be government demonstration projects. That is indicative of the commitment Cypress was ready to make. PowerLight was the largest PV installer at that time. They would later figure decisively in SunPower's future.

13. Goodbye Concentrators

We also planned to install a large, 20 kW concentrating dish on the Cypress campus as a test of our concentrator concept. There emerged a problem, however. The dish would require a large set back covered with only gravel to guard against fire caused by reflected light in the event that the dish was not pointed at the sun.

That was a problem in that the dish was to be in the parking lot. The dish was also rather large, in fact as tall as the three-story Cypress headquarters building. It would be a visible landmark for much of northern San Jose, California. This is to be compared to the non-concentrating system on the already roof that provided 15 times the power, and was invisible from the ground. I had a dream one day, or rather a nightmare, in which we had installed the dish and we were showing it to the local press. A reporter asked me what portion of the building's electrical consumption was supplied by the dish. I was forced to admit that this giant landmark only supplied a tiny fraction of the needed power. He then looked at me like I was crazy. It was embarrassing. Bob Lorenzini, Bill Mulligan met on May 1, 2001 to talk about the fate of concentrators. Our one-sun project Mercury was going very well. We all looked at each other and simultaneously said "let's cancel the concentrator project and focus our effort exclusively on one-sun cells." We later discussed the decision with TJ, and he agreed — luckily. We had thrown away the radiator 20 years earlier when we abandoned TPV, now we had thrown away the concentrator too. It was a propitious decision that proved correct. Putting the cells directly in the sun, without the complication of concentrating mirrors and tracking equipment, was the way to go.

SunPower was now getting serious. Charlie Gay, who had at one time been president of the then largest PV company, ARCO Solar, and at another time the director of NREL, recognized the value of our high efficiency approach. He came on board, helped put our strategy together, and sell the vision. We also needed PV marketing experience and knowhow. Luckily, with Charlie's help we convinced Peter Aschenbrenner, then Senior VP of Global Operations at AstroPower, to join. Charlie and Peter in turn convinced the rest of us that we could indeed manufacture and sell PV our high efficiency modules on a large scale. The team was coming together nicely.

14. Becoming a Manufacturer

Now we had to figure out how to transfer our one-sun cell design into high volume manufacturing. The cells would have to be made at low cost and with high manufacturing yield. Cypress helped here by installing our pilot production line in an existing Cypress factory in Round Rock, Texas. We worked closely with experienced manufacturing personnel. The idea was to quickly inoculate our team with a manufacturing mentality, and train them in the latest in manufacturing methodology. Many of our engineers moved to Texas temporarily to participate in this project, making big sacrifices in order that we succeed. By the spring of 2003 we had cells from the pilot line that met our requirement of 20% conversion efficiency. SunPower announced its new high efficiency PV module in May of 2003. Now we needed a factory in which to make them. This factory would have to be more than 20 times as large as our pilot line. We had met the goals set out by Cypress, and they purchased the remaining outstanding shares of SunPower early in 2004. We were now a wholly owned subsidiary of Cypress. They did promise to take us public if we succeeded, however. Cypress would eventually invest $150 million turning SunPower into a manufacturer. In retrospect, we were lucky none of the VC's did invest because there is no way they would have been able to bring this much money to the task, let alone provide the manufacturing experience.

It was also time to bring in a manager with experience in large, fast growing enterprises. We courted TJ's preferred candidate, Tom Werner, who was running another Cypress subsidiary at the time. Tom had the perfect background having had high level management positions at companies like GE and 3-Comm. Tom was unfamiliar with PV and a little skeptical. We sent him to a conference in Osaka, Japan where he saw first-hand how dynamic and interesting the PV industry had become. He was hooked. Tom became CEO, and I President and CTO, now able to concentrate on our technology.

As a newcomer to the mainstream flat-plate PV module business, we knew that we would have to gather every advantage we could muster in order to compete against the established players. By this time, Sharp in Japan had become the industry leader, partly as a response to the Japanese rooftop incentive program. They, and the other leading companies, were much larger than us. Cypress felt that we needed more than just our efficiency advantage, so we looked to a low cost manufacturing region in which to locate. We chose the Philippines because Cypress already had a factory in the Philippines, and was familiar with operating there. We found a suitable factory that had previously been used to assemble disk drives, bought it, and began converting it into a high efficiency solar cell factory. This was a bold move. At that time nearly all PV production was located either in Europe, Japan or the US. Cypress, as part of the semiconductor industry which long before had moved much production to lower cost regions, was quite familiar with how to do it, whereas the existing PV industry was not. Today, of course, things have changed, and there is lots of PV manufacturing throughout Asia.

Once again our engineering team moved, this time to the Philippines, some for up to a year. Cypress donated much of their manufacturing management talent to the cause. Some moved to the Philippines permanently. The integration of one company into another is always fraught with difficulty; however in this case things went remarkably well. I think this was in large part because the SunPower team really wanted to succeed in changing the world, and was psychologically prepared to do what was necessary. The effort and sacrifice put in by SunPower people was again enormous and key. Conversely, the support we received throughout Cypress grew and grew as people became excited about the new challenge of making solar cells in high volume. There are similarities between chip and solar cell manufacturing because many of the basic processes are similar. But there is a big difference in terms of volume. Our first factory was to have a production capacity of 100 MW per year. That translates in to 33 million, 3-watt solar cells, or about 100 000 cells per day. A typical chip factory might not start that many wafers in a year. The challenges created by this volume were daunting and exciting. After much hard work, the factory was ready to start producing in the fall of 2004.

We began shipping product late in 2004 and the response was amazing. The demand for high efficiency modules proved much large than we anticipated even in our dreams. We rapidly moved to the number one spot for residential installations in California, something that would have been unthinkable if we built standard PV modules. Then at Christmas a huge problem unfolded. Reports started coming in that our systems in the field had begun to lose power.

15. Polarization

We had no idea what the problem was. I put together a team to focus intently on the issue. It was scary indeed. $150 million invested and perhaps it was all going to crash down around us. After working around the clock for several months, we had identified the cause of the problem, and found a solution. Very high efficiency cells are susceptible to a phenomenon we discovered, which we named polarization. It had heretofore not been observed because it only occurs when modules are installed in a large system that generates high voltages. The incredible thing was that it only occurred when our high efficiency modules are operating at positive voltages compared to earth, and not when negative. The normal procedure was to have the negative terminal grounded, some leaving the modules at a positive voltage. We found that he whole problem is avoided by simply grounding the positive terminal instead. More amazingly, the power loss in modules could be recovered in the existing systems by simply changing the ground connection. Needless to say, we were overjoyed by this finding. Teams went into the field and fixed all the existing installations. The systems recovered, and no problem has since developed. It was a scary event, and now we were back in action. But I could feel the wind of the bullet as it went by.

16. IPO

Big changes were afoot in the PV business as the decade progressed. The market had grown dramatically in the latter half of the 1990s driven by the Japanese rooftop incentives. The mantel would soon pass to Germany, which passed a far reaching PV incentive program called the Renewable Energy Law in 2000. Companies were expanding production at a rapid pace to serve the new demand in Germany. In order to grow into a major manufacturer we would need more capital than Cypress could supply alone. It was time to look to the public markets. There had been very little investor interest in PV in the US, and only two solar initial public offerings (IPOs) in the last 20 years. The situation in Germany was better. A handful PV companies had gone public on the German stock exchanges, starting with

Solarworld in 1999. Entrepreneurs had discovered that it was now possible to make money manufacturing solar cells because the manufacturing scale had become large enough that costs were coming down quickly. So when we began talking with investment bankers in 2005 there was lots of skepticism, but certainly a glimmer of curiosity and hope. We decided to go public on the larger US stock exchange, NASDAQ. No one was prepared, however, for the huge reception we received in November of 2005 when we began our investor road show in Europe. Each presentation was delivered to a standing room only crowd. It was clear we could have sold all the offered shares in the first morning of the first day. And there was a week of presentations left. There had developed a large pent up demand for solar stock that seemed to have gone unnoticed by the investment banks. They noticed it now. The IPO was a huge success, with the share price 41 percent on the first day of trading.

A flurry of PV IPOs was to follow. In a short period of time everything had changed for the industry. Lots of capital was now available to start-up PV companies, fueling rapid growth in manufacturing capacity and market size.[11] The older established players, which were largely subsidiaries of large corporations, found it hard to play this new game due to their slow decision making process. For example, as the industry grew there developed a shortage of highly refined silicon. The existing silicon refiners were reluctant to add capacity because of perceived uncertainty in whether the new PV market would be lasting. New companies, flush with capital, were capable of funding this expansion in the form of down payments. The large multinational corporations balked at these prepayments, and soon found themselves without sufficient silicon. By the end of the decade, most of the industry leaders had been in the market for less than ten years.

17. PowerLight

One of SunPower's largest customers was PowerLight. PowerLight had pioneered a new way to install PV panels on flat commercial roofs that involved no roof penetrations. Prior to their invention, the common wisdom was that panels should be mounted on steel frames that tilted the panels south to receive the most sun. Installing the frames required many roof penetrations for anchoring, a tedious and time consuming task that was disliked by building owners due to the potential for roof leaks. Tom Dinwiddie, PowerLight's founder, discovered a way to interlock panels that simply lay on the roof such that wind could not blow them off. This resulted in vastly reduced installation cost and time, and catapulted PowerLight to the PV largest installer in the world. From this beginning, PowerLight became a leader in ground mounted power plants that use trackers to

---

11 SunPower alone has raised $1 billion in the public markets.

Figure 5. Large SunPower PV power plant using tracking structures to follow the sun.

follow the sun. Tracking the sun results in more energy delivered throughout the day. PowerLight loved our panels because they could get 50% more power off of the same installation. We liked PowerLight's innovative products and understanding of PV customers, so we decided to merge with them. SunPower thereby became in integrated company covering the PV value chain all the way from growing silicon crystals to installing systems at customer sites. Figure 5 shows one of SunPower's large-scale PV power plants.

18. Epilog

Today, 25 years after our fledgling start as Eos, SunPower is a billion dollar company with 5000 employees and a global presence. We are proud to be part of the dynamic PV industry that is helping the world wean itself off of fossil fuel. The next decades will see continued dramatic changes in our industry as we grow to become a major energy source. Some think PV is too expensive to be taken seriously. This is not the case. For example, SunPower, as well as our competitors, routinely bids large power plant projects that guarantee an electricity cost of less than the so-called market price referent in California, which is around 12 cents per kilowatt hour. Peaking power often costs utilities well in excess of this. Sometimes it is hard to imagine where that old 200 times to expensive PV cost all went. Of course, lots of hard work was needed to get this cost down. There is another factor at work, however. We discussed how silicon cost has decreased over time. Other major components of a PV system have also. Steel and glass cost about the same today as they did in 1975 — without adjusting for inflation. There has been a four-fold loss in dollar value since then, so in inflation adjusted dollars they cost 25% as much as in 1975. Commodities have their own learning curve cost reduction too, at least until they become scarce. On the other hand, energy cost has gone up in inflation adjusted dollars, nearly doubling in fact. Rising fossil fuel cost is simply a consequence that it is starting to become scarce. It is nice to be in an industry where ones cost is coming down and ones competitor's is going up.

Some argue that even if PV did become cost competitive, it will remain too small an energy source to matter. Nonsense. It is true that PV generated only about 0.2% of the worlds electric energy last year. But our industry has grown 200 fold in the last twenty years, and we expect to grow anther 200 in the next twenty years. Simple math indicates that we will then be quite significant. And PV can support such growth. For example, SunPower can install a power plant at a rate of 1 MW per day using a small field crew (Fig. 5). That translates into 300 MW per year, or 3 GW in ten years. Such a plant will produce as much annual energy as a large nuclear power plant, which would take at least as long to permit and build. One small crew installing the equivalent of a large nuclear plant. We can convert our energy source to solar. It can be done, and it must be done.

It has been an exciting and memorable journey for me. I have enjoyed countless friendships among the many wonderful people who have been attracted to our industry. It is a privilege to have had the opportunity to be a part of it, and a real pleasure and honor to have been recognized by my friends and peers through awards such as the Cherry Award of the IEEE and the Becquerel Prize from the European Communities. I look forward very much to watching the next generation of PV workers carry the ball to the next step of a solar powered world.

Richard M. Swanson

Richard Swanson was born in Davenport, Iowa in 1945. He received his BSEE and MSEE from Ohio State University in 1969 and then began dissertation research as a National Science Foundation Fellow at Stanford University. His paper, "Ion-implanted Complementary Transistors in Low-voltage Circuits," jointly authored with Prof. James D. Meindl, was awarded classic paper status by the

Journal of Solid State Circuits in 2003 as one of the ten most cited papers since the inception of the journal in 1966. In 1976, he joined the Electrical Engineering faculty. His research investigated the semiconductor properties of silicon relevant for better understanding the operation of silicon solar cells. These studies have helped pave the way for steady improvement in silicon solar cell performance.

In 1991, Dr. Swanson resigned from his faculty position to devote full time to SunPower Corporation, a company he founded to develop and commercialize cost-effective photovoltaic power systems. In 2005, SunPower became publicly listed on the NASDAQ Stock Exchange under the symbol SPWR. Today, SunPower is a leading manufacturer and installer of photovoltaic panels.

Dr. Swanson has received widespread recognition for his work. In 2002, he was awarded the William R. Cherry award by the IEEE for outstanding contributions to the photovoltaic field, and in 2006 the Becquerel Prize in Photovoltaics from the European Communities. He was elected a Fellow of the IEEE in 2008 and a member of the National Academy of Engineering in 2009. He received the 2009 Economist Magazine Energy Innovator Award.

Chapter 39

Terrestrial Photovoltaic Industry — The Beginning

Peter F. Varadi

P/V Enterprises, Inc. Chevy Chase, MD, USA

The successful utilization of Photovoltaic (PV) cells to supply the electric needs of satellites used for various purposes gave the idea and an assurance that PV could also be used successfully for large scale terrestrial applications. The obvious problem was that at that time the price of PV cells made for the satellites was about $300/Wp.[1] By 1972, several of the US government organizations, which had needs for telecommunication in remote areas, decided to try to use PV systems to power those equipment. The two US manufacturers of solar cells and panels for space use, Heliotek (Spectrolab) and Centralab, both located in California, were contacted to supply PV cells connected in a module to provide 12 V dc power. Obviously the price of the space solar cells at that time was still over $100/Wp and therefore to use them for these terrestrial applications was not feasible, only PV cells which were rejects for space could be used. The amount of these rejects were not much and only a very few experimental telecommunication systems were made.

1972 was also the year that scientists and academics started to discuss the utilization of PV for terrestrial purposes. In May 1972, the 9th IEEE PV Specialist conference was the first to include a session on terrestrial PV. In the same year, the Institute of Energy Conversion was established at the University of Delaware (USA) for the research of thin film PV.

Dr. Joseph Lindmayer the head of the Physics laboratory of the Communications Satellite Corporation's (COMSAT) Research Center in Clarksburg, Maryland was successfully working to improve the efficiency of PV solar cells for satellite use, which was an extremely important project for COMSAT as the

---

[1] J. J. Loferski, *Progress in Photovoltaic Research and Applications*, Vol. 1, 67–78 (1993).

transmission capacity of the communication satellites depended on the amount of electricity generated by their PV solar modules. Lindmayer developed the so-called "violet cells".[2] This was considered a breakthrough because the "violet cells" were able to increase the efficiency, to convert light to electricity of PV solar cells by 50%, compared to the conventional solar cells produced at that time.

Lindmayer and his colleague Dr. Peter F. Varadi, who was the head of the Material Science Department of COMSAT Research Center, realized that PV could also be used for terrestrial applications, but for this application the PV cells had to be considerably less expensive than the PV used for space applications. They realized this would require a technology very different from the technology utilized to produce solar cells and solar modules for space applications.

After COMSAT's management did not approve any expenditures to develop technology to produce inexpensive PV cells for terrestrial applications, Lindmayer and Varadi — two Hungarian refugees — decided on December 31, 1972 to leave their very good jobs and start a company, which they named SOLAREX, devoted entirely to develop low price PV products to be utilized for terrestrial applications, to provide electricity to power electrical equipments and sell these at a price, which would be attractive to users, for telecommunication and for other electrical equipment.

In 1973, two important conferences were focusing on the possibility utilizing PV for terrestrial purposes. One was "The Sun in the Service of Mankind" held in July in Paris. The other, the so-called "Cherry Hill Conference" was held in October 1973. This conference by coincidence was held at the beginning of the "oil embargo" and it brought the US government's attention to the possibility of the utilization of PV to avert future energy crisis. While these conferences were important events directing the spotlight to the utilization of PV as an alternative to the nuclear and fossil fuel energies, but they did not tried to solve the issue of the commercialization of PV for terrestrial purposes.

When Lindmayer and Varadi after the first euphoria — that they were now entrepreneurs and pioneers in an extremely interesting field — faded away they reviewed the situation, and realized that while PV in the future will be surely the alternative to the fossil fuel and the nuclear energy sources, but that would be only in a very distant future. When they opened the doors of Solarex on August 1, 1973, they opened the door to a totally unexplored new field of science, technology and business.

Lindmayer concentrated on the technology. His idea was to reduce the cost of the machinery needed to produce PV cells and modules, to eliminate vacuum and batch processing, which was essential for space solar cells and that all the production should be under atmospheric pressure utilizing off the shelf equipments, production bands and automation. Also the technology to be developed should be

---

[2] J. Lindmayer and J. F. Allison, The violet cell an improved silicon cell. *COMSAT Tech. Rev.* 3, 1 *1973) also: *Electronics*, May 22, 1972, p. 30.

able to produce good quality Silicon (Si) solar cells even from the cheapest low quality wafers. He also had ideas to develop new and less expensive technology to produce Si wafers.

Varadi realized, that if he would still be a scientist, which he was one month before Solarex was started, he would also give papers, predicting the future possibility of the $0.5–1/Wp PV cells and its utilization to replace nuclear and fossil fuel to power the utilities. But life looked entirely different, when he had to worry to pay the employees and obviously himself and Joseph and the suppliers and try to achieve that they should produce at least 10 000 to 50 000 Wp in the first year at a sales price of about $ 10–30/Wp.

This price was realistic, but it was obviously not for the utility market and therefore his idea was to develop a commercial and diversified market for PV, a market, where PV would be an inexpensive solution compared to what the customer was using at that time. He also believed to rely as little as possible on government-supported business. He believed, that if Solarex had hundred commercial customers and lost one, that would be only 1% of the business, but if the only customer was the government and due to changes in its priorities and the support was cut, 100% of the business would be lost and that would be the end of Solarex. Therefore Solarex was concentrating to develop the commercial applications of PV.

This idea was very interesting, but in reality the issue, to find hundred customers, was extremely complicated. PV modules could be used for a great variety of applications. The first question was to find out what these applications were. What nobody predicted happened in October 1973, shortly after Solarex started the production. This was the oil embargo.[3] That was a wake-up call for the people in the USA and provided for the tiny PV industry — if one could call it an "industry" — an incredible free advertising and exposure. Solarex was located in the suburbs of Washington DC and was visited daily by reporters and TV crews.

The effect of this free advertising was so great, that when Solarex's sales person paid to put a "bingo card"[4] in a magazine, the post office had to send a small truck carrying at least four large bags filled with "bingo cards". Obviously this should have solved the issue to find applications for PV, but as it turned out some of the applications where for a housewife paying the family house's electric bill got the idea of mounting solar modules on the roof of their house, which would provide them with free electricity from the sun. She probably would have gotten a heard attack if somebody would have called her and tell her what the price

---

[3] Daniel Yergin, *The Prize*, Simon & Schuster, 1991.

[4] "Bingo card" was called the post card used by magazines to be detached by the reader and mailed to the company advertising the product, to obtain information about their product. Obviously the company had to pay the magazine to include the Bingo card, as well pay the postage on the card they received.

would be. But there were many applications, in which PV even with those prices would be cost effective. The problem was which of the many applications would lead to open up a market for PV? Varadi used to say: "Many people can make solar cells, but few people know where to sell them."

When Dr. Wolfgang Palz, at that time the chief for Energy Development at the French National Space Agency in Paris, visited Solarex in those days and got into conversation with Joseph about large PV systems for utilities, Varadi used to say: "Where is the purchase order?"

Without purchase orders for utility scale PV systems, the business started to develop. The United States had several organizations, which were taking care of vast areas where nobody was living, but communication was still needed. US Forest Service, Bureau of Land Management, Weather bureau, and in many states, e.g. Arizona, the police. PV was tried and was performing very well and these organizations used it. The first mass produced PV system utilized for these applications was the Solarex H and HP type series (6 to 10 Wp) (Fig. 1) of which many thousand units were fabricated and installed world wide from 1974 to 1979.

For navigation aids PV offered big advantages. Canada was the very first in the world to start a program to convert all of their navigational aids to solar power. Mr. Sunny Leung was the architect of the Canadian Coast Guard Program. The very first solar powered lighthouse in the world was the one at False Duck, Ontario (Fig. 2). This lighthouse was commissioned in 1981. It was a hybrid system with two diesels and had 22 pc. of 30 Wp Solarex modules (Fig. 3) The load was reduced in 1995 and the diesels were removed and 18 out of the 22 modules

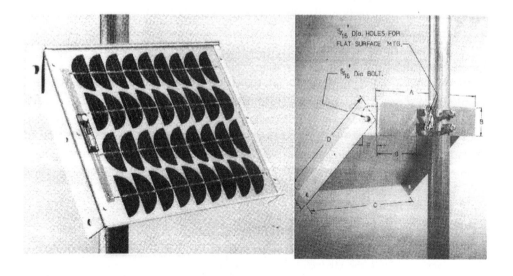

Figure 1. The first mass produced PV system (1974–1978).

Figure 2. The first Solar Powered Lighthouse: Falls Duck, Ontario, Canada.

Figure 3. Solar modules — Falls Duck, Ontario, Canada.

were transferred to other locations. Presently, after 28 years of service, the original four modules are still operating the lighthouse.

The northern-most navigational aid installed by the Canadian Coast Guard, powered by PV is at Resolute Bay, 150 km from the magnetic North Pole of the

Figure 4. Navigational aid near the Magnetic North Pole, Canada.

Earth (Fig. 4). This navigational aid was also installed in 1981 and is still in operation.[5]

The introduction of PV to power the US Coast Guard's navigational aids was technically more complicated because of the environmental conditions: the relatively warm salt ocean waters, atmospheric conditions, high humidity and strong UV radiation, as failure was not an option, the PV modules required very protective encapsulation for the solar cells and their interconnections. The US Coast Guard under the direction of captain Lloyd Lomer, conducted for many years testing and based on that, established a testing procedure and came up with a specification to use only PV modules fulfilling that specification and test procedure (Fig. 5). The US Coast Guard converted in the early 1980's all of the navigational aids to be powered by PV. President Ronald Reagan commended Lomer for "saving a substantial amount of the taxpayer's money through your initiative and managerial effectiveness as project manager for the conversion of aids to navigation from battery to solar photovoltaic power."[6] The savings utilizing PV to charge the batteries in Navigational Aids with PV for the Canadian and the US Coast Guards were substantial. It eliminated the need to send ships to periodically replace batteries.

Utilizing Arco Solar cells and modules two manufacturers in the Houston and New Orleans area, Automatic Power Corporation and Tideland Signal equipped large number of oil platforms with PV navigational aids.

---

[5] Peter F. Varadi: CanSIA Solar Conference, November 2006; Ottawa, ON, Canada.

[6] http://www.californiasolarcenter.org/history_pv.html.

Figure 5. US Coast Guard module.

The Suez Canal was reopened in June 1975. The navigational aids used at that time were replaced in 1982 by 628 modern buoys produced by the Resinex Corporation of Italy. All of them were equipped with PV powered lights, (Fig. 6). The utilized PV system, modules and charge controllers were supplied by Solarex Corporation, providing also engineering supervision of the installed PV systems.

This is also a good example that even at the beginning, when PV was still expensive there was a market for it, where PV was less expensive than the alternative. The success of utilizing PV for navigational aids is only one example.

Other important market areas developed where PV was very useful. Obviously telecommunication was one of the major area, but cathodic protection, water pumping and others including homes in developed countries e.g. in Europe or America, which had no access to the electric grid but needed electricity for lighting and for radio/TV reception.

This expanding PV production was many times affected by the availability of its basic material, the Si wafer. PV cells could use Si wafers the quality of which was not satisfactory to produce semiconductor devices. The availability of these relatively low cost Si wafers depended on the requirement for semiconductor devices. If the demand was high, the available not satisfactory wafers were also high, if the demand for semiconductor devices was low, there was a shortage of the so-called "solar grade" wafers and their price obviously increased. Lindmayer and Varadi decided, that their expanding PV business could not rely on the unpredictable availability and price of their most important material, the Si wafer. One alternative was to buy crystal pullers and fabricate their own wafers, but that would have required a very large amount of capital, which was not available to

Figure 6. Suez Canal — solar powered navigational aid — 1982.

Solarex and would not yield relative inexpensive wafers. Therefore Lindmayer started to develop one of his original ideas, to simply melt the Si material and cast it in a crucible. After the Si solidified in a block, it can be made into wafers. Lindmayer developed a very simple casting technique, which could also use relatively inexpensive Si and also developed the technology to produce PV solar cells from the wafers produced by casting.

The process of producing Si wafers utilizing the casting technique was extremely successful and today close to half of the world's production of PV solar cells use these wafers which are called "multicrystalline" wafers to differentiate them from wafers produced in crystal pullers which are called "single crystal" wafers. Solarex made the World's first large PV installations utilizing these multicrystalline wafers. The first was in 1982, a 200 kWp roof installed on Solarex's factory in Frederick, MD, USA (Fig. 7). It was called the "Solar Breeder", because the electricity produced by solar cells was used to produce solar cells. The next one was a 300 kWp multicrystalline roof installed on the roof of Georgetown University's Intercultural Center in Washington DC in 1983 (Fig. 8). Both of these systems are still in operation.

Arco Solar with ARCO financing built the first one megawatt PV power plant for utility grid support in 1982. The system was located adjacent to the Southern California Edison substation in the Mojave Desert community of Hesperia in Southern California.

The October 23–25, 1973 Cherry Hill conference, had two important effects on the development of PV:

It called on the attention of the US Government to the possibilities in PV for the Research Applied to National Needs (NSF/RANN) program. As a result, the

Figure 7.   Solarex's 200 kWp factory roof ("The Solar Breeder") — 1982.

Figure 8.   Georgetown University, Intercultural Center, Washington, DC: 300 kWp building integrated PV — 1983.

US government recognized the need to expand research and development activities in alternative forms of energy, which included PV and established in 1975 the Energy Research and Development Administration (ERDA). The program received bigger recognition, when in 1977 the US Government consolidated the federal energy policy and ERDA was integrated with the Federal Energy Administration and with other federal energy functions and created a Cabinet level U.S. Department of Energy (DOE). The result was that during the presidency of Jimmy Carter (1977–1981) substantial money was earmarked for PV R&D and for demonstration project.

The other and for the future success of the utilization of PV for terrestrial applications an extremely important result was, that in January, 1975 the US government initiated a terrestrial PV research and development project, assigned to the Jet Propulsion Laboratory (JPL), which was patterned after the recommendations of the Cherry Hill Conference. At JPL, John V. Goldsmith who headed JPL's space oriented PV activities became the Technical Manager also for the terrestrial PV program, the aim of which was to help the terrestrial PV industry to reduce prices and produce reliable PV modules. The JPL program was extended to Si material, crystal growth of Si, encapsulation of solar cells and also fabrication methods of PV modules. Goldsmith, who spent many years working in the field of space oriented PV systems, knew very well, that one of the most important issue was to achieve excellent quality PV modules.

To achieve excellent quality PV modules, utilizing the experiences from the space programs, JPL established a test specification for terrestrial PV modules and was buying blocks of modules from manufacturers who could qualify their products. The first block (Block I: 1975 to 1976) purchased 54 kW off the shelf modules, to establish what is available. The specification was simple, it required verifying the electrical test per manufacturers ratings and environmental tests limited to: temperature cycle and humidity soak. Five companies participated in Block I: M7 International, Sensor Technology, Solarex, Solar Power and Spectrolab. Even these simple tests required several design improvements during production.

In Block II modules totaling 127 kW were purchased. The design (e.g. the requirements of interconnect and terminal redundancy) and testing specifications (thermal and humidity cycle, structural loading) were extended. In this program Sensor Technology, Solarex, Solar Power Corporation and Spectrolab participated.

Included in the Block II program was also the introduction of a quality management program. At that time the ISO quality management system did not exist (it was introduced only in 1987). JPL introduced a Quality Management System, which was very similar to what ISO introduced ten years later. The manufacturer had to prepare a Quality Assurance Plan, including inspection criteria. The program had to be approved by JPL personnel. Manufacturers had also to agree, that a JPL inspector could be in residence and observe the execution of the QA program.

These two blocks were followed by three more block buys ending with Block V. The test specifications were tightened in each of them, based on the experiences in the previous Block buys. It was expected, that PV modules produced, which passed JPL's Block V tests would be able to have a 20 year warranty.

In retrospect, the JPL Block I — V program bought a significant amount of PV modules (in the Block III program in 1978, 30 to 50 kWp was purchased from each of five manufacturers, which at that time was a large quantity). This attracted more companies e.g. General Electric, Motorola. Spire and Arco Solar to participate, and the manufacturers were eager to qualify even if they had to comply with rigorous specifications and their enforcement. This resulted that the quality and reliability of the PV modules were vastly improved. Without this program the expected failures would have destroyed the image and usefulness of PV. The JPL Block program provided manufacturers a means to evaluate quality and the resulting expected life of their PV modules and they were able to offer a warranty long enough to make PV competitive with other electric generation systems.

JPL's final, the Block V test specification became the PV industry's standard. Later it was somewhat revised in the USA and became IEEE1262 and also by the European Union's Solar Test Center, JRC-Ispra (JRC-Ispra 503: Qualification Test Procedures for Crystalline Silicon PV modules). Finally these were merged and became the presently used International Electrotechnical Commission (IEC) standard (IEC 61215 Crystalline silicon terrestrial photovoltaic modules — Design qualification and type approval).

In spite of the participation of several companies in the "block buys", in the late 1970s only a few meaningful terrestrial PV manufacturers remained in the USA: Solarex, Inc., was privately owned, Solar Power Corporation started by Mr. Elliott Berman was supported, and was in 1975 taken over entirely by Exxon. When Bill Yerkes, President of Spectralab, left Spectralab in 1975, started Solar Technology International (STI). STI was acquired in 1977 by Atlantic Richfield (ARCO) and was renamed to ARCO Solar. Mobil in 1974 established Mobil-Tyco to produce thin sheet silicon wafers to produce solar cells.

After 1976, the situation changed. The interest to use PV became global. Besides of these US manufacturers of PV there was still not many in Europe. Philips had a small operation in Paris and sold PV modules in Australia. Philips however discontinued this venture, when they encountered technical problems with their product. Another entry in France was Photowatt, which participated in 1980/81 in the JPL Block IV program with a 38 W module. In the UK BP acquired a small company and was involved in successfully installing PV systems in many countries of the world.

The demand was worldwide and Solarex expanded outside of the USA. In Europe it established in Switzerland a manufacturing operation for multicrystalline silicon produced by the casting process and module assembly and started to manufacture flash simulators. It formed joint ventures in France and Italy to manufacture solar cells and also modules. It started a factory in Australia, also for the manufacturing of PV modules and later also solar cells. It acquired the company of its representative in the UK and manufactured electronics especially for cathodic protection systems. It also established a factory in Hong Kong and manufactured small PV modules and novelty items powered by PV cells.

ARCO Solar under the direction of Mr. Charles Gay developed its worldwide sales of its PV products. It established joint ventures in Europe with Siemens and in Japan with SHOWA Shell. ARCO Solar's sales was also expanded to the southern hemisphere, which only had access to diesel generator sets.

The beginning of the 1980s brought a fundamental change for PV in the US. In 1980, President Carter, under whose presidency from 1977 until 1981, the US Government which had exhibited great support for PV, lost the election. The new administration under President Reagan who took office in January 1981 cut the budget and the support for PV was drastically reduced. This had a great effect on PV companies, whose primary customer was the US Government. Two companies were not much affected, because of they developed a global commercial market for PV. One was Solarex, the other was ARCO Solar.

In case of Solarex, Varadi believed in expanding the commercial market and minimize the government dependence, furthermore he did not believe, that the large utility market was imminent. He believed also, that Lindmayer was an extremely competent scientist with lots of practical ideas and his research and development staff was extremely good to work out his ideas. An additional pillar to support Solarex's success was that John V. Goldsmith joined Solarex and brought his expertise of running large programs in the field of PV. Solarex therefore could be and became very diversified. It had a large international presence, having factories on four continents. Solarex started to design and install large PV systems, also developed a sizeable consumer business, manufacturing PV products for this purpose in Hong Kong and selling them in catalogues and establishing Renewable Energy oriented stores under the name of "Energy Sciences" in the best shopping malls in the Washington — Baltimore area.

Solarex formed a corrosion protection division under the direction of Mr. Ramon Dominguez, selling turnkey cathodic protection systems powered by PV. Under the direction of Goldsmith, it developed a business to make PV solar cells and panels for space application.

Solarex was also involved in producing miniature PV arrays to power electronic watches and calculators. To their big surprise, this business was lost suddenly because some Japanese companies started to manufacture the thin film a-Si (amorphous silicon) mini PV arrays. For calculators, this was not only cheaper, but its spectral response was also better. While the crystalline Si PV array worked better in a bar at candlelight, the a-Si worked better in an office illuminated by fluorescent light.

That was the first time thin film PV was used in quantity. The idea of utilizing thin film PV solar cells started with selenium cells, but the efficiency was very low. In the early 1970s, there were attempts to use the CdS/Cu_2S thin film PV cell. A factory was built by Shell in Delaware (USA), but it turned out that the product was extremely moisture-sensitive and would have needed moisture impermeable hermetic sealed encapsulation. Many other thin film structures were experimented with, a-Si, CIGS (Cupper-Indium-Gallium di-Selenides), CdTe (Cadmium Tellurides). The first successful system was a-Si (amorphous silicon) developed by the group headed by Dr. David Carlson at the RCA Research Laboratories, in Princeton, NJ. When the government support started to decline, RCA did not want to support further the a-Si development work. Solarex realizing that the a-Si is a viable PV system acquired from RCA the a-Si patents, know-how and equipment and hired all of the scientists working on the subject and opened an R&D and manufacturing facility in Newtown PA, near Princeton.

ARCO Solar on the other hand introduced also a-Si modules and started to develop the CIGS system and set up a small facility for thin film PV modules. Interestingly the CdTe idea was dormant for 25 years, when it was developed and became an extremely important factor in the PV business.

In September 1983, AMOCO acquired Solarex. Solar Power Corporation, which was relying heavily on government programs, by being involved in several demonstration projects, became the casualty of the Reagan era and EXXON announced that it will discontinue its operation and Solarex acquired the Solar Power Corporation assets on June 15, 1984.

About ten years after the beginning of the terrestrial PV business with these consolidations only two major PV manufacturers, Solarex and ARCO Solar, both owned by a major oil company, remained.

Chapter 40

Solar Power in Geneva, Switzerland

Philippe Verburgh

SIG-GE, Switzerland

1. A First-Class Solar Potential

The Geneva region benefits from a substantial solar resources. A study conducted on behalf of the Geneva State Council found that all public buildings had a total roof area suitable for PV installations of 0.7 km2; it could produce 50 million kWh per year. If one adds all private, commercial and residential buildings in Geneva, one finds an outstanding solar resource in the State. The Geneva Industrial Services SIG develop this potential on behalf of the State Council and together with the Geneva people by means of an innovative and ambitious strategy.

2. The "5 MW Solar" Project and the "SIG Vitale Range"

Since 2004, SIG has embarked on the "5 MW Solar" project and committed itself to grow solar production in Geneva ten-fold within five years. This objective was already achieved after three years whilst most people had considered it unrealistic at the time it was started. SIG has set up a voluntary support policy with the products "Green SIG Vitale" and "Sun SIG Vitale". This pioneer initiative in Switzerland was followed by Zurich in 2006 and Bern in 2007.

As of end 2009, 15% of all power consumers in Geneva buy "green electricity" adding up to 180 million kWh annually; a part of it was produced from PV. The commitment of Geneva's SIG was recognized internationally in 2008 when it was awarded the Prize from the Swiss Solar Agency.

Power for the World by W. Palz
Copyright © 2011 by Pan Stanford Publishing Pte Ltd
www.panstanford.com
978-981-4303-37-8

3. A Sunny Future for Geneva

As the objectives of the "5 MW Solar" project had been more than met, SIG has adopted one even more ambitious goal: 8 MW of PV capacity by 2010 and 15 MW by 2015. The new objectives are supported by the following strategic initiatives:

- Promotion of voluntary incentives through "SIG Vitale",
- Participation in the Swiss federal "Feed-in Tariff" initiative,
- Support for local solar generation by fixing attractive tariffs (for the solar electricity produced locally).

The Geneva Canton has the ambition to pursue these efforts even though they amount to a small impact on the planet. But they confirm Geneva's leadership for solar development in Switzerland.

Philippe Verburgh

Mr. Philippe Verburgh joined SIG (Geneva Industrial Services) in January 2001. He led the the Energy Department (Gas and Electricity) as Director for over eight years. Since April 1, 2009, he has taken on the position of Director of the Customer Department, leading sales, marketing, communication, supply of gas and electricity as well as the development of renewable energies.

He graduated from the Ecole Polytechnique Fédérale de Lausanne (EPFL), in Switzerland, with a Ph. D in Technical Sciences.

Mr. Verburgh started his career as Teaching Assistant with Professeur Neirynck, at the EPFL. He then moved on to ELCA Informatique as IT Engineer, before joining EOS as Head of IT Services first, and then Vice President in charge of IT and sales from 1985 to 2000.

Before starting at SIG in Geneva, M. Verburgh also managed the sales department at Avenis Trading.

Chapter 41

Early PV Markets and Solar Solutions in South Asia

Neville Williams

Founder, Solar Electric Light Company, Co-Founder, SELCO-INDIA,
Founder and Former Chairman, Standard Solar Inc., USA

"True, they don't have any electricity and solar could provide it," John Corsi, CEO of Solarex Corporation told me in 1990, "but they also don't have any money." We were talking about the two billion people in the world without access to electricity. The head of America's largest solar company, later absorbed into BP Solar, headquartered near my home in Maryland, expressed doubts about my idea of starting a non-profit organization to bring solar lighting to rural people in the developing world. This could be the world's largest potential market for solar photovoltaics (PV), I told him.

He wished me luck, and said "At least your organization and our company will have something in common: we're both non-profit!" That year I launched the Solar Electric Light Fund (SELF, for energy self-reliance). John Corsi donated $5000. For the next decade developing countries became the world's largest and most exciting market for solar power. And *yes*, ostensibly "poor" rural people *could afford solar*, we discovered, provided they had access to credit financing.

But I'm getting ahead of the story. When I joined the US Department of Energy in 1979 to help promote President Carter's nascent solar energy program, it was clear that PV, enjoying a vast R&D budget, was the sexiest and most promising new energy technology. Ten years later, after PV module prices fell from $90 a Watt to $5 a Watt (today we're at $2 a Watt) , it seemed to me that solar PV would be a great human development tool, bringing clean, affordable light and — and also power for black & white televisions — to some of the millions of families who lived their lives in the dark, with only candles and kerosene for illumination (and kerosene won't run a radio, TV or a fan).

Power for the World by W. Palz
Copyright © 2011 by Pan Stanford Publishing Pte Ltd
www.panstanford.com
978-981-4303-37-8

From US charitable foundations we raised money to launch solar "seed" projects, initially in South Asia, to try to make solar technology "sustainable," a new word back then. Solar could also address two problems: the lack of basic household electricity, and carbon emissions from millions of kerosene lamps. For Western donors, it was an environmental issue — bringing clean power to the developing world — but for families without electricity it was a development issue. Solar lights improved farmers lives. Clean was beside the point.

We began our work in Sri Lanka, working with the extraordinary NGO, Sarvodaya, as our partner. That worked so well that it fostered the growth of numerous local solar installation companies, which led to more funding, and eventually to a $40 million World Bank solar project. The result has been the deployment of over 100 000 of rural solar home lighting systems island wide in lieu of extending the country's power-constrained grid.

SELF expanded its program to include India, Vietnam, Nepal, China, South Africa and a half dozen other countries. In each country, we addressed the affordability problem by setting up revolving funds, allowing for a three-year credit purchase of solar home systems. Later, we figured out how to involve third-party rural financial institutions. We found that with credit, even very low-income families could and would buy 35 Wp solar home systems.

The first thing we learned was that "rural poor" was a misnomer. Most of the poor I saw in 11 countries lived in the cities. Our beneficiaries were cash crop farmers with nice four room houses on land they owned, but far from the electric grid. Our technicians wired them like a normal house, with wall-mounted light switches and extremely bright DC compact fluorescent light fixtures, and solar

Figure 1. Kathmandu's Centre For Renewable Energy and SELF's Neville Williams garlanded by villagers during initial solar electrification planning visit to Pulimarang, Ghorka District, Nepal, 1995. Photo courtesy of Neville Williams.

panels mounted on poles. We served the "richest of the poor." There are hundreds of millions of such rural families without electricity (later, other NGO's would begin to bring light to the poorest of the poor using various Chinese-made solar lanterns and flashlights).

Lesson two was that solar technology already existed in most countries. Solar panels needed to be imported, except in China and India where they have been made domestically for over 20 years. Light fixtures, charge controllers, and well-made deep cycle batteries (some of the best were made in Bangladesh) could all be procured locally or regionally. And talented entrepreneurs devoted to building solar businesses always emerged wherever we went. In Nepal, I worked with a local solar enterprise to electrify the country's first all-solar village in the shadow of Annapurna. The project was dedicated by the Prime Minister himself, who helicoptered in to Pulimarang, near Ghorka, one day and pronounced that solar would be good for all of Nepal's remote villages. Today, a dozen solar companies serve this huge off-grid market so people living on their high mountain ridges can have electric light and television and not feel abandoned by the modern world.

But SELF could only do so much as an NGO. The Rockefeller Foundation and Rockefeller Brothers Fund urged us to consider "commercializing" some of SELF's activities, to make solar "projects" into actual businesses. SELF started in China, launching a Sino-American solar joint business venture that was responsible for China's first exemplar "solar village", Magiacha, in a remote corner of Gansu Province.

But it was in India that this epiphany blossomed. I had hired a young man, Harish Hande, then working on a PhD in solar engineering at the University

Figure 2. Neville Williams and Village Development Committee chief Dak Bahadur with solar powered homes in Pulimarang, Ghorka District, Nepal. Photo courtesy of Petra Schweizer.

of Massachusetts, to investigate the prospects for SELF in India, which the World Bank had proposed we do (with no funds ever to follow, thank you). Hande decided Karnataka was the best bet, a huge prosperous southern state with a hopeless utility and no money to extend the grid. But we could not figure out what SELF could do to effect solar rural electrification. So, Harish and I decided to launch a brand new solar business in India and call it the Solar Electric Light Company (SELCO). It soon made sense for me to leave SELF and charter a US corporation to raise money for solar enterprises in several countries. With capital from patient European investors who shared our vision we were able to fund SELCO-India, and launch SELCO subsidiaries in Vietnam and Sri Lanka.

Harish found a powerful regional banker, Mr. K. M. Udupa, to be on SELCO-India's board, and with his help we ended up in 2002 with some 475 South Indian rural banks making solar loans to our customers. I traveled around Karnataka giving talks about solar power to village bankers. Solar loans, surprising the skeptics, had the highest recovery rates. Tata BP Solar, India's largest manufacturer of solar cells and modules, offered SELCO special pricing and support. Managing director Hande learned how to assemble reliable four-light solar home systems employing DC lighting fixtures and charge controllers made in SELCO's own woman-run workshop in Mangalore. SELCO systems featured top quality deep-cycle batteries, and 30 to 40 Watt peak Tata BP Solar panels. Technicians managed to carry the smaller systems on motorbikes, and could install one in half a day. The company also bought a fleet of jeeps and minivans.

Bangalore-based SELCO-India grew to 165 employees with 27 "solar service centers" located in towns across the state marketing and selling solar home systems. Shell Solar also got into the game. Today, competitors abound, but SELCO remains the well-regarded leader. Last year, Dr. Harish Hande received the Entrepreneur of the Year Award at the World Economic Forum in Davos. He used to tell me, "I wake up every morning amazed that we are making money selling solar energy to poor people."

SELCO was eventually restructured and its headquarters moved from the Washington suburbs to Bangalore. Today, SELCO has sold and installed over 110000 solar home lighting systems in South India. Europe's GoodEnergies is among the consortium of SELCO's recent investors who believe social enterprises that marry "profit and purpose" can fulfill the critical mission of bringing clean, reliable, affordable solar power to millions of people who have no other hope of enjoying electricity and electric light. SELCO and the many entrepreneurial solar companies flourishing across South Asia are proving it can be done.

Neville Williams

Neville Williams, a former journalist and long time solar entrepreneur, is the author of Chasing the Sun: Solar Adventures Around The World (New Society Publishers, 2005). In 2005 he brought what he learned in South Asia back to America and founded Standard Solar in Maryland, which is today the largest installer of residential and small commercial solarelectric systems in the mid-Atlantic region. (See: www.self.org, www.selco-india.com, www.standardsolar.com.)